Solid Waste Management

Scrivener Publishing
100 Cummings Center, Suite 541J
Beverly, MA 01915-6106

Publishers at Scrivener
Martin Scrivener (martin@scrivenerpublishing.com)
Phillip Carmical (pcarmical@scrivenerpublishing.com)

Solid Waste Management

Challenges, Sustainability and Advancements

Edited by

Priyanka Singh

Pooja Agarwal

and

V. Vivekanand

Scrivener Publishing

WILEY

This edition first published 2026 by John Wiley & Sons, Inc., 111 River Street, Hoboken, NJ 07030, USA and Scrivener Publishing LLC, 100 Cummings Center, Suite 541J, Beverly, MA 01915, USA

For more information about Scrivener publications please visit www.scrivenerpublishing.com.

Wiley Global Headquarters
111 River Street, Hoboken, NJ 07030, USA

For details of our global editorial offices, customer services, and more information about Wiley products visit us at www.wiley.com.

The manufacturer's authorized representative according to the EU General Product Safety Regulation is Wiley-VCH GmbH, Boschstr. 12, 69469 Weinheim, Germany, e-mail: Product_Safety@wiley.com.

Library of Congress Cataloging-in-Publication Data

ISBN 9781394267934

Front cover barrels image courtesy of Wikimedia Commons
Recycling Cover Image: Generated with AI using Adobe Firefly
Cover design by Russell Richardson

Set in size of 11pt and Minion Pro by Manila Typesetting Company, Makati, Philippines

Printed in the USA

10 9 8 7 6 5 4 3 2 1

Contents

List of Figures

List of Tables

Preface

Solid waste products have been continuously generated due to rapid industrialization activities and population growth. These solid waste products mainly include fruit-vegetable waste residue, glass, plastics, metallic electronic items, agricultural waste products. These solid waste products severely affect the environment and health of societal life. The major issues for handling of healthcare waste items have been brought by COVID-19 pandemic situation in the whole world. Policymakers have adopted safety measures for combating these contaminations and applied policies for managing healthcare waste. Therefore, an efficient management system for these solid waste products required appropriate transportation, infrastructure and disposal facilities. The conventional disposal techniques have encountered many challenges of inadequate landfill disposal, inadequate management of hazardous solid waste (like electronic, biomedical, radioactive) and operational inefficiencies of policies. These challenges of waste management systems have been overcome using advance technologies based on predictive models of artificial intelligence (AI) and geographic information system (GIS). Remote sensing along with GIS system has efficiently predicted the nature and trend of waste generation areas using data captured by aerial photography, videography and integrated sensors

This book has compiled all information in the field of solid waste management to provide extensive knowledge to all stakeholders (policy-makers, waste management organizations, researchers, academicians). Chapter 1 explored the several forms of solid waste and challenge due to trash from cities, businesses, electronics, and construction projects. Chapter 2 has evaluated pertinent organizations and policies for waste management in the world, and provide integrated sustainable management as a solution to the shortcomings. Chapter 3 has investigated legislative framework for biomedical waste management, legislative actions for COVID-19-based biomedical waste (C-BMW), the ability of treatment and disposal facilities to handle challenges, potential management strategies, and perspectives in India. Chapter 4 and Chapter 5 has provided a detailed overview on the use of artificial intelligence, GIS and remote sensing technique for efficient waste management to encourage sustainable development. Chapter 6 has highlighted effective sustainable role of informal waste recycling and modern waste management in developing countries. It described different approaches based on enzymatic, fermentation and valorisation processes for bioconversion of biodegradable waste product into value added industrial products. Chapter 7 has discussed role of cutting-edge technology, interdisciplinary collaborations, and sustainable practices to address the technological, economic, and environmental concerns associated with producing biofuel from biowastes. Chapter 8 has dispensed the planning, extensive research, and collaborative efforts to unleash the full economic potential of halophile-driven biowaste-to-energy projects. Chapter 9 has provided comprehensive information for bioconversion of at waste into a resource for value addition. Waste valorization imbibed the natural recycling principles of zero waste, loop closing, and underlines the importance of sustainable and environmentally friendly alternatives- like fertilizer, wastewater treatment, and biofuel to reduce, reuse, and recycle the waste into composting, anaerobic digestion and vermicomposting. Chapter 10 has highlighted the role of nanotechnology for sustainable waste valorization in agriculture. The transformative potential of Nanotechnology in converting agricultural waste into valuable resources, promoting ecological balance, and advancing the sustainability of modern farming practices have been also elucidated. Chapter 11 has described conventional methods as well as the use of nanomaterials with their regeneration methods for heavy metal removal from wastewater. Chapter 12 has provided information for different regulatory concern including Extended Producer Responsibility programs and "zero waste" goals align with sustainable practices

All chapters have provided comprehensive knowledge of role of advance technology for solid waste management, their appropriate disposal strategies and valorisation methods for their bioconversion into value added products.

Abbreviations

- 3Rs – Reduce, reuse and recycle
- 4-chloro-2-methylphenoxyacetic acid (MCPA)
- Adaptive and Interactive Modelling Systems (AIMS)
- Adaptive network fuzzy inference system (ANFIS)
- Aerosol Optical Depth (AOD)
- Air cathode microbial fuel cell (ACMFC)
- Air quality index (AQI)
- Anaerobic digestion (AD)
- Analytic Hierarchy Process (AHP)
- Arbuscular mycorrhizal fungus (AMF)
- Artificial intelligence (AI)
- Artificial Neural Networks (ANN)
- Best Environmental Management Practices (BEMPs)
- Bioactive compounds (BACs)
- Biomedical waste (BMW)
- Building Information Modeling (BIM)
- Carbon dots (CDs)
- Carbon nanofibers (CNFs)
- Carbon-based materials (CBMs)
- Carbon-nitrogen (C/N) ratio
- Cellulose nanocrystals (CNCs)
- Cellulose nanofibers (CNFs)
- Central Pollution Control Board (CPCB)
- Clean Water Act (CWA)
- Combined Heat and Power (CHP)
- Common Bio-Medical Waste Treatment and Disposal Facilities (CBWTF)
- Community-based organizations (CBOs)
- Comprehensive Environmental Response, Compensation, and Liability Act (CERCLA)
- Conducting polymers (CPs)

- Constructive solid geometry (CSG)
- Consumer Social Responsibility (CSR)
- Convolutional neural networks (CNN)
- COVID-19-based biomedical waste (C-BMW)
- Debris from Construction and Demolition (C&D)
- Deep eutectic solvents (DESs)
- Dimethyl Ether (DME)
- Dimethyl sulfoxide (DMSO)
- Electrocoagulation (EC)
- Electronic health records (EHRs)
- Enhanced Landfill Mining (ELFM)
- Enhanced Waste Management (EWM)
- Environmental Management (EM)
- Environmental Management Bureau (EMB)
- Environmental Protection Agency (EPA)
- Extended Producer Responsibility (EPR)
- Final disposal site (FDS)
- Final Disposal Sites (FDSs)
- Flexible organic light-emitting diodes (FOLEDs)
- Food waste (FW)
- Fruits and vegetable wastes (FVWs)
- Geographic Information System (GIS)
- Geographic Information Technology (GIT)
- Global Navigation Satellite System (GNSS)
- Global positioning system (GPS)
- Graphene oxide (GO)
- Graphene oxide (GOx)
- Graphene oxide nanoparticle (GO-NP)
- Graphene quantum dots (GQDs)
- Greenhouse gas (GHG)
- Health care facilities (HCFs)
- Heavy metals (HMs)
- Hydroelectric Power Plant (HPP)
- Photovoltaic (PV)
- Hydrothermal Carbonization (HTC)
- Hydrothermal Gasification (HTG)
- Hydrothermal Liquefaction (HTL)
- Integrated Pest Management (IPM)
- Integrated remote sensing (IRS)
- Integrated Sustainable Waste Management (ISWM)
- Internal combustion (IC)

- International Atomic Energy Agency (IAEA)
- Internet of Things (IoT)
- Land surface temperature (LST)
- Land use/land cover (LULC)
- Landfill suitability index (LSI)
- Life Cycle Assessment (LCA)
- Machine learning (ML)
- Material recovery facilities (MRFs)
- Medium-sized enterprises (MSMEs)
- Metal nanoparticles (MNPs)
- Metal-organic frameworks (MOFs)
- Microbial fuel cells (MFC)
- Multi-criteria decision-making (MCDM)
- Multi-Criteria Evaluation (MCE)
- Multiwall carbon nanotubes (MWCNT)
- Municipal Solid Waste (MSW)
- Nano-biochar (nano-BC)
- Nanoparticles (NPs)
- Nitrogen-phosphorus-potassium (NPK)
- Non-air flow box (NAB)
- Non-Governmental Organization (NGO)
- Nuclear Regulatory Commission (NRC)
- Object-Based Image Analysis (OBIA)
- Organization for Economic Cooperation and Development (OECD)
- Particulate Matter (PM)
- Personal protection equipment (PPE)
- Plasma Pyrolysis Technology (PPT)
- Poly (butylene succinate-co-butylene adipate) (PBSA)
- Polyaniline (PANI)
- Polycaprolactone (PCL)
- Polychlorinated biphenyls (PCBs)
- Polyethersulfone (PES)
- Polyethylene terephthalate (PET)
- Polyhydroxyalkanoates (PHA)
- Polyhydroxybutyrate (PHB)
- Polylactic acid (PLA)
- Product-as-a-service (PaaS)
- Public-private partnerships (PPPs)
- Radio-frequency identification (RFID)
- Random Forest (RF)

- Random Tree (RT)
- Rapid solid liquid dynamic (RSLD)
- Refuse-derived fuel (RDF)
- Residence time (RT)
- Resource Conservation and Recovery Act (RCRA)
- Rice husk (RHSil)
- Silica nanoparticles (SiNPs)
- Solid waste management (SWM)
- South African National Income Dynamics Study (SA-NIDS)
- Sustainable Development Goals (SDGs)
- Spatial analysis using GIS
- Spatial multiple criteria evaluation (SMCE)
- Sulfamethoxazole (SMX)
- Supercritical water gasification (SCWG)
- Support Vector Machine (SVM)
- Sustainable Integrated Waste Management (SIWM)
- Sustainable solid waste management (SSWM)
- Sustainable waste management's (SWM)
- Swachh Bharat Mission (SBM)
- Tetracycline (TC)
- Tons per day (TPD)
- Total organic carbon (TOC)
- Transfer Stations (TSs)
- Ultraviolet (UV)
- Unified State Ecological Security Office (USEPA)
- United Nations Environment Program (UNEP)
- Unmanned aerial vehicle (UAV)
- Urban heat islands (UHIs)
- Urban local bodies (ULBs)
- Urban Waste Expertise Programme (UWEP)
- Vehicle Routing Problem (VRP)
- Volatile fatty acids (VFA)
- Volatile organic compounds (VOC)
- Waste-To-Energy (WTE)
- Waste-to-Energy (WtE)
- Waste-to-Product (WtP)
- Web Ontology Language (OWL)
- Weighted linear combination (WLC)
- World Trade Organization (WTO)
- Zeolites (ZLs)

1

Towards a Circular Future: Sustainable Management of Solid Waste Products

Deep Gupta[1]* and Sudhir Kumar Gaur[2]

[1]School of Smart Agriculture Science, COER University Roorkee, Vardhaman Puram, Roorkee, Uttarakhand, India
[2]School of Business Studies, COER University Roorkee, Vardhaman Puram, Roorkee, Uttarakhand, India

Abstract

Solid waste management has an impact on the environment, society, and the economy, making it a critical global concern. The notion of sustainable solid waste management is introduced in this chapter, emphasizing the necessity for all-encompassing plans to deal with the rising amounts of trash produced globally. It lays the groundwork for realizing how crucial it is to implement sustainable practices in order to reduce solid waste's negative environmental effects. Effective systems for the collection and transportation of waste take front stage, highlighting environmentally friendly practices including source segregation and community involvement. This content also touches on resource recovery and waste minimization, including case studies of effective waste reduction and recycling initiatives. Examined are cutting-edge resource recovery technologies that support the ideas of a circular economy, such as composting and waste-to-energy procedures.

The importance of public awareness campaigns, community involvement, and legislative initiatives in promoting sustainable waste management practices is acknowledged in the chapter's conclusion. For legislators, companies, and communities attempting to align waste management procedures with more general sustainability goals, it offers a succinct yet thorough reference. The urgent need for practical solutions to the growing problems associated with trash generation is examined in this chapter on the sustainable management of solid waste products. It navigates the present trash production trends, highlighting the variety of items thrown worldwide and the need for customized treatment strategies. The chapter

**Corresponding author*: deepgupta154@gmail.com

Priyanka Singh, Pooja Agarwal and V. Vivekanand (eds.) Solid Waste Management: Challenges, Sustainability and Advancements, (1–48) © 2026 Scrivener Publishing LLC

explores several forms of solid waste, emphasizing the special difficulties that come with trash from cities, businesses, electronics, and construction projects.

Keywords: Solid waste, biological waste, medical waste, sustainable management

1.1 Introduction

The management of solid waste products is a major concern in the modern world, intertwined with public health, environmental sustainability, and socioeconomic well-being. Solid waste generation has reached previously unheard-of levels due to changing consumption habits and growing global populations, making immediate action to lessen its negative effects necessary. An overview of sustainable solid waste management is given in this chapter, along with a discussion of key concepts, challenges, and workable solutions for this pressing issue. Solid waste is a diverse range of items that are dumped from homes, businesses, industries, and organizations. It presents a multitude of challenges to societies across the globe [1]. The traditional "take, make, dispose" linear approach has resulted in the accumulation of enormous amounts of garbage, placing a burden on natural resources, contaminating ecosystems, and posing a health risk to humans. Furthermore, these issues are made worse by inadequate waste management infrastructure, especially in developing nations where informal disposal methods are common.

The circular economy, which promotes closed-loop systems where resources are reused, recycled, or repurposed, resource efficiency, and minimizing waste output, is central to the conversation on sustainable solid waste management. A comprehensive strategy including stakeholders from many sectors—from communities and consumers to government organizations and business sectors—is needed to fully embrace this paradigm shift.

The enormous volume and variety of trash produced makes managing solid waste one of the biggest difficulties. Only municipal solid trash has a diverse range of items, such as organic garbage, paper, plastics, metals, and hazardous compounds, all of which require specific treatment and disposal techniques. Thus, efficient waste categorization and segregation become essential processes for maximizing resource recovery and reducing contamination to the environment. The need of establishing sustainable waste management procedures is further highlighted by the negative environmental effects of inappropriate garbage disposal, including soil and water contamination, air pollution, and habitat damage. Landfills, which were once the most popular method of getting rid of waste, are dangerous for the environment because of things like leachate contamination and greenhouse gas emissions [2]. Because they provide advantages for the environment and the economy, alternative waste treatment methods like composting, anaerobic digestion, and waste-to-energy conversion are becoming more and more important to investigate. Furthermore, the social dimensions of solid waste management cannot be overlooked, as marginalized communities often bear the brunt of inadequate waste infrastructure and environmental degradation. Fair access to waste management services and substantial local community participation in decision-making processes are required by environmental justice principles. Fostering equitable and sustainable waste management systems also requires empowering informal trash workers, who are vital to recycling and rubbish collection in many situations.

The shift to sustainable waste management needs strong legislative frameworks, effective enforcement strategies, and financial incentives to encourage ethical trash management. For example, extended producer responsibility (EPR) programmes encourage producers to manage the end-of-life of their products, which in turn promotes eco-design and waste reduction initiatives. Likewise, market-driven tools like pay-as-you-throw plans and landfill levies can internalize the costs associated with disposing of rubbish while encouraging recycling and waste avoidance. The goal of environmental sustainability and human well-being necessitates the sustainable management of solid waste products. A coordinated effort involving technology innovation, policy reform, community participation, and behavioral change is needed to address this complicated situation. Societies may reduce the environmental effects of waste generation, preserve finite resources, and create resilient, inclusive communities for future generations by embracing the circular economy's tenets and using integrated waste management techniques.

1.1.1 Importance of Sustainable Managing Solid Waste Products

In order to address the urgent environmental, social, and economic issues that contemporary societies face, sustainable management of solid waste products is essential. For the reasons listed below, this project is extremely important.

1. **Environmental Conservation:** Natural resources and ecosystems are seriously threatened by improper solid waste management. Landfills release pollutants and greenhouse gases into the environment, which leads to contamination

of the soil and water. Recycling and composting are examples of sustainable waste management techniques that lessen pollution, ease the load on landfills, and support biodiversity preservation.

2. **Resource Conservation:** Valuable items that can be recovered and repurposed are frequently found in solid trash. Resource efficiency is increased by using sustainable waste management techniques to collect and recycle resources like plastics, paper, and metals. By doing this, the extraction of virgin resources is decreased, energy is preserved, and the negative effects of resource extraction and processing on the environment are lessened.
3. **Mitigation of Climate Change:** Methane, a powerful greenhouse gas that contributes to climate change, is primarily obtained from landfills. Methane emissions can be greatly decreased by removing organic waste from landfills and utilizing anaerobic digestion or composting. Furthermore, waste-to-energy technologies have the ability to utilize waste's energy potential, replacing energy sources based on fossil fuels and reducing greenhouse gas emissions.
4. **Public Health and Safety:** Inadequate waste management techniques put communities' health at danger; especially in underdeveloped nations where burning and open dumping of waste are commonplace. Garbage disposed of inappropriately can expose people to viruses and dangerous substances that can cause respiratory disorders, waterborne infections, and other health risks. Sustainable waste management contributes to the protection of public health and guarantees a secure living environment for all through appropriate collection, treatment, and disposal of waste.
5. **Economic Benefits:** By reducing costs, fostering innovation, and creating jobs, the transition to sustainable waste management offers economic benefits. The recycling and resource recovery sectors boost the economy and create job opportunities. Adopting circular economy concepts, such as product stewardship and extended producer responsibility, also encourages eco-design and builds a stronger, more resilient economy.
6. **Social Equity:** By guaranteeing fair access to garbage collection and disposal services, especially in underserved communities, sustainable waste management promotes

social equity. By addressing inequities in waste management infrastructure and services, community engagement in waste management programmes empowers local stakeholders, develops social cohesion, and advances environmental justice.

To mitigate environmental degradation, conserve resources, safeguard public health, promote social fairness, and support economic progress, solid waste products must be managed sustainably. Societies can move towards a circular economy model that strikes a balance between socioeconomic development and environmental stewardship, guaranteeing a more robust and sustainable future for future generations, by adopting sustainable waste management techniques.

1.1.2 Importance of Switching to a Circular Economy Strategy

Making the shift to a circular economy strategy is crucial for tackling urgent global issues and promoting sustainable development in terms of the economy, environment, and society [3]. A circular economy paradigm requires rethinking resource utilization, minimizing waste generation, and optimizing resource efficiency across the product lifespan in order to transition from a linear "take-make-dispose" approach. The following crucial elements can help to clarify the significance of this shift.

1. **Resource Security and Conservation:** In a linear economy, resources are taken out, turned into goods, and then thrown away as trash, which causes the depletion of natural resources and harm to the environment. A circular economy, on the other hand, uses techniques like reuse, repair, remanufacturing, and recycling to try and prolong the life of resources [4]. A circular economy ensures that vital raw materials will be available for future generations by reducing reliance on virgin materials, conserving finite resources, and improving resource security through closing the loop on resource flows.
2. **Environmental Sustainability:** The linear economy model is intrinsically unsustainable, leading to pollution, habitat destruction, and the depletion of natural resources. A circular economy reduces its bad effects on the environment by using low energy, generating less waste, and emitting fewer

greenhouse gases during the extraction, production, and disposal of resources [5]. A circular economy reduces environmental degradation, helps to moderate climate change, and conserves biodiversity by encouraging eco-design, the use of renewable energy, and sustainable consumption practices.

3. **Economic Competitiveness and Resilience:** By promoting innovation, opening up new commercial avenues, and increasing resource productivity, a circular economy generates economic benefits. Businesses may lower production costs, improve supply chains, and extract value from waste streams by designing products with durability, recyclability, and resource efficiency in mind. Moreover, circular business models can minimize environmental impacts and create revenue streams through the use of sharing platforms and product-as-a-service. By encouraging a more diversified, sustainable economy and lowering reliance on erratic commodities markets, adopting the concepts of the circular economy increases economic resilience.
4. **Well-being and Social Inclusion:** By encouraging inclusive growth, job creation, and community resilience, a circular economy puts social equity first. A circular economy encourages cooperation, social cohesion, and democratic decision-making by including stakeholders from all points of the value chain, such as workers, consumers, and local communities. Sustainable production and consumption methods also improve public health and well-being by lowering chemical exposure, enhancing the quality of the air and water, and fostering better living conditions for all.
5. **Infrastructure and Supply Chain Resilience:** The COVID-19 pandemic highlighted how susceptible linear supply systems are to shocks and disturbances. By decentralizing production and distribution networks, decreasing reliance on single-use materials, and diversifying supply sources, the shift to a circular economy fosters resilience. A circular economy decreases emissions associated with transportation, promotes local manufacturing, repairs, and recycling of infrastructure, and increases community resilience to shocks and crises from the outside world.

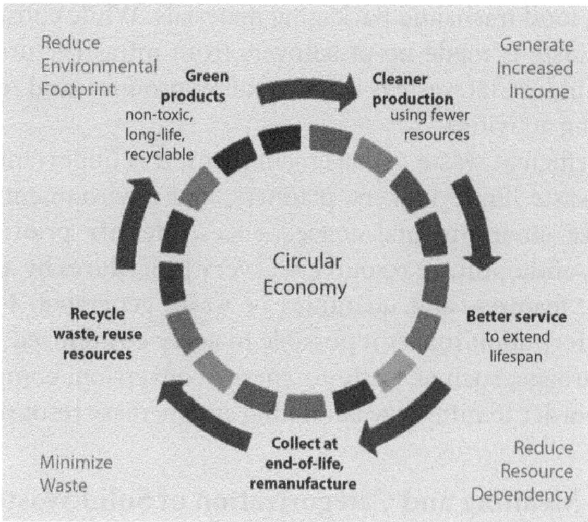

Figure 1.1 Circular economy with 3R's (reduce, reuse, recycle). Source Insight IAS [6].

Achieving sustainable development and tackling global issues like resource scarcity, environmental degradation, and socioeconomic injustice require a shift to a circular economy model. A circular economy provides a way forward for a future that is more resilient, sustainable, and inclusive for both people and the environment by boosting resource efficiency, closing the loop on resource flows, and encouraging innovation (Figure 1.1). The adoption of circular economy concepts necessitates cooperation and coordinated efforts from various sectors, governments, corporations, and civil society; nonetheless, the advantages greatly exceed the difficulties, opening the door to a more prosperous and sustainable future for future generations.

1.2 Comprehending Solid Waste

Solid waste comprises materials that have been dumped from a variety of sources, such as homes, businesses, industries, and institutions, and are no longer needed or valuable. This garbage comprises many different types of materials, such as glass, metal, plastics, paper, organic matter, and hazardous compounds [7].

Geographical location, socioeconomic class, and consumption habits are only a few examples of the variables that might affect the composition of solid waste materials. Municipal solid waste (MSW) is mostly produced by household businesses at homes. It usually consists of plastics, paper goods, food trash, and packaging materials. While construction and demolition waste is made up of leftovers from infrastructure and building projects, industrial waste is made up of by-products and residues from manufacturing activities.

To create efficient waste management plans, it is important to comprehend solid waste. Policymakers, planners, and environmental managers can minimize environmental consequences, identify priority areas for intervention, and optimize resource recovery procedures by analyzing the composition, features, and quantities of waste generated. Furthermore, waste characterization makes it possible to apply customized methods for managing garbage, such as waste-to-energy conversion, composting, and recycling, in order to minimize landfilling and increase resource efficiency.

1.2.1 The Meaning and Categorization of Solid Waste

Any undesired or discarded materials that need to be disposed of because they can no longer be used for their intended function are mentioned as solid waste. All such materials may come from institutional, commercial, industrial, agricultural, or residential activity, among other sources. Solid waste is usually classified based on its characteristics, composition, and source [8]. The following are the main classifications of solid waste.

1. **Municipal Solid Trash/Waste (MSW):** This kind of waste consists of materials generated at institutions, public areas, businesses, and residences. Commonplace materials such as food scraps, paper, glass, plastic, metal, yard garbage, and various household objects are included in it. One of the most prevalent kinds of solid waste is MSW, which is frequently used as the foundation for municipal waste management initiatives.
2. **Waste from Industry:** By-products, leftovers, and wasted materials produced during manufacturing, processing, or industrial processes are referred to as industrial waste. Materials like chemicals, solvents, sludge, metals, and packaging materials can all fall under this category of garbage. Industrial trash needs to be handled and disposed of carefully to avoid contaminating the environment because it may include hazardous materials.
3. **Debris from Construction and Demolition (C&D):** C&D garbage is the leftover material from excavation, demolition,

renovation, and construction projects. Typical components of construction and demolition trash are metals, plastics, wood, gypsum, asphalt, and concrete or bricks. Appropriate C&D waste management is necessary to promote the recycling and reuse of valuable resources while minimizing the effects on the environment.

4. **Hazardous Waste:** Because of its toxic, corrosive, combustible, or reactive qualities, hazardous waste puts both human health and the environment in serious danger. Garbage products like chemicals, insecticides, solvents, batteries, electronic garbage (e-waste), and medical waste fall under this category. Certain procedures must be followed while handling, treating, and disposing of hazardous waste in order to ensure safety and avoid contamination.
5. **Biodegradable Waste:** This category includes organic wastes that are able to break down spontaneously over time through biological processes. Paper goods, yard garbage, food waste, and other organic materials fall under this category. Anaerobic digestion and composting are two effective ways to handle biodegradable waste that can help cut greenhouse gas emissions and restore essential minerals for healthy soil.
6. **Special Waste:** Waste that doesn't cleanly fall into any of the other categories is referred to as special waste. Items like tyres, appliances, electronics, batteries, fluorescent lights, and hazardous trash from the home may fall under this category. Special trash frequently needs specific procedures for collection, recycling, or disposal in order to guarantee effective management and reduce environmental effects.

For the purpose of promoting resource conservation, implementing efficient waste management techniques, and reducing environmental contamination, it is imperative to comprehend how solid waste is classified. Communities can minimize landfilling, maximize recycling and resource recovery, and safeguard the environment and public health by properly classifying and managing solid waste [9].

1.2.1.1 Classification of Solid Wastes

The biological and inorganic waste products generated by a society, such as product packaging, paint cans, batteries, furniture, clothing, bottles, kitchen trash, paper, appliances, and so on, are referred to as solid wastes. These waste products usually have no value for the original user or users. Therefore, solid wastes include both a more homogeneous collection of wastes from industry, agriculture, and mining as well as a diverse mass of garbage from the urban society. Wastes may obtain enormous value in another location, even when they are of little or no value to the person wanting to dispose of them [10]. Strategy and operation of the practical elements related to solid waste management therefore require information of the origins and forms of solid wastes as well as evidence on their composition and the place from where they are produced and disposed of.

1. Classification according to sources

(i) Residential waste: This category includes wastes from homes, apartments, and the like, such as leftover surplus food, peels of vegetables, plastic, clothing, and ashes.
(ii) Commercial: This category includes garbage from supermarkets, eateries, lodging facilities, hotels, motels, auto repair companies, and hospitals that include ashes, glass, metal, and leftover food.
(iii) Institutional: Paper, plastic, glassware, and other items made by educational, administrative, and public facilities, such as schools, colleges, offices, jails, etc., make up the majority of this.
(iv) Municipal: This category comprises waste products from several municipal operations, such as street cleaning, landscaping, building debris, dust, and leaves and other organic materials.
(v) Industrial: This category mostly includes wastes resulting from industrial processes, such as ashes, demolition and building wastes, hazardous wastes, etc.
(vi) Agricultural: Decomposing fruits and vegetables, garbage, agricultural waste, and other materials from orchards, vineyards, farms, and other agricultural locations are included in this.
(vii) Open areas: This category comprises trash from places like roads, parks, playgrounds, open spaces, alleyways, parks, and recreational areas.

2. Classification according to kind

The following is a classification of wastes according to their types, or their physical, chemical, and biological characteristics.

(i) Trash: Animal and vegetable wastes that come from handling, selling, storing, preparing, cooking, and serving food. These wastes' garbage comprises putrescible (rotting) organic debris, which attracts rats and other vermin and emits an unpleasant stench. As such, it needs extra care when handling, storing, and disposing of it.

(ii) Ashes and remaining residues: These are materials left over after wood burning, and other combustible materials such as coal, charcoal, coke in homes, businesses, and small industrial facilities for warmth and cooking. These are considered as industrial wastes when formed in considerable quantities in factories and in some power plants. Tiny fragments of glass and metal are frequently combined with cinders, clinker, and fine powdered residue to form ashes. Ash and waste residues are desirable in landfills because they are nearly completely inorganic.

(iii) Combustible and non-flammable wastes: All wastes that come from businesses, homes, institutions, and other sources; they do not contain food wastes or other materials that quickly decompose. Combustible materials often include things like fabric, paper, cardboard, rubber, garden plant trimmings, etc., whereas non-combustible resources include things like glass, dishes, tin and aluminium cans, metal dirt, ferrous and non-ferrous materials.

(iv) Bulky wastes: These consist of major appliances from the home such as refrigerators, washers, couch, crate, car tyres, wood, trees, and branches. These domestic wastes need a unique collection system because they can't fit in standard storage bins.

(v) Street wastes: These wastes, which contain plastics, waste paper, cardboard, dirt, leaves, and other vegetative substance, are gathered from streets, pathways, alleys, parks, and unoccupied plots. In fact, there is a serious and pervasive problem with littering in public areas in many nations, including India. This threat needs to be adequately addressed by a solid waste management system.

(vi) Dead animals: Dead animals in relation to municipal garbage are those that pass away naturally or unintentionally while travelling. Please take note that carcasses and animal parts from slaughterhouses are not included in this category as they are considered industrial trash. There are two categories of dead animals: large and little. Dogs, cats, rabbits, rodents, and other animals of similar size are classified as small animals, and horses, cows, goats, sheep, and pigs are classified as large animals. This distinction is made because large animals need specialized tools to be lifted and handled when removed. Dead animals can be dangerous to public health if they are not removed quickly because, as they decompose, they appeal to flies and other vermin. Their existence in public places is particularly unpleasant from the aesthetic point of view as well.

(vii) Automobiles that have been abandoned: Vehicles left on roadways and in other public areas include cars, lorries, and trailers. Abandoned cars, however, are very varied in value to collectors and have a large scrap value for their metal.

(viii) Wastes from concrete structure and demolition: This category includes wastes produced during the building, renovation, and demolition of homes, businesses, and other structures. They are mostly made up of soil, rocks, concrete, bricks, timber, plumbing, roofing, electrical, and heating systems, as well as a portion of the general municipal waste stream.

(ix) Farm wastes: The wastes come from a variety of agriculture activities and operations, including at the time of planting, harvesting, milk yield, raising animals for slaughter, and running feedlots. Animal waste disposal has grown to be a serious issue in many places, particularly with regard to feedlots, poultry farms, and dairy farms.

(x) Dangerous wastes: Wastes of consumer origin, industrial, institutional, or that pose a risk to human health and the environment now or in the future are classified as hazardous wastes. This is because of their ignitability, corrosiveness, reactivity, toxicity, and other physical, chemical, biological, or radioactive properties. Be aware that hazardous wastes, either gaseous or liquid, may occasionally be the active agents.

Despite being contained in solid containers; these are nonetheless categorized as solid wastes. Empty paint, solvent, and pesticide containers are common examples of hazardous wastes. These materials are often combined with municipal trash and end up in the urban waste stream. Certain hazardous wastes have the potential to ignite landfill fires and explode in incinerators. Additional materials that need particular processing include radioactive waste and hospital-generated pathological waste. Hazardous wastes should be set aside then gathered, transported, and finally disposed of individually, ideally after necessary action to make them not harmful, according to effective management methods.

(xi) Sewage wastes: Sewage wastes are the solid leftovers from the treatment of sewage. The majority of them are organic, having been produced by treating organic sludge that was isolated from both raw and treated sewage. Grit and eggshells are examples of the inorganic portion of raw sewage that is detached at the first initial treatment because they can contain pathogens and putrescible organic waste that needs to be buried right away. Although treated and dewatered sludge is often not cost-effective, it can be beneficial as a soil conditioner. Thus, unless definite strategies are established for its proper disposal, solid sludge ends up in the municipal garbage stream.

(xii) Mining wastes: Wastes from mining are produced when mineral resources are extracted. They comprise elements that need to be removed in order to reach the mineral resource, such as waste rock and topsoil. Moreover, other waste materials are discharged during or after the processing of mineral ores, including final slags, mine water, mine tailings, mining water treatment sludge, gaseous wastes, etc. Since some of these wastes are inert, the ecosystem is not thought to be threatened by them. But other fractions produced by the mining of non-ferrous metals include significant amounts of hazardous materials like heavy metals. Following extraction and subsequent mineral processing, these metals and metal compounds have a tendency to become more chemically accessible, which can lead to the formation of acid or alkaline drainage. Therefore, mine wastes requires careful characterization to prevent and minimize air, water, and soil contamination.

(xiii) Biodegradable and non-biodegradable wastes: Materials made of organic matter, such as leftover food, fruit and vegetable peels, paper, textile, wood, etc., produced by a variety of domestic and commercial processes, are primarily referred to as biodegradable wastes. Microorganisms break down these wastes into simpler molecules from more complex ones. Wastes that are not biodegradable are made up of recyclable and inorganic elements including metal, glass, plastic, and cans.

(xiv) Medical waste: Waste produced during healthcare activities that can endanger public health or the environment because of its hazardous nature is referred to as biomedical waste, healthcare waste, or hazardous waste. Medical waste is usually classified based on its properties and possible dangers. The primary categories of medical waste are as follows:

a) Infectious waste: Materials tainted with pathogens—bacteria, viruses, parasites, or fungi—that could infect humans are classified as infectious or biohazardous waste. Tissues from surgeries or autopsies, bandages, swabs, cultures, blood products, and discarded gloves are all included in this category. In order to stop infectious diseases from spreading, infectious waste must be handled and disposed of properly.

b) Sharps waste: Needles, syringes, scalpels, lancets, and broken glass are examples of objects that might cause puncture or laceration injuries. These things are referred to as sharps waste. Sharps waste presents a danger of needle stick injuries that can spread blood-borne infections like HIV, hepatitis B, and hepatitis C. It may also be contaminated with infectious agents. To safeguard waste handlers, the public, and healthcare professionals, safe disposal of sharps waste is essential.

c) Pathological waste: Human organs, body parts, tissues, and fluids removed during surgery, autopsy, or other medical treatments are all considered pathological waste. Animal carcasses and tissues from veterinary clinics and research labs are also included in this category. In order to avoid environmental contamination and threats to public health, pathological waste must be handled, disposed of, and contained properly as it may contain infectious pathogens.

d) Pharmaceutical waste: Medication, drug, vaccine, and pharmaceutical product waste includes those that are contaminated, expired, or not used. Drug vials, packaging supplies, and pharmaceutical containers are also included in this category. Pharmaceutical waste should be disposed of properly to prevent pollution of the environment, drug contamination of waterways, harm to aquatic life, and harm to human health. To properly manage pharmaceutical waste, safe disposal techniques like drug take-back programs or incineration are crucial.

e) Chemical waste: Hazardous chemicals, solvents, reagents, disinfectants, and cleaning agents used in healthcare institutions are all included in the category of chemical waste. Chemically polluted laboratory waste, disinfectants, and chemicals used in laboratories are also included in this category. Inappropriate handling, storage, or discarding of chemical waste can have a destructive impact on both the environment and humans. Safe handling of chemical waste requires proper labeling, segregation, and disposal in accordance with legal requirements.

f) Radioactive waste: Materials contaminated with radioactive isotopes from radiation therapy, nuclear medicine, diagnostic imaging, and research are involved in the category of radioactive waste. This group includes contaminated equipment, sealed radiation sources, radioactive sources, and protective clothing. Specialized techniques are required for the management, storage, and disposal of radioactive waste in order to reduce the hazard of radiation exposure and avoid contaminating the environment.

g) Electronic waste: Medical equipment, such as monitors, scanners, and diagnostic machines, as well as electronic devices utilized for patient records, communication, and administration, are examples of the electronic source waste products produced by hospitals. Lead, mercury, and cadmium are among the dangerous compounds found in these objects, and their safe disposal is necessary to avoid contaminating the environment and endangering public health.

In order to implement appropriate handling, storage, transportation, treatment, and disposal techniques, healthcare facilities, waste management businesses, and regulatory agencies must have a thorough understanding of the classification of medical waste. The environment, healthcare personnel, patients, waste handlers, and the general public are all protected when medical waste is managed properly.

1.2.1.2 The Need for Sustainable Solid Waste Product Management

Addressing the complex issues brought on by growing urbanization, population expansion, and consumption patterns requires the sustainable management of solid waste products. Inadequate waste management techniques result in financial losses, hazards to public health, and environmental deterioration [11]. The conventional method of disposing of waste landfills emits greenhouse gases, contaminates water and soil, and takes up precious land. Furthermore, incorrect handling of non-biodegradable and hazardous waste exacerbates pollution of the environment and jeopardizes the stability of ecosystems.

Viable solutions to reduce these negative effects are provided by sustainable waste management, which encourages the circular economy, pollution avoidance, and resource conservation. Societies may reduce waste creation, enhance resource recovery, and lessen their dependency on landfilling by using integrated waste management methods including 3Rs, specially recycling, composting and energy generation from waste. Sustainable waste management techniques also support socioeconomic development, public health protection, and the mitigation of climate change, ensuring a more resilient and just future for all. The distinct and vital characteristics of this waste stream make sustainable treatment of solid waste products, especially medical waste, necessary. If not correctly managed, medical waste which includes infectious materials, sharps, and pharmaceuticals poses serious threats to human health and the environment.

Inappropriate disposal can damage ecosystems, contaminate water sources, and promote the spread of infectious illnesses. Sustainable management techniques, like appropriate disposal, treatment, and segregation, are crucial for maintaining environmental sustainability, public health, and regulatory compliance. Furthermore, implementing sustainable methods for managing medical waste can support a circular economy, lower greenhouse gas emissions, and save resources. Therefore, it is critical to prioritize sustainable medical waste management in order to reduce dangers, protect public health, and promote environmental sustainability. Below is additional information regarding the sustainable treatment of solid waste products.

1.3 Products Derived from Diverse Metabolic Processes in Live Organisms

Biological solid waste products are produced as by-products. These waste products include a diverse range of materials, such as metabolic wastes and cellular detritus. Comprehending these waste products is essential for managing the environment as well as for understanding physiological processes in living things. Any undesired leftover material from hospital operations or related activities is referred to as waste. Any compact or liquid category waste, along with its vessels and any transitional product, that is collected during the diagnosis, treatment, or immunization of humans or animals in related research, or in the manufacture or testing, is referred to as bio-medical waste [12].

Contagious and potentially less contagious wastes more sharps, pharmaceutical waste, pathological wastes, wastes of radioactive materials, chemical wastes, general wastes, and pressurized containers are the eight primary categories into which this hospital solid waste belongs. Due to the widespread usage of both medical and non-medical disposals, hospital waste and the percentage of infection waste are unquestionably higher in India than one would anticipate. The majority of garbage produced in hospitals, like food waste, is just as safe as regular municipal waste. Consequently, hospital waste needs to be disposed of appropriately by being divided into risk and non-risk categories. The necessity to accommodate the growing population has led to the rapid growth of hospitals in both the community and private sectors during the past century. Additionally, the introduction and adoption of "disposable" medical equipment has increased the amount of trash generated in hospitals today.

Hospital waste generation in India is estimated to be between 1.59 and 2.2 kilogram per day per bed, with 10-15% of the trash being biomedical. According to a survey conducted in Bangalore, the amount of solid waste produced daily per bed in hospitals and nursing homes typically ranges from ½ to 4 kg. ½–2 kg per day for government hospitals, and ½–1 kg per day for commercial institutions [13]. In Bangalore, hospital waste generation amounts to approximately 40 tonnes per day. A study was carried out at ten major hospitals in Delhi, Mumbai, Kolkata, and Nagpur between 1993 and 1996 to determine the average composition of waste. It is estimated that 20% of hospital waste in affluent nations comes from biomedical waste. According to a research study done in Delhi, there were approximately 40,000 hospital beds in the city, and the amount of hospital waste produced was 1.5 kg/day/bed. Compared to other Indian states, Kerala has excellent public health and healthcare facilities. Kerala is currently dealing with the issue of securely disposing of garbage produced by medical facilities due to the growth in healthcare services. The majority of hospitals are located in constrained regions of metropolitan centers. It is exceedingly challenging to dispose of waste produced by daily activities in this environment safely. The various kinds of biological solid waste products, along with their sources, contents, and importance are listed below.

1.3.1 Seven Waste Categories Used to Define Infectious Trash

a) Cultures and stocks: Cultures and stocks include biological waste from biological manufacturing, live and attenuated vaccines, agents that are infectious to humans and related biologicals, and anything that is used to mix, transmit, or contain agents. Petri dishes, pipettes, pipette tips, microliter plates, disposable loops, eppendorfs, and toothpicks are a few examples of items that fall under this category [14]. Cellular debris is one of the main sources of biological solid waste products. There is a continual turnover of cells, with new cells replacing older or damaged ones. Cellular debris is produced as a result, which includes pieces of dead cells, organelles, and extracellular matrix elements. Usually, specialized cells like macrophages phagocytose these waste products, or epithelial tissues use shedding mechanisms to get rid of them.

b) Blood products, human blood, and bodily fluids that are infectious: Serum, plasma, and other blood products, as well as non-glass containers containing such discarded fluids, fall under this category. Blood that is not contained by a disposable item or is clearly dripping is also included. Any material that contains visible blood, semen, vaginal secretions, synovial fluid, cerebral fluid, peritoneal fluid, or pericardial fluid is also included. Sharps are defined as glass containers that hold such waste fluids. Blood products do not include intravenous bags that were not filled with blood or blood products. Dialysates are not regarded as bodily fluids like blood.

c) Sharps: The EPA and DEP classify certain items as regulated medical waste, including needles, scalpel blades, hypodermic needles, syringes (with or without attached needles), and needles with attached tubing regardless of contact with infectious organisms. Additional sharp objects include razor blades, blood vials, test tubes, pipette tips, pasteur pipettes, disposable pipettes, and other broken and unbroken glass waste (such as cover slips and microscope slides) that may have come into contact with infectious materials [15]. Additionally, items that can rip or puncture autoclave bags are also considered sharp objects.
d) Research animal waste: All animals that were exposed to infectious agents for experiment and research or for testing purpose may leave behind contaminated remains, infected body parts, and bedding. A contracted vendor picks up and disposes of animal corpses and infected body parts that were not purposefully exposed to infectious pathogens during research or testing.
e) Isolated waste: biological waste and rejected materials tainted with bodily fluids from animals or people that are quarantined due to suspected highly contagious disease infections (biosafety category 4 agent).

1.3.2 The Effects of Biological Solid Waste Products on the Environment

Products made from biological solid waste have a big impact on sustainability and environmental health. Ecosystem waste buildup can result in pollution, nutritional imbalances, and habitat destruction. For instance, too much nitrogenous waste from agricultural runoff can cause water bodies to become eutrophic, which can result in algae blooms and oxygen shortage. Moreover, biogeochemical cycles and ecosystem dynamics may be impacted by microbial metabolites generated as waste products. For example, methane produced by anaerobic microbial activities is a factor in both climate change and greenhouse gas emissions [16].

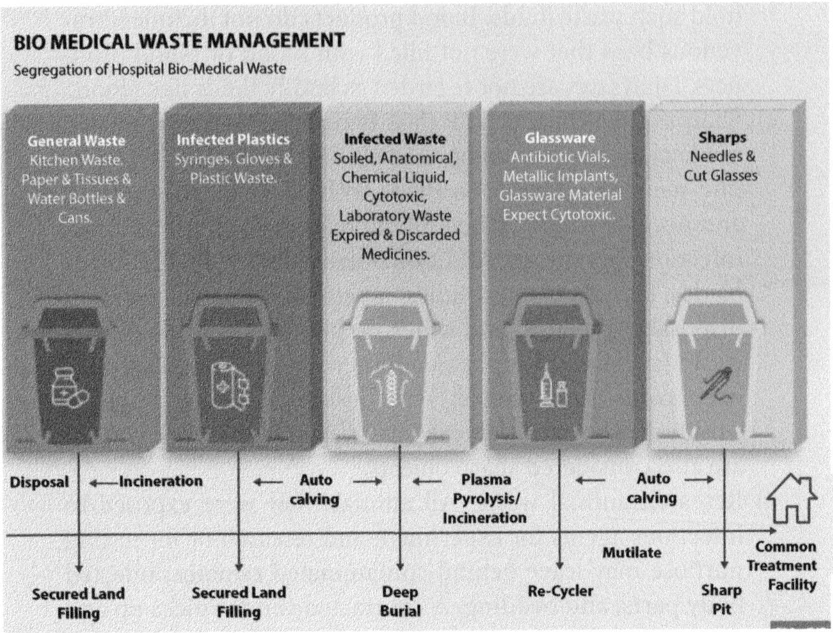

Figure 1.2 Segregation of bio-medical waste [17].

Biotechnological applications: Biosolid waste products have potential uses in biotechnology, despite their detrimental effects on the environment. For instance, fermentation and bioremediation techniques can be used to use certain microbial metabolites to produce biofuels, bioplastics, and other useful substances. Solid waste products derived from biological activities are an essential component of live organisms' physiological functions. Comprehending the diverse kinds, sources, and consequences of these refuse materials is crucial for biotechnological advancements, ecological management, and our general understanding of ecosystem functioning. We can reduce the negative consequences of biological solid waste products and maximize their potential for sustainable development by properly researching and managing them. Figure 1.2 shows segregation of bio-medical waste.

1.3.3 Steps to Effective Biomedical Waste Management

Reducing environmental pollution and safeguarding public health depend on efficient biomedical waste management. The following are essential actions to guarantee appropriate handling of biomedical waste:

a) Segregation: Sorting trash according to its source is the first stage in the management of biomedical waste. Different sorts of garbage, including infectious, sharps, chemical, radioactive, and general waste, should be separated. Table 1.1 shows the waste category according to trash nature. Each form of garbage is managed and disposed of properly thanks to this segregation.
b) Packaging: Biomedical waste needs to be carefully packaged after it has been separated to avoid contamination, leaks, or spills during handling and transportation. Sharps waste should be deposited in leak-proof, puncture-resistant

Table 1.1 Classification of wastes based on trash nature.

Category	Waste content
Cate. 1	Human Organs and Body Parts: Human tissues and anatomical waste.
Cate. 2	Animal Waste: Any kind of tissue, organ, corpse, bleeding portion, etc., that is produced by various health sectors.
Cate. 3	Biotechnology and Microbiology Waste: Remaining materials from microorganism specimens used in research, stocks, or laboratory cultures.
Cate. 4	Waste Sharps: Blades, scalpels, glass, syringes, needles, etc.
Cate. 5	Cytotoxic Pharmaceuticals and Abandoned Medications: Antiquated, tainted, and discarded medications.
Cate. 6	Solid Waste: Cotton stained with blood, dressings, solid plaster molds, etc.
Cate. 7	Nature's Plasticity Solid Waste: Catheters, IV sets, tubing, and PPE kits.
Cate. 8	Waste produced by washing, cleaning, and other disinfection processes in laboratories is referred to as liquid waste.
Cate. 9	Burning Ash: Ash resulting from burning any biomedical waste.
Cate. 10	Chemical Waste: Substances utilized in the creation of living things.

Table 1.2 Waste category based on color coded labels containers.

Kinds of collecting container	Waste category
Simple Plastic bag	Category 1,2
Sterilized/Disinfected container/Plastic bag	Category 3,6
Perforated proof container/Plastic bag	Category 4,7
Special thick Plastic bag	Category 5,9,10

containers; infectious waste, on the other hand, needs to be disposed of in legally required color-coded bags or containers (Table 1.2).

c) Labeling: The kind of waste, the generation date, and any applicable hazard signals or warnings must all be prominently displayed on the labels of all biomedical waste containers. Identification and safe handling of the garbage are made easier with proper labeling.
d) Storage: Appropriate storage spaces that are safe, well-ventilated, and off-limits to unauthorized individuals are where biomedical waste should be kept. To avoid spills or leaks, make sure storage locations have the proper containment mechanisms in place.
e) Transportation: Moving waste within 48 hours of generation to a location outside of health clinics for final treatment.
f) Treatment of waste: Garbage treatment refers to properly disposing of garbage after thoroughly cleaning and disinfecting highly contaminated and contagious items. The most crucial stage in trash management is waste treatment. The following methods can be used to treat biological waste: land disposal, microwave radiation, chemical disinfection, incineration, and wet and dry thermal treatment.
g) Storage and consolidation of biomedical waste: The gathering and keeping of various waste types in distinct containers inside sizable dustbins (as mentioned in Tables 1.3 and 1.4).
h) Disposal: Lastly, it is important to dispose of treated biological waste in keeping with local laws and policies. This could entail deep burial, landfilling, or other permitted techniques. Verify the license status and environmental safety of disposal sites.

Table 1.3 Different categories of wastes based on handled/treated differently.

Dustbin color coding	Waste kinds/category	Handling of waste
Color Yellow	Type/Category 1,2	Incineration/burning or burying deeply
Color Red	Type/Category 3,6	Burning, microwaving, and autoclaving/incineration
Color Blue	Type/Category 4,7	Chemical sterilization, dry and wet heat treatment, incineration
Color Black	Type/Category 5,9,10	Incineration/demolition/disposal in a municipal landfill, and destruction

Table 1.4 Waste category based on color coding of dustbins.

Different color coding for dustbins	According to waste kind/category
Color Yellow	Kind/Category 1,2
Color Red	Kind/Category 3,6
Color Blue	Kind/Category 4,7
Color Black	Kind/Category 5,9,10

The following are waste treatment methods:

1. **Incineration:** For the majority of hazardous medical wastes, incineration was once the recommended treatment and is still in use today. This method, which consists of oxidizing ardours at a high temperature, will reduce organic and combustible waste into incombustible material and have a significant negative impact on waste volume, dimensions, and mass. Usually, wastes that cannot be reused as recycled, repurposed, or dumped in a landfill are treated using this procedure [18]. This process can be used to treat wastes that typically fall into categories 1, 2, and 3, which include biological wastes

from laboratories and anatomical wastes from humans and animals that have combustible or flammable matter content more than 60%, moisture below 30%, non-combustible solid content below 5%, and non-combustible fine particles below 20%. The different types of incineration technologies are as under:

- **Single-chamber kilns/furnaces with static gates:** These furnaces can handle wastes with low heating values, such as 2000–3500 Kcal/kg.
- **Double-chamber pyrolytic incinerator:** This device is specifically made to burn garbage related to infectious diseases in healthcare settings. A double-chamber pyrolytic incinerator can be used to treat wastes with a high heating value, or more than 3500 Kcal/kg.
- **High-temperature rotary kilns:** This technology can be applied to the breakdown of heat-resistant compounds and genotoxic substances found in medical waste.

2. **Chemical disinfection:** In order to eliminate or deactivate the pathogens present in trash classified as categories 4 and 7, such as used sharp needles, scalpels, syringes, glass, blades, PPE kits, catheters, etc., chemicals are added to the wastes. Usually, the procedure is done on hospital property. In order to minimize any negative visual effects on disposal and lower the volume of trash, the wastes are first ground into a fine powder in order to maximize the amount of contact between the waste and the disinfectant by expanding the surface area and removing any enclosed spaces. This also makes any body parts unrecognizable. Shredded paper is typically dampened with water to avoid overheating and to make contact with the disinfectant easier later on.

Commonly used are rotating blade shredders, which have two wheels rotating in opposite directions with blades connected. The shredder may deteriorate if there is an excessive number of sharps in the garbage. Waste can be compacted after disinfection and disassembled to reduce its initial volume by 60–70%. Below is a discussion of several typical compounds used as disinfectants.

Formaldehyde: When combined with steam at 80°C for 45 minutes, it is used to disinfectant all microorganisms, including single cell bacteria, viruses, and bacterial spores in both the dry and solid wastes.
Ethylene oxide or oxirane: All microorganisms, including bacteria, viruses, and bacterial spores, are likewise rendered inactive after 4–12 hours at 60–80% humidity and 37–55°C in solid wastes.
Glutaraldehyde: Both bacteria and parasite eggs can be disinfected by it. It is used with an acetate buffer in a 2% aqueous solution. Medical equipment can be disinfected in five minutes, and spores can be killed in ten minutes.
Sodium hypochlorite: In order to eradicate harmful bacteria, viruses, and minor spores, it is extensively utilized as chemical treatment for wastewater treatment, sanitation purpose and drinking water purification.

Commercially available are a number of self-contained waste treatment systems designed especially for biomedical waste that use chemical disinfection. Some systems are simple to use since they have air filtering systems and are completely automated. It also has less of an effect on the environment as a result. While some of these commercial systems generally use the thermal process based on wet or dry chemical fumigation, the majority of them pulverize the waste. While some may treat pathological waste, it cannot be used to treat cytotoxic or chemical waste (Figure 1.3 shows the municipal waste handling techniques).

Self-contained chemical disinfection systems are one type of commercial chemical disinfection treatment where trash is treated with peroxide, then shred and oxidized by burning lime before being encapsulated in a siliceous mass. Following that, the trash has been processed to the point where it may be disposed of in landfills without requiring special handling. The amount of waste is cut by 80% in this method.

3. **Wet thermal treatment:** Using this technique, steam at a high temperature and pressure is applied to the shredded pathogenic trash. If the right combination of temperature and contact duration is used, it inactivates the majority of bacteria. 99.9% of bacteria and bacterial spores must be inactivated at a minimum temperature of 1210C. In order to further cleanse biological wastes, heat treatment is an option after chemical disinfection. Prior to being treated for sharps, the wastes must be shredded, and in order to maximize disinfection effectiveness, grinding or crushing is advised.
4. **Dry thermal treatment:** This is an additional form of thermal treatment where the wastes are shredded into particles

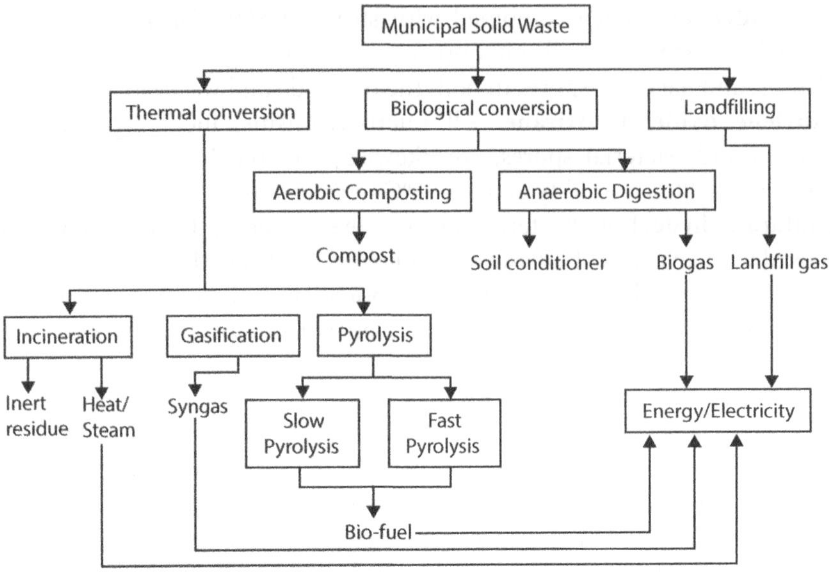

Figure 1.3 Municipal solid waste handling techniques. (Source: Ahmad [20]).

the size of 25 mm and then heated to 1100C–1400C by oil circulation for 20 minutes. Wastes are decreased by 20–35% in weight and by 80% in volume. This method works well for processing sharps and infectious waste, but it shouldn't be utilized for processing radioactive, cytotoxic, or pathological waste.

5. **Microwave irradiation:** Microwaves with a wavelength of 12.24 cm and a frequency of approximately 2450 MHz are capable of destroying the majority of bacteria. By using a microwave to quickly heat the water in the waste, the pathogenic components are destroyed through heat conduction. This procedure can be used to handle biomedical wastes that fall into categories 3 and 6, which include unused samples from pathology labs, specimens utilized in research labs, cotton tainted with blood, bandages, etc.
6. **Inertization:** Waste must be mixed with cement and other materials before it may migrate into groundwater or surface water as part of the inertization process [19]. It is particularly appropriate for burning ashes with a high metal content and pharmaceutical waste.

The packaging for pharmaceutical waste should be removed before adding a solution of water, lime, and cement to initiate the inertization process. Cubes or pellets are made on-site using a homogeneous combination made up of 65% pharmaceutical waste, 15% lime, 15% cement, and 5% water. This mixture can then be transported to a suitable storage place. The procedure can be carried out using quite simple equipment and is moderately priced.

7. **Land disposal:** It is also safe to dispose of garbage that falls into categories 1 and 2 on land. Waste can be disposed of on land using a variety of techniques, as listed below:
   Municipal garbage disposal: Landfill use must be considered a valid disposal option if a medical authority truly lacks the resources to remediate wastes prior to disposal. There are two ways to dispose of garbage on land.
   Open dump: The careless and dispersed disposal of waste materials at a location results in air pollution, fire hazards, an increased risk of disease transmission, and unrestricted access for animals and scavengers. It is not appropriate to place waste on or near open dumps. There is a clear risk that infectious diseases may infect humans or animals, and there is also a chance that the disease will spread later on through wounds, ingestion, inhalation, or indirect pathways such as the food chain or pathogenic host species.
   Sanitary landfill: For the harmless disposal of biomedical waste, it is strongly advised to use isolation of wastes from the environment, suitable engineering work before the site is ready to receive wastes, workers on-site to supervise operation, orderly deposit, and daily coverage of waste.
   Encapsulation: Biomedical waste that is already pre-treated is better suited for municipal landfill disposal than untreated garbage. Encapsulation is one type of pre-treatment solution that entails packing waste into decided containers with addition of an immobilizing substance, and sealing of containers. Sharps and chemical or pharmaceutical residues are placed within metallic drums or high-density polyethylene cubic boxes used in the process [21]. After that, a substance, such as clay, bituminous sand, cement mortar, or plastic foam was poured into the boxes or containers. The containers must be sealed and disposed of in landfills once the medium has dried.

This method is safe and significantly less expensive. Because of this, it is suitable for facilities with restricted funding and facilities for disposing of sharp objects and chemical or pharmaceutical wastes. For non-sharp infectious trash, encapsulation alone is not advised; however, it can be used in conjunction with burning of such garbage. The primary benefit of this procedure is its high efficacy in mitigating the possibility of biomedical waste being accessed by scavengers.

Safe burial on hospital grounds: When it comes to healthcare facilities with minimal biomedical waste management programs, especially those in isolated areas, temporary refugee settlements, or areas with extreme hardship, safe burial of waste on hospital grounds might be the only practical choice available at the time [22]. But hospital administration should nevertheless set certain fundamental guidelines:

- If available, the burial site should be coated with a material of low permeability, such as clay, to prevent polluting of any shallow groundwater that may eventually reach surrounding wells.
- Access to the disposal site should be limited to authorized personnel only.
- Only biological waste that poses a risk should be buried. The available space would soon fill up if regular hospital garbage were also buried there.
- To prevent major environmental pollution issues, it is not advisable to bury large amounts of chemical waste (more than 1 kg) all at once.
- The burial site ought to be maintained like a landfill, with an earthen layer covering each layer of waste to keep rodents and insects away and to prevent odors.

1.4 Hospital Radioactive Waste Products

Because radioactive waste can be hazardous to human health and the environment, hospitals produce a large amount of medical waste that presents special issues. This waste comes from a variety of medical processes, including radioactive material research, diagnosis, and therapy [23]. In order to reduce dangers to human health and the environment, it is imperative that we comprehend the nature of radioactive medical waste and put proper treatment systems into place.

Nature of Radioactive Medical Waste: The term "radioactive medical waste" refers to a wide variety of substances, such as the following:

- **Diagnostic Materials:** radioactive isotopes used in diagnostic imaging, such as iodine-131 for thyroid imaging and technetium-99m for nuclear medicine studies.
- **Therapeutic Materials:** Radioactive materials utilized in cancer therapies, such as radiation therapy's usage of cesium-137 and cobalt-60.
- **Laboratory Waste:** Radioisotopes, which are used to mark molecules in tests, are examples of radioactive materials employed in biomedical research. Equipment that has been contaminated includes gloves, protective clothing, syringes, and other items that come into touch with radioactive materials while being used in medical procedures.

Techniques for Handling Radioactive Medical Waste: Radioactive medical waste must be managed effectively using a variety of treatment techniques to lower its radioactivity and guarantee safe disposal [24]. Among these techniques as mentioned in Figure 1.4, are:

a) **Segregation and Collection:** To avoid combining radioactive waste with other forms of medical waste, hospitals must segregate it at the site of formation. For collecting, specific containers with the relevant radiation symbols labeled on them should be utilized.
b) **Storage:** In order to protect workers and the environment from radiation exposure, radioactive waste must be stored in safe buildings with radiation shielding. During storage, short-lived radioactive elements may decay to safe levels, lowering their potential for hazard.
c) **Decay:** Over time, some radioactive isotopes naturally decay and stop being radioactive. Before disposing of some waste items, hospitals can lower their radioactivity by using decay storage.
d) **Incineration:** For the treatment of solid radioactive waste, including contaminated items like lab coats, gloves, and equipment, incineration is a popular choice. Emission controls and monitoring systems used in specialized incinerators guarantee full combustion and reduce the number of radioactive particles released into the atmosphere.

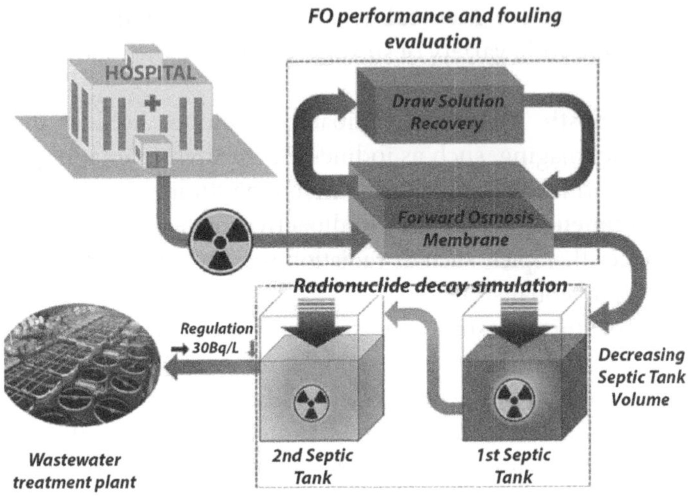

Figure 1.4 Medical radioactive liquid waste treatment. (Source: Lee [25]).

e) **Chemical Treatment:** Radioactive material-contaminated surfaces and equipment can be cleaned using chemical methods. Radioactive particles can be safely removed from surfaces by using solutions that contain chelating agents or detergents, allowing for reuse or disposal.

f) **Solidification:** Radiation-sensitive liquid waste, such as radioactive solutions and lab reagents, can be cemented or absorbent-material-solidified. The radioactive pollutants become immobile by solidification, allowing for safe handling and disposal.

g) **Encapsulation:** Long-lived radioactive waste can be encased in sturdy materials like concrete or lead containers. Examples of such materials are sealed radioactive devices and expended radioactive sources. By isolating the radioactive substance, encapsulation lowers the possibility of environmental contamination.

h) **Transportation and Disposal:** After being processed, radioactive medical waste needs to be transferred to disposal sites that are licensed and abide by regulations. Deep geological disposal, burial in approved repositories, or sanctioned landfills with designed barriers to stop radioactive materials from seeping into the environment are some examples of disposal techniques.

Regulatory Considerations: National and international agencies have established strict regulatory frameworks for the management of radioactive medical waste. Standards and guidelines for the safe management, treatment, and disposal of radioactive waste are defined by regulatory bodies like the International Atomic Energy Agency (IAEA) and the Nuclear Regulatory Commission (NRC) in the United States.

The regulations governing the licensing, storage, transportation, and disposal of radioactive materials must be followed by hospitals and other healthcare facilities [26]. By adhering to these rules, the environment and public health are shielded from the risks posed by radioactive waste. Because radioactive waste can pose threats to human health and the environment, hospitals generate special issues in this regard. Segregation, storage, treatment, and disposal techniques that are specific to the type and concentration of radiation are all necessary for the efficient management of radioactive medical waste. In order to guarantee the safe treatment and final disposal of radioactive waste and reduce its negative effects on the environment and public health, regulatory compliance is crucial. Hospitals may contribute to a safer healthcare environment and reduce the risks connected with radioactive medical waste by putting the right precautions in place.

1.5 Electronic Source Waste Products

The hospital electronic waste products, commonly referred to as medical e-waste, present serious threats to human health and the environment if not correctly managed. These wastes comprise a variety of electronic gadgets, including computers, printers, monitors, medical equipment, batteries, and other electronics utilized in healthcare institutions. Because of their makeup and functionality, medical e-waste needs specific treatment techniques to reduce potential risks. The nature of medical e-waste will be examined, and treatment options to guarantee secure disposal and reduce environmental effect will be covered [27].

Nature of Medical E-Waste: A wide variety of electronic equipment and devices used in healthcare facilities are included in the category of medical e-waste. These gadgets frequently include dangerous compounds including polychlorinated biphenyls (PCBs) and brominated flame retardants, along with heavy metals like lead, mercury, cadmium, and arsenic. These materials provide medical e-waste the potential to be hazardous to both human health and the environment if improperly managed.

Sensitive patient data kept in electronic health records (EHRs) or other digital systems may also be present in medical e-waste. The security and privacy of patients could be jeopardized by improper disposal of these devices, underscoring the significance of safe data destruction in e-waste management.

Treatment Methods for Medical E-Waste: Several treatment techniques are used in the effective management of medical e-waste with the goals of lowering environmental contamination and safeguarding public health. Usually, these techniques consist of:

- **Segregation and Collection:** In order to distinguish medical e-waste from other waste streams, healthcare facilities need to put in place appropriate policies for collection and segregation. Within the facility, designated collecting sites should be set up to guarantee the safe handling and storage of electronic devices that are about to be disposed of.
- **Inventory and Assessment:** Healthcare facilities can determine the kinds and amounts of electronic devices that need to be disposed of by performing a thorough inventory and evaluation of medical e-waste. This data is necessary for choosing the best treatment strategies and creating an effective waste management plan.
- **Reuse and Recycling:** To reduce waste production and save resources, healthcare facilities should give priority to the reuse and recycling of electronic devices. Refurbished and repurposed gadgets can be given to charitable groups for use in impoverished communities, or they can be used within the institution. Recycling initiatives can also lower the need for virgin resources by removing useful components from outdated gadgets that can be used to create new goods.
- **Decontamination:** Medical e-waste, especially equipment utilized in patient care areas, may include biohazards or infectious organisms. As a result, before disposing of any material, comprehensive decontamination methods need to be used to get rid of any microbiological or biological pollutants. Depending on the type of electronic device and how it will be used, decontamination techniques may involve chemical disinfection, autoclaving, or sterilization.

Secure Data Destruction: Ensuring patient privacy is crucial when discarding electronic devices that hold confidential medical data. In order to guarantee that all stored data is completely removed from electronic devices before disposal, healthcare facilities should put safe data destruction procedures in place. Techniques such as data wiping, degaussing, or physical destruction can be used to stop unwanted access or data breaches.
Hazardous Waste Management: Medical e-waste is frequently categorized as hazardous waste since electronic gadgets contain dangerous materials, and therefore needs to be handled as such. To avoid contaminating the environment and incurring fines, healthcare facilities should abide by local laws controlling the processing, storing, transporting, and disposing of hazardous waste. Hiring certified waste management firms guarantees that hazardous medical e-waste is properly treated and disposed of in accordance with legal specifications.
Environmental Monitoring: To identify potential concerns and put remedial measures in place as needed, regular monitoring and assessment of the environmental impacts connected with medical e-waste management is crucial [28]. To test the quality of the air and water, sample soil, and conduct ecological evaluations, monitoring programs can be used to examine the efficacy of waste management techniques and lessen any negative environmental effects.

To sum up, in order to reduce environmental pollution, safeguard public health, and guarantee regulatory compliance, medical e-waste must be managed properly. Healthcare facilities can successfully reduce the dangers connected with disposing of medical e-waste by putting in place procedures for segregation, collection, reuse, recycling, decontamination, secure data deletion, hazardous waste management, and environmental monitoring. To effectively solve the issues raised by medical e-waste and advance a safer and healthier environment for all, healthcare providers, waste management firms, government agencies, and other stakeholders must work together.

1.6 Global Trends in the Formation and Management of Solid Waste

These trends are still changing as a result of a number of reasons, including changes in consumer habits, urbanization, economic development, and population expansion [29]. According to the most recent data available, the following patterns are seen:

a) **Rapid Increase in Waste Generation:** The World Bank projects that by 2050, the amount of solid trash generated worldwide will have increased by 70% from 2016 levels. This indicates a continued global trend of rising garbage output. This tendency is mostly driven by the rapid urbanization, industrialization, and population increase that occur in developing nations. Waste production rises as a result of rising consumption rates brought about by an increase in the number of people moving to cities and leading contemporary lifestyles.
b) **Changing Composition of Waste Streams:** As a result of evolving technology and shifting consumption habits, the composition of solid waste streams is changing. Paper, plastics, metals, and biological trash are all important components of waste, but the amount of electronic waste, or "e-waste," is increasing quickly. Furthermore, the widespread use of single-use plastics and packaging materials exacerbates waste management issues and raises environmental concerns.
c) **Pressure on Landfills:** The increasing volume of solid waste is placing pressure on conventional waste disposal options like landfilling. There is an increasing awareness of the concerns to public health, environmental degradation, and land scarcity that come with landfills, especially in highly populated urban areas. It is getting harder and harder to find good locations for new landfills, which makes using alternate waste management techniques necessary.
d) **Emphasis on Waste Reduction and Recycling:** To lessen the negative effects of trash generation on the environment, there is an increasing focus on resource recovery, recycling, and waste reduction. Policies and programs are being implemented by businesses, communities, and governments to encourage the circular economy, reduce waste, and boost recycling rates. Resource recovery from waste streams is greatly aided by material recovery facilities (MRFs) and advanced recycling technology.
e) **Extended Producer Responsibility (EPR):** Extended Producer Responsibility (EPR) frameworks are being

adopted by numerous nations in order to hold manufacturers responsible for the product life cycle management. Producers are required by EPR to be in charge of product and packaging material collection, recycling, and appropriate disposal. EPR encourages the creation of recycling infrastructure and eco-friendly product design by placing the cost of waste management on producers.

f) **Circular Economy Initiatives:** The circular economy is becoming more and more popular as a sustainable method of waste reduction and resource management. Through increasing resource efficiency, encouraging product reuse, repair, and recycling, and lowering dependency on finite resources, the circular economy projects seek to minimize waste production. Establishing closed-loop systems that preserve resource value and reduce environmental effect requires cooperation between different industries and value chains.

In conclusion, the need for comprehensive and sustainable waste management policies is highlighted by global trends in the development and handling of solid waste. A more sustainable future can be achieved by stakeholders by giving priority to resource recovery, recycling, waste reduction, and the adoption of circular economy principles. This will help them manage the issues faced by rising waste volumes. Governments, businesses, communities, and individuals must work together to manage garbage effectively in order to protect the environment and conserve resources.

1.7 Challenges in Solid Waste Management

Inadequate solid waste management procedures present serious obstacles with far-reaching environmental, social, and economic ramifications. Solid waste management is a key global issue [30]. The environmental effects of inappropriate waste disposal and the social and economic ramifications of insufficient waste management are the two main difficulties in solid waste management that will be discussed in this article.

Environmental Impacts of Improper Waste Disposal: Ecosystems, water quality, air pollution, and general environmental health are all negatively impacted by improper trash disposal. Among the major effects on the environment are:

- **Land Pollution:** Landfills are the most popular way to dispose of waste in the world, yet leachate and dangerous chemicals from them can contaminate groundwater and soil. Methane, a strong greenhouse gas that contributes to climate change, is produced when organic waste breaks down. Long-term soil pollution is another consequence of inappropriately disposing of non-biodegradable items like plastics.
- **Water Contamination:** Landfill leachate has the potential to contaminate surface and groundwater bodies with dangerous materials. Solid waste contains chemicals and pollutants that can seep into water sources and endanger aquatic ecosystems as well as human health. Additionally upsetting ecosystems, contaminated water can have an impact on biodiversity and fauna.
- **Air Pollution:** In many areas with poor waste management infrastructure, open burning of rubbish is a frequent practice that discharges dangerous pollutants into the air. Burning rubber, plastics, and other materials release particulate matter, dioxins, and furans, among other harmful gasses that pollute the air and cause respiratory issues in the people around. Moreover, methane emissions from organic waste that decomposes in landfills worsen air quality problems and fuel global warming.
- **Habitat Destruction:** Waste that is improperly disposed of in natural habitats—such as rivers, woods, and oceans—can damage species and disturb ecosystems. Particularly, marine pollution from plastic trash presents a serious risk to marine life, as millions of marine creatures perish every year as a result of ingesting or becoming entangled in plastic debris.
- **Soil Degradation:** Soil degradation and fertility loss can result from inappropriate landfill management practices or the unapproved disposal of garbage. Waste-derived chemical pollutants have the power to change the composition of soil, making it less conducive to plant growth and increased agricultural yield. Food security and safety are impacted by soil pollution because contaminated crops can accumulate toxins.

- **Ocean Pollution:** Marine life and ocean ecosystems are seriously threatened by marine debris, particularly plastic garbage. Marine animals that consume or become entangled in plastic pollution suffer injuries, asphyxia, and even death. Furthermore, pollution and invasive species can be carried by plastic waste, causing ecological disruptions in large maritime habitats.

Waste reduction, recycling, and sustainable disposal techniques are the main focus of comprehensive waste management methods that address the environmental effects of incorrect garbage disposal. Communities can reduce environmental pollution and preserve natural resources for future generations by putting in place efficient waste management rules and making infrastructure investments.

Social and Economic Implications of Inadequate Waste Management: The public's health, people's ability to support themselves, and economic growth are all negatively impacted by inadequate waste management procedures [31]. Among the principal difficulties are:

- **Public Health Risks:** Communities are exposed to a range of health risks as a result of improper garbage disposal, such as skin infections, respiratory ailments, and infectious diseases. Open waste disposal and burning increase the risk of vector-borne illnesses including cholera, dengue fever, and malaria by providing breeding habitats for disease-carrying insects like rodents and mosquitoes.
- **Impact on Livelihoods:** Inadequate waste management techniques can seriously harm people's ability to make a living, especially for underprivileged groups that engage in recycling and trash collecting on the streets. Waste pickers are vulnerable to health concerns due to exposure to hazardous waste materials and pollutants, and their vulnerability is further increased by limited access to protective gear and treatment. Moreover, contamination of agricultural land and water supplies by waste disposal site pollution can have a negative impact on farmers' income and productivity.
- **Economic Costs:** Governments and communities bear substantial financial expenditures as a result of inadequate waste management. Municipal budgets are strained by the direct expenses of trash collection, transportation, and disposal,

especially in low-income communities with little resources. Furthermore, the indirect costs associated with environmental remediation, medical expenses, and lost production as a result of disease place further strain on economies and impede efforts toward sustainable development.

- **Social Inequities:** Vulnerable groups and marginalized communities are disproportionately affected by inadequate waste management, which exacerbates social injustices. The majority of waste-related pollution and health hazards are concentrated in poorer neighborhoods and informal settlements, which exacerbates social inequities and sustains environmental injustices. Communities that are already marginalized by poverty and environmental degradation are further exacerbated by a lack of access to adequate garbage services and infrastructure.

Comprehensive plans that give priority to resource recovery, recycling, trash reduction, and sustainable waste management techniques are needed to address the environmental, social, and economic issues brought on by insufficient waste management. Governments, communities, corporations, and civil society organizations must work together to design and carry out waste management programs and policies that safeguard public health, advance environmental sustainability, and encourage inclusive and equitable development.

Technological and Infrastructural Limitations: Restrictions in technology and infrastructure provide major obstacles to the efficient handling of solid waste [32]. Among these restrictions are:

- **Lack of Infrastructure:** Many areas lack the necessary infrastructure for waste management, including collection systems, treatment facilities, and disposal locations. This is especially true in low- and middle-income nations. The lack of access to basic trash services leads to a lot of unlawful dumping, open burning, and careless garbage disposal, which worsens the threats to public health and the environment.
- **Resource Constraints:** In many areas, the construction and upkeep of waste management infrastructure is impeded by financial and human resource constraints. Governments'

capacity to invest in trash infrastructure and execute sustainable waste management methods is hampered by a lack of finance, limited workforce, and conflicting objectives.

- **Technological Barriers:** The effective management of solid waste is hindered by technological impediments, such as obsolete equipment and inadequate waste processing technology. Ineffective garbage collection and transportation systems cause operational and logistical problems, which raise expenses and decrease the efficacy of waste management initiatives.
- **Capacity Building Needs:** Developing technical know-how and expanding institutional capacity are necessary for putting waste management plans into practice. Many communities lack the resources, expertise, and know-how needed to create and operate environmentally friendly waste management systems. Initiatives aimed at increasing capacity, such as technical support, knowledge exchanges, and training courses, are essential for enabling regional players and creating robust waste management systems.
- **Integration of Informal Sector:** In areas with no formal infrastructure, informal rubbish pickers are especially important to waste management systems. Regulating, organizing, and recognizing the contributions of informal trash workers present difficulties when integrating them into regular waste management procedures. Providing chances for underprivileged populations to earn a living, formalizing and bolstering the informal trash sector can improve the effectiveness and inclusivity of waste management systems.

Governments, business associations, and international organizations must work together to address infrastructure and technology constraints. Overcoming these obstacles and creating sustainable waste management systems that safeguard the public's health, conserve natural resources, and advance socioeconomic development will require investing in cutting-edge waste management technologies, modernizing the infrastructure already in place, and bolstering institutional capacity. Through the prioritization of innovation, collaboration, and capacity building, communities may surmount technological and infrastructural obstacles and shift towards waste management techniques that are more resilient and sustainable.

1.8 Emerging Trends and Technologies in Solid Waste Management

The following new trends and technology are influencing how waste management will develop in the future as the world looks for more sustainable and effective ways to handle solid waste [33].

a) **Advanced Recycling Technologies:** Waste sorting and recycling operations are becoming more accurate and efficient due to advancements in recycling technology including artificial intelligence (AI), robotics, and sensor-based sorting systems. By increasing recovery rates and decreasing contamination in recycling streams, these technologies make it possible to automatically sort recyclable materials.
b) **Waste-to-Energy (WtE) Systems:** Technologies for converting waste into energy, such as pyrolysis, gasification, and incineration, are becoming more and more popular as practical substitutes for conventional landfill disposal. Waste-to-energy (WtE) systems reduce the amount of garbage transported to landfills by converting solid waste into energy, such as heat or electricity, and producing renewable energy sources in the process.
c) **Decentralized Waste Management Solutions:** Local communities can manage their garbage more sustainably when they use decentralized waste management techniques including on-site waste treatment systems, small-scale biogas production, and community-based composting. These decentralized approaches lower transportation expenses, lessen their negative effects on the environment, and strengthen community resistance to waste management issues.
d) **Smart Waste Management Systems:** Real-time monitoring, optimization, and predictive maintenance of garbage collection and disposal processes are made possible by the integration of Internet of Things (IoT) technology and data analytics into waste management systems. Intelligent waste management systems lower expenses, increase operational effectiveness, and help waste management authorities make data-driven decisions.
e) **Circular Economy Initiatives:** Innovation in waste management techniques is being fueled by the shift to a circular

economy, which places an emphasis on closed-loop material flows, resource efficiency, and waste prevention. Initiatives for a circular economy encourage product redesign, remanufacturing, and extended producer responsibility (EPR) in order to maximize resource recovery and reduce waste generation over the course of a product's existence.

f) **Biodegradable Packaging Materials:** Packaging waste has a negative environmental impact, but there are sustainable alternatives to conventional plastics thanks to the development of biodegradable and compostable packaging materials. Deriving from renewable sources like plant-based polymers, biodegradable packaging materials naturally break down in the environment, minimizing pollution and preserving limited resources.

g) **Block Chain Technology for Supply Chain Transparency:** A transparent and safe platform for tracking and validating waste management procedures, from collection and sorting to recycling and disposal, is offered by block chain technology [34]. Supply chain management systems that are built on blocks promote sustainable practices and build stakeholder trust by improving waste management operations' traceability, accountability, and integrity.

1.9 Policy Recommendations for Fostering a Circular Economy

The following policy proposals might be taken into consideration by policymakers in order to improve sustainable waste management practices and encourage a circular economy [35]:

- **Implement Extended Producer Responsibility (EPR) Programs:** Pass laws requiring manufacturers to manage a product's whole life cycle, including collection, recycling, and appropriate disposal. Eco-design is encouraged, resource efficiency is promoted, and the environmental costs of product usage are internalized through EPR programs.
- **Introduce Economic Instruments:** Use financial tools to internalize the externalities of waste generation and encourage sustainable production and consumption practices, such

as landfill levies, deposit-refund programs, and green procurement regulations. Businesses and consumers are encouraged to embrace the circular economy and minimize waste generation through the provision of financial incentives.

- **Strengthen Waste Regulations:** To encourage recycling, trash prevention, and environmentally friendly waste management techniques, rules and enforcement systems should be strengthened. Set minimum recycling requirements, outlaw single-use plastics, and impose severe fines on unauthorized dumping and degradation of the environment.
- **Foster Collaboration and Innovation**: Encourage cooperation between civil society, academia, business, and government to promote innovation and information exchange in circular economy projects. Encourage the creation of new technology, business plans, and best practices for sustainable waste management through research and development.
- **Promote Green Public Procurement:** Green public procurement rules should be used to incentivize public and private entities to prefer environmentally friendly goods and services. In order to encourage supply chains to embrace the concepts of the circular economy, procurement criteria ought to take the environment, resource efficiency, and recyclability into account.
- **Raise Awareness and Education:** To increase public understanding of the advantages of a circular economy, as well as the significance of sustainable consumption and waste avoidance, start public awareness campaigns and educational programs [36]. To encourage behavior change and a sustainable culture, offer businesses, consumers, and waste management professionals' educational programs and technical support.

1.10 Strategies for Promoting Sustainable Consumption and Waste Prevention

The following tactics can be put into practice by stakeholders to encourage trash reduction and sustainable consumption [37]:

Educate Consumers: Inform customers about how their consumption patterns affect the environment so they may make wise decisions. To promote responsible consumption, educate people about sustainable alternatives including reusable items, minimum packaging, and eco-friendly materials.
Encourage Product Design for Durability and Repair: To increase a product's lifespan and decrease the need for replacement, manufacturers should be encouraged to design their goods with longevity, reparability, and durability in mind. Encourage the availability of product warranties, repair services, and spare parts to assist customers in caring for and mending their belongings.
Promote Sharing and Collaborative Consumption: Encourage cooperative consumption models, including tool libraries, community gardening, and car sharing, to boost the sharing economy and cut down on waste production and resource use. To encourage resource sharing and build community resilience, support collaborative consumption projects and peer-to-peer sharing systems.
Implement Waste Reduction Strategies: To reduce waste creation at the source, put waste reduction measures into practice, such as packaging optimization, material substitution, and source reduction [38]. Urge companies to implement zero-waste programs, lean manufacturing techniques, and product redesign tactics in order to cut waste production and boost resource efficiency.

- **Support Circular Business Models:** Encourage companies to implement circular business models to reduce waste and conserve resources. Examples of these models include product-as-a-service (PaaS), leasing, and take-back schemes. Give companies converting to sustainable business practices and the concepts of the circular economy financial incentives, technical support, and regulatory support.
- **Facilitate Community Engagement:** Utilize outreach activities, community-based education, and participatory techniques to involve the local community in trash reduction and sustainable consumption projects. Encourage a sense of ownership and responsibility for trash management by empowering community organizations, schools, and grassroots initiatives to spearhead waste reduction campaigns, recycling drives, and community clean-up events.
- **Promote Green Procurement Practices:** Encourage both public and private companies to use sustainable supply chain management techniques and green procurement policies

to give preference to eco-friendly goods and services. To encourage sustainable production and consumption practices, set up procurement rules that give preference to recyclable materials, eco-label certified items, and products with low environmental impact.

Stakeholders may reduce resource consumption, encourage sustainable consumption and waste avoidance, and accelerate the shift to a circular economy that maximizes resource efficiency and reduces environmental consequences by putting these methods and ideas into practice. Governments, corporations, civil society organizations, and consumers must work together to promote systemic change and build a more sustainable future for future generations.

Conclusion

Sustainable solid waste management, particularly with regard to biomedical waste, is essential to preserving public health, preserving the environment, and advancing sustainable development in the transition to a circular economy. Through the adoption of circular economy ideas and the implementation of creative waste management techniques, stakeholders may effectively tackle the obstacles presented by solid waste and simultaneously seize possibilities for resource conservation and recovery.

In conclusion, creating a circular future that puts public health, environmental preservation, and resource efficiency first requires the sustainable management of solid waste, including biomedical waste. Through the adoption of circular economy ideas, strategic investments in cutting-edge technologies, and the establishment of strong regulatory frameworks, stakeholders have the power to revolutionize solid waste management into a robust, sustainable process. Governments, manufacturers, healthcare providers, and civil society must work together to promote systemic change and build a future where resources are maximized, waste is reduced, and communities live in harmony with the environment. By working together, we can create a circular future in which garbage, instead of being a problem, is recovered, reused, and regenerated for future generations.

References

1. Ajibade, F.O., Adelodun, B., Lasisi, K.H., Fadare, O.O., Ajibade, T.F., Nwogwu, N.A., Sulaymon, I.D., Ugya, A.Y., Wang, H.C., Wang, A., Environmental pollution and their socioeconomic impacts, in: *Microbe Mediated Remediation of Environmental Contaminants*, pp. 321–354, Woodhead Publishing, 2021.
2. Abubakar, I.R., Maniruzzaman, K.M., Dano, U.L., AlShihri, F.S., AlShammari, M.S., Ahmed, S.M.S., Al-Gehlani, W.A.G., Alrawaf, T.I., Environmental sustainability impacts of solid waste management practices in the global South. *Int. J. Environ. Res. Public Health*, 19, 19, 12717, 2022.
3. Ferasso, M., Beliaeva, T., Kraus, S., Clauss, T., Ribeiro-Soriano, D., D. Circular economy business models: The state of research and avenues ahead. *Bus. Strategy Environ.*, *29*, 8, 3006–3024, 2020.
4. Tuntova, A., Depletion of natural resources as a threat to environmental security and a factor in international relations. *Polit. Secur.*, 6, 2, 3–10, 2022.
5. Gallucci, T., Lagioia, G., Dimitrova, V., Marinov, S., Amicarelli, V., Vassileva, B., Pashova, S., Boshnakov, P., Palamarova, P., Ivanov, Y., *Theory and Practice of Circular Economy*, p. 306, Direct Services, Sofia, 2019.
6. Insight IAS, https://www.insightsonindia.com/2023/07/26/editorial-analysis-moving-away-from-the-take-make-dispose-model/#.
8. Iqbal, L.X. and Chen, G.H., Municipal solid waste: Review of best practices in application of life cycle assessment and sustainable management techniques. *Sci. Total Environ.*, 729, 138622, 2020.
9. Deus, R. M., Bezerra, B. S., & Battistelle, R. A. G. Solid waste indicators and their implications for management practice. *International Journal of Environmental Science and Technology*, 16, 2, 1129–1144, 2019.
10. Cowger, W., Steinmetz, Z., Gray, A., Munno, K., Lynch, J., Hapich, H., Primpke, S., De Frond, H., Rochman, C., Herodotou, O., Microplastic spectral classification needs an open source community: open specy to the rescue! *Anal. Chem.*, 93, 21, 7543–7548, 2021.
11. Das, S., Lee, S.H., Kumar, P., Kim, K.H., Lee, S.S., Bhattacharya, S.S., Solid waste management: Scope and the challenge of sustainability. *J. Clean. Prod.*, 228, 658–678, 2019.
12. Vakilchap, F., Mousavi, S.M., Baniasadi, M., Farnaud, S., Development and evolution of biocyanidation in metal recovery from solid waste: a review. *Rev. Environ. Sci. Bio/Technol.*, 19, 509–530, 2020.
13. Sahoo, S., Rathod, W., Vardikar, H., Biswal, M., Mohanty, S., Nayak, S.K., Biomedical waste plastic: bacteria, disinfection and recycling technologies—a comprehensive review. *Int. J. Environ. Sci. Technol.*, 21, 1, 1141–1158, 2024.
14. Reinhardt, P.A., *Infectious and Medical Waste Management*, CRC Press, 2018.
15. Buchan, B.W., Relich, R.F., Mahlen, S.D., The American Society for Microbiology, Medical College of Wisconsin, Eskenazi Health, Wisconsin

Diagnostic Laboratories, Indiana University Health, Indiana University School of Medicine, & Sanford Health Bismark. *Interim Clinical Laboratory Guideline for Biological Safety*, 2019.
16. Scaglia, B. and Adani, F., An index for quantifying the aerobic reactivity of municipal solid wastes and derived waste products. *Sci. Total Environ.*, 394, 1, 183–191, 2008.
17. Alur, M., How to prevent biomedical waste from contaminating our air and water. *Citizen Matters*, 2022. https://citizenmatters.in/be-the-prescription-for-change-initiative-for-managing-household-biomedical-waste/.
18. Hong, J., Zhan, S., Yu, Z., Hong, J., Qi, C., Life-cycle environmental and economic assessment of medical waste treatment. *J. Clean. Prod.*, 174, 65–73, 2018.
19. Mehrotra, A., Inertization, Utilization, and Safe Disposal of Incineration Residues. Doctor of Engineering (D Eng), Doctoral Project, Civil & Environmental Engineering, Old Dominion University, DOI: 10.25777/0d5d-3w81, https://digitalcommons.odu.edu/cee_deng_projects/1, 2017.
20. Ahmad, M., Beddu, S., Itam, Z.B., Alanimi, F.B.I., State of the art compendium of macro and micro energies. *Adv. Sci. Technol. Res. J.*, 13, 1, 88–109, 2019.
21. Singh, A.P., Saxena, R., Saxena, S., An Elaborated Study on Bio-Medical Waste and Its Successful Administration. *Int. J. Recent Sci. Res.*, 13, 8, 2121–2126, 2022.
22. Annapoorani, S.G., Sustainable management of biomedical waste during COVID-19 pandemic, in: *COVID-19: Sustainable Waste Management and Air Emission*, pp. 159–195, Springer Singapore, Singapore, 2021.
23. Komilis, D., Fouki, A., Papadopoulos, D., Hazardous medical waste generation rates of different categories of health-care facilities. *Waste Manage. (Oxford)*, 32, 7, 1434–1441, 2012.
24. Capoor, M.R. and Bhowmik, K.T., Current perspectives on biomedical waste management: Rules, conventions and treatment technologies. *Indian J. Med. Microbiol.*, 35, 2, 157–164, 2017.
25. Lee, S., Kim, Y., Park, J., Shon, H.K., Hong, S., Treatment of medical radioactive liquid waste using Forward Osmosis (FO) membrane process. *J. Membr. Sci.*, 556, 238–247, 2018.
26. Thompson, B.G.J., The role of performance assessment in the regulation of underground disposal of radioactive wastes: An international perspective. *Risk Anal.*, 19, 5, 809–846, 1999.
27. Alabi, O.A., Adeoluwa, Y.M., Huo, X., Xu, X., Bakare, A.A., Environmental contamination and public health effects of electronic waste: an overview. *J. Environ. Health Sci. Eng.*, 19, 1209–1227, 2021.
28. Arya, S. and Kumar, S., E-waste in India at a glance: Current trends, regulations, challenges and management strategies. *J. Clean. Prod.*, 271, 122707, 2020.

29. Kumar, A. and Agrawal, A., Recent trends in solid waste management status, challenges, and potential for the future Indian cities – A review. *Curr. Res. Environ. Sustain.*, 2, 100011, 2020.
30. Owusu, G., Social effects of poor sanitation and waste management on poor urban communities: a neighborhood-specific study of Sabon Zongo, Accra. *J. Urban.*, 3, 2, 145–160, 2010.
31. Sawant, A.D., Warrier, S., Pawar, G., Socio-Economic Implications of Waste Management, in: *Urban Growth and Environmental Issues in India*, pp. 239–252, 2021.
32. Zohoori, M. and Ghani, A., Municipal solid waste management challenges and problems for cities in low-income and developing countries. *Int. J. Sci. Eng. Appl.*, 6, 2, 39–48, 2017.
33. Batista, M., Caiado, R.G.G., Quelhas, O.L.G., Lima, G.B.A., Leal Filho, W., Yparraguirre, I.T.R., A framework for sustainable and integrated municipal solid waste management: Barriers and critical factors to developing countries. *J. Clean. Prod.*, 312, 127516, 2021.
34. Scott, I.J., de Castro Neto, M., Pinheiro, F.L., Bringing trust and transparency to the opaque world of waste management with blockchain: a Polkadot parathread application. *Comput. Ind. Eng.*, 109347, 2023.
35. Ngan, S.L., How, B.S., Teng, S.Y., Promentilla, M.A.B., Yatim, P., Er, A.C., Lam, H.L., Prioritization of sustainability indicators for promoting the circular economy: The case of developing countries. *Renew. Sustain. Energy Rev.*, 111, 314–331, 2019.
36. Lorek, S. and Spangenberg, J.H., Sustainable consumption within a sustainable economy–beyond green growth and green economies. *J. Clean. Prod.*, 63, 33–44, 2014.
37. Schröder, P., Vergragt, P., Brown, H.S., Dendler, L., Gorenflo, N., Matus, K., Quist, J., Rupprecht, C.D., Tukker, A., Wennersten, R., Advancing sustainable consumption and production in cities – A transdisciplinary research and stakeholder engagement framework to address consumption-based emissions and impacts. *J. Clean. Prod.*, 213, 114–125, 2019.
38. Fercoq, A., Lamouri, S., Carbone, V., Lean/Green integration focused on waste reduction techniques. *J. Clean. Prod.*, 137, 567–578, 2016.

2

Sustainable Integrated Waste Management

Paridhi, Sameer and Subhalaxmi Pradhan*

Division of Chemistry, School of Basic Sciences, Galgotias University, Greater Noida, UP, India

Abstract

Sustainable Integrated Waste Management (SIWM) has always been essential to human civilizations. A society's characteristics should be taken into consideration while developing SIWM strategies. According to worldwide patterns, the systems are focused on sustainability challenges, primarily by integrating 3R (reduce, reuse, and recycle) technology. However, the kind and extent of advancements toward sustainability differ and are contingent upon a nation's economic standing. Rich nations with ample resources, such as South Korea, Japan and USA can afford to invest more in 3R technology. The majority of current initiatives center on "Zero Waste" and/or "Zero Landfilling," which is undoubtedly costly for countries with poorer fundamentals like those of India, southern African countries or southeast Asian countries like Indonesia. It is necessary to evaluate the practical requirements of SIWM systems in these economically weaker countries. Also, due to the COVID-19 pandemic–induced catastrophe, waste creation has altered internationally in almost every industry, necessitating extra care. SIWM not only offers novel approaches to address contemporary waste management problems, but it also suggests several important prospective suggestions to comprehensively reduce the likelihood of pandemics in the future. Hence, the purpose of this chapter is to evaluate pertinent organizations and policies, identify the underlying barriers to waste management in the world, and provide integrated sustainable management as a solution to the shortcomings.

Keywords: Waste, sustainability, 3R polices, covid-19, zero waste technology

**Corresponding author*: subhalaxmi.pradhan@galgotiasuniversity.edu.in

Priyanka Singh, Pooja Agarwal and V. Vivekanand (eds.) Solid Waste Management: Challenges, Sustainability and Advancements, (49–92) © 2026 Scrivener Publishing LLC

2.1 Introduction

In the contemporary world, waste management has emerged as a crucial worldwide issue, as resource depletion and environmental degradation are worsened by growing populations and patterns of consumption. The customary "take, make, dispose" linear approach is no longer practical, which leads to landfills overflowing to their full capacity, air, water, and land pollution, and the depletion of scarce natural resources. In order to tackle all the complex problems of social, economic and environmental implications related to waste management challenges, the concept of Sustainable Integrated Waste Management (SIWM) was introduced as a holistic approach. An integrated system would lessen the environmental impact and enable more efficient financing of waste management techniques from a financial standpoint. Waste management is a public service that requires outside funding and cost recovery, not an economic activity meant to generate revenue, as the World Bank emphasized. Therefore, external finance will serve as sustainable waste management's mainstay [1, 2].

In large metropolitan and urban cities, which have a fast population boom and are experiencing high economic and social growth, effective municipal solid waste (MSW) management poses a global threat. Sustainable integrated waste management mainly refers to the integration of effective planning and technological advancements covering all the sources and aspects with the emphasis on resource efficiency, in the management of waste generation by the industrialized countries to protect the environment as well as the societies and human health [2, 3]. Additionally, the physical aspects of garbage collection, disposal, recycling, and governance such as fiscal sustainability and proactive policies in developing nations are included in sustainable integrated waste management.

Additionally, waste is thought of as a "second-generation" pollution issue, one that gradually reduces to other issues. By placing stringent limits on conventional pollutants' emissions, European nations have made significant strides toward lowering emissions into the air and water throughout the 1970s. Nonetheless, environmental rules have dealt with air and water pollution issues independently due to their single-media mandate. As a result, the least managed kind of pollution and the least regulated environmental medium are now experiencing pollution difficulties. Solid waste from manufacturing processes has frequently grown as a result of the use of emission control systems. Furthermore, there has been a rise in the concentration of dangerous materials in solid waste.

Sustainable integrated waste management also prioritizes resource recovery methods such as waste-to-energy systems, in addition to recycling and composting. By converting waste materials into heat, electricity, or biofuels, waste-to-energy technologies such as anaerobic digestion and incineration with energy recovery provide a sustainable energy source while lowering the amount of garbage dumped in landfills. The traditional waste management facilities are underway which are mainly associated with incineration, which is the collection and blending of all the waste materials from the upstream sources for the energy recovery and the reduction in the volume of the municipal solid waste for it to be landfilled. Although organic wastes are converted to valuable products like compost and thermal energy through recovery centers, their operation capacities are limited. A more significant issue is the acute shortage of applications for nutrient-rich digestate (used as pond fertilizer in aquaculture and agriculture) and compost (used in agriculture). This implies that even if it takes energy for the conversion of all these products, a large number of these by-products will wind up in the landfill somehow. To increase the chances for employing recycled goods, such as feeds, fertilizer, fuel, etc., we should revitalize local agricultural and aquaculture operations, with an emphasis on the reuse and recycling of food waste [72].

Sustainable integrated waste management gives recycling and composting first priority when it comes to trash generation in order to keep materials out of landfills and incinerators. Recycling is the process of turning waste materials into new goods, which uses less energy and conserves natural resources rather than making things from raw materials. Contrarily, composting is the process of breaking down organic waste into a nutrient-rich material that may be added to soil to improve its quality and boost agricultural output [4].

Likewise, sustainable intertwined waste operation promotes the relinquishment of ingenious technologies and practices for managing dangerous non-recyclable waste aqueducts. This includes styles similar to chemical treatment, thermal transformation, and tip mining, which aim to minimize the environmental and health pitfalls associated with certain manners of waste while maximizing resource reclamation.

A key component of sustainable integrated waste management is its acknowledgement of the value of community involvement and engagement in decision-making. Involving stakeholders at all levels, government departments, corporations, nonprofits, and private citizens, it promotes the adoption of sustainable practices and a sense of shared accountability for waste management [5].

Global social and economic institutions are experiencing an unprecedented level of turmoil due to the COVID-19 pandemic. Every industry has undergone transformation, be it the production and distribution of hygiene goods, the implementation of social distancing programs, or the restructuring of the public health system. Numerous direct and indirect environmental effects of COVID-19 have also been reported. However, there are reasons to be concerned since unfavorable consequences linked to a rise in solid waste generation and a decline in recycling initiatives may have medium- or long-term implications caused by the pandemic.

Municipal solid waste (MSW) management has emerged as one of the most difficult environmental issues throughout this epidemic. This widespread problem has also altered the ways of waste generation and has been an issue for the sanitary workers and the policymakers as well, for the proper disposal of all the waste generated. In many developing nations, the management of trash is not sustainable, which makes them more vulnerable to the spread of COVID-19. Which also leaves a chance for the contamination of the normal waste due to the improper collection and disposal of the biologically contaminated waste, which in turn increases the risk of transmission of the virus. Thus, the proper handling and the disposal of this waste is a critical step in the effective treatment during the pandemic [6, 7].

To put it briefly, the COVID-19 epidemic has put a strain on supply chain management (SWM) worldwide. This is due to the intricacies of the supply chain bottleneck in the production, distribution, use, and disposal of personal protective equipment. The main risk of spreading COVID-19 in developing and metropolitan cities is improper waste management and a lack of personal protective equipment (PPE) kits. Improper solid waste management puts the lives of the informal worker community, i.e., the waste workers, at a higher risk, and they in turn can be the carriers of the virus and spread it into the community due to a lack of knowledge. The entire community would be at increased risk as a result of inappropriate treatment of coronavirus-infected household waste generated by patients getting care at home, municipal staff, and other formal workers. They carry out informal recycling, garbage collection, sorting, and classification for the general public. Therefore, in order to control risk and waste uprightly, the current study examines how MSW operations are managed in industrialized and developing nations during a pandemic [8, 9].

The sustainable integrated waste management strategy acknowledges that waste management cannot be seen in isolation; rather, it must be included in more general sustainable development objectives, such as addressing climate change, protecting biodiversity, and promoting social justice. To provide creative solutions suited to regional settings and difficulties, it encourages cooperation among stakeholders across sectors and scales, including governments, businesses, communities, and academic institutions. It takes a comprehensive and integrated approach, providing a way forward for a more resilient and sustainable future in which garbage is seen not as a burden but as a precious resource that has to be managed carefully for the benefit of both present and future generations. The current chapter examines the complex aspects of SIWM and the tactics required to move toward a circular economy that protects the environment and advances human well-being.

2.2 Sustainable Waste Management in Practices

Sustainable waste management means reducing the amount of trash produced, optimizing recycling and reuse, and making sure that garbage that cannot be avoided is disposed of properly and sustainably.

Reducing greenhouse gas emissions, avoiding landfill expenses, encouraging resource repurposing, generating revenue for waste management operators, lowering health risks associated with haphazard scavenging, and generally enhancing environmental effects and public awareness are all advantages of sustainable waste management. These are a few advantages that come with using sustainable waste management techniques. Communities may lessen the demand for raw material extraction and save natural resources by putting into practice sustainable waste management techniques. These techniques can also aid in the growth of a circular economy, which minimizes waste production and maximizes resource efficiency by continually reusing items. Reducing the quantity of trash produced in the first place is one of the main objectives of sustainable waste management. This may be accomplished by taking steps like redesigning products, encouraging the usage of reusable and durable items, and motivating customers to make environmentally friendly purchases. Furthermore, educating people about garbage sorting and segregation as well as promoting the value of waste reduction may assist individuals and communities in making wise decisions. It can be practiced as the three Rs, as described below.

2.2.1 The Three Rs – Reduce, Reuse and Recycle

2.2.1.1 Reduce

Reducing trash creation at its source is the first step in waste management. This entails consuming as little as possible, selecting goods with little packaging, and implementing habits that reduce waste production in general. Reducing waste output results in less energy being used in waste disposal procedures and less resources being used in manufacturing. This has a direct impact on sustainability and environmental protection.

The reduction in waste production starts with every individual person. If each and every household contributes to the cause or takes the initiative, the problem might be controlled. For waste reduction, consumers can take these precautionary measures:

1. Buying only what a person requires and not what they want: People tend to upgrade their electronic equipment to the latest model even though the current one is in functional condition. This not only increases the carbon footprint but also causes excess and irrelevant consumption of raw materials and sources.
2. Buying high-quality materials: Even though buying goods made from higher quality material is a bit expensive they tend to last longer and are more sustainable. Using disposable goods of low quality not only causes excess consumption of raw materials but also leads to production of excess solid waste.
3. Use of minimum packaging: Packaging materials like polythene bags, plastic wrappers, cardboard boxes and Styrofoam peanuts all end up in a landfill, and being non-biodegradable in nature they are very hard to dispose of. It is advisable to use jute bags or cloth bags for more sustainability.
4. Buying from local markets or vendors: Buying from large cooperative companies not only feeds into the stream of consumerism but also causes unwanted waste production and consumption of excess raw materials. Buying from locals not only sustains raw materials but also reduces material consumption during the transportation of goods [10–12].

2.2.1.2 Reuse

Another efficient strategy to reduce waste is to reuse products. Products can be reused numerous times before being disposed of, as opposed to being thrown away after a single usage. To increase the lifespan of products, this may entail fixing, remodeling, or reusing them. Reuse saves resources by reducing the amount of garbage that is dumped in landfills and by postponing the need to create new items.

Reusing begins with the idea that the things we consume on a daily basis may be turned into resources instead of waste. After all, one's definition of waste is subjective. What's trash to one is treasure to another. When we take a close look at the items we discard, we might start to recognize them as resources that can be utilized again to meet requirements and find solutions to issues in daily life. However, the majority of us haven't even started to utilize the riches found in our waste. Once you've decided to recycle garbage, you may start coming up with ideas and organizing a brainstorm. Reusing meets the human need for creativity while saving money and conserving resources. Reuse of materials and goods can be done through some of these examples:

1. Containers or packaging boxes can be reused at home for storage or for the use of school projects.
2. Outgrown clothes or footwear can be donated to the less fortunate.
3. Reusing plastic shopping bags for storage or for carrying belongings.
4. Old bed sheets or towels can be cut into small pieces and used as cleaning cloth or dusters.
5. Old furniture can be repolished and maintained to extend its usability lifespan or can be donated to charities or homeless shelters.
6. Broken appliances can be donated to vocational centers or schools for the pupil to practice repairing them for their future job profiles.
7. Old books, magazines and newspapers can be donated to the local library.
8. Packing materials like cardboard boxes, plastic wrappers, Styrofoam peanuts, and silica gel packets can be stored and reused for storing household materials and future packing.
9. Microwaveable containers provided by food takeout companies can be washed and reused at home for storing food.
10. Old tires can also be used in gardens as a low fencing or for a makeshift swing.

2.2.1.3 Recycle

Recycling is the process of gathering garbage, processing it, and making new things out of it. It is an essential part of waste management since it contributes to energy conservation, resource preservation, and a reduction in pollution from raw material extraction and manufacture. Recycling helps create a sustainable economy where resources are used more responsibly and effectively by keeping items out of landfills and reusing them in the manufacturing cycle.

The main benefits of recycling are:

1. It generates another working industry: For all the recycled goods collected, the industries would have to come up with plans for their uses and application. As more recycling plants are established, more products can be made to meet the public requirement.
2. It creates more job opportunities: Authorities have established collaborative relationships with handicapped workshop operators, managed job-training partnerships, or found alternative ways to employ jobless workers through recycling initiatives.
3. It causes cost avoidance: The notion that recycling ought to be profitable has hindered it for a long time. For certain recyclables, it could be the case, but not for others. Recycling, on the other hand, needs to be considered an affordable disposal solution. Generally speaking, it needs less government funding than incineration or landfilling. It preserves natural resources and aids in environmental protection.

The collection of recyclable goods is also done through various methods and plans for the sustainable waste management processes and their effective implementation:

1. It is mandatory for households to segregate recyclables from their trash for curbside pickup. Garbage is to be deposited in regular bins, and clean recyclables may need to be placed in specific containers. Both are set out for pickup by different trucks at the curb.
2. People can deposit their old glass, metal, plastic, and paper at a designated recycling drop-off location. Drop-off sites are one of the simplest ways to gather recyclable items. Usually situated close to a busy area, including parking lots and supermarket entrances, these locations offer easy access.
3. Aluminum and other metals, glass, plastic, paper, and occasionally batteries and other commodities are purchased by

buyback centers which are mainly run by large manufacturing or recycling industries [10–12].

2.3 Composting

The process of biologically breaking down and stabilizing organic substrates in a way that promotes the development of thermophilic temperatures due to heat produced by the organism is known as composting. The end product is sufficiently stable to be used and stored on land without causing harm to the environment. Protozoa, fungus, and bacteria are examples of microorganisms that participate in the biological breakdown process [18]. Mineralization and humification are the two biological reaction processes that make up decomposition.

2.3.1 Mineralization

As a result of microbes' metabolic processes, easily fermentable organic materials like sugars and amino acids are broken down during mineralization, releasing water, carbon dioxide, and heat. Mineralization mainly results in the formation of organic residuals which are partially stabilized. The biochemical reactions taking place in the mineralization process are:

- Carbohydrates —> Sugars —> Organic acids —> CO_2 + H_2O + Organic Matter
- Proteins —> Peptides —> Amino Acids —> NH_4 compounds —> N_2 or ammonia

The mineralization process mainly occurs through the following steps or processes:

1. Initial decomposition: The breakdown of organic materials by bacteria, fungus, and actinomycetes is the first step in the composting process. The organic components are broken down into simpler chemicals by these microbes when they feed on them.
2. Mineralization of organic matter: In this phase, nutrients are released when organic compounds break down even further into simpler forms. As a result of this breakdown, organic nitrogen found in proteins and other organic compounds becomes ammonium (NH_4^+), a kind of inorganic nitrogen.

Likewise, organic sulfur turns into sulfate (SO_4^{-2}) and organic phosphorus turns into inorganic phosphate (PO_4^{-3}).

3. Microbial action: Numerous microbes are essential to the process of mineralization. For example, ammonification—the process by which organic nitrogen is changed into ammonium—is carried out by bacteria. The ammonium is then further converted by other microorganisms, such as nitrifying bacteria, producing nitrate (NO_3^-), a type of nitrogen that plants may easily absorb.
4. Maintenance of optimum temperature and humidity: The compost pile's moisture content and temperature are two variables that affect how quickly minerals are mineralized. The mineralization process proceeds more quickly in ideal microbial activity conditions, which are normally found between 120°F and 160°F (49°C and 71°C). Sufficient moisture levels are necessary for the growth and efficient breakdown of microorganisms.
5. The carbon-to-nitrogen (C/N) ratio in the compost pile: It's important to consider the composting materials' carbon-to-nitrogen ratio. Optimizing the C/N ratio to a ratio of 25:1 to 30:1 guarantees effective breakdown and mineralization. Mineralization may be slowed down if nitrogen becomes limited due to an excessive carbon content (high C/N ratio).
6. Duration of the mineralization process: Over the course of the composting process, mineralization occurs. The process of compost maturation and stabilization persists long after the active composting stage has ended [13, 14].

2.3.2 Humification

The conversion of organic matter into humus, a stable, black, nutrient-rich substance that is good for soil health, is referred to as the "humification" process in composting. As organic materials break down in the compost pile, microbes produce humus.

The humification process for the compost takes place in the following steps:

1. Initial decomposition: The compost pile is broken down into simpler compounds by microbes like bacteria, fungus, and

actinomycetes when organic materials like kitchen scraps, yard trash, or manure are added.

2. Chemical reactions occurring: Complex organic molecules undergo a variety of chemical processes, such as oxidation, hydrolysis, and acid-base interactions, which lead to their breakdown into simpler compounds during decomposition. As a result, nutrients including potassium, phosphorus, and nitrogen are released.
3. Action of microorganisms: The bacteria, fungi and microbes breakdown all the organic matter by producing enzymes which help in the formation of the humus, these are always engaged in breaking down the matter which further enriches the soil or compost with nutrients.
4. Stabilization of the soil structure: Humus enhances soil structure by strengthening the soil's capacity to hold onto moisture and nutrients and encouraging aeration and drainage. Additionally, humus enhances soil fertility by binding soil particles to form aggregates.

2.4 Waste Sorting and Division

Sorting and dividing waste into different categories in order to make recycling and proper disposal easier is known as waste segregation. Reducing general trash output, identifying goods that can be reused, and putting away items that can be recycled are all made easier with waste sorting. Furthermore, though, there is a moral obligation to take responsibility for the garbage that is produced and how it is handled.

The generated waste will wind up intermingled in landfills just as it was mixed in your bins if adequate segregation isn't followed. Food scraps, paper, and liquid waste are among the waste materials that can combine and break down to release hazardous gases into the sky and runoff into the soil.

Waste segregation, when carried out correctly, may lower the amount of garbage that needs to be managed, improving the efficiency of collection and disposal in the end. Furthermore, separating garbage at the source can facilitate processing and handling, improve the possibility of recovering resources, encourage recycling and reuse, and save operating expenses [15].

Based on its biological, physical, and chemical characteristics, waste is identified and categorized into the following groups for the purpose of waste segregation at the source:

1. Dry waste: It mainly refers to everything that is not regarded as filthy or damp. Materials that are not recyclable as well as recyclable ones are included. Bottles, cans, clothes, plastic, wood, glass, metals, and paper are examples of dry waste materials.
2. Wet waste: It includes all organic materials, such as food items, soiled food wrappers, personal hygiene items, yard waste, tissues, and paper towels, as well as any other contaminated material that could contaminate recycling.
3. Sanitary waste: It refers to any waste, whether liquid or solid, that comes exclusively from people and their actions. Items from medical waste may also be included.
4. Hazardous waste: It refers to all materials, goods, and by-products that are flammable, poisonous, corrosive, or reactive.
5. E-waste: It refers to garbage that is electronic in nature.
6. Inert waste: It refers to waste materials that neither readily degrade nor are reactive to chemicals or biological agents [16, 17].

Promotion of recycling and waste segregation techniques necessitates appropriate rules and regulations as well as strong enforcement of them. Scholarly recommendations for policies that incorporate financial incentives for recycling and reuse would also be appropriate for this goal. As a result, policy evaluations are the first step in implementing changes to the management cycle; these include choices for formalizing and promoting recycling practices as well as offering financial incentives. Some of the key components of an efficient waste management strategy include formalization, appropriate coordination of all parties at all levels, and curriculums that are specifically designed to improve the quality of education [18].

2.5 Extended Producer Responsibility (EPR)

Extended producer responsibility, or EPR, is a strategy for environmental protection that aims to reduce the overall environmental impact of a product and its packaging by making sure that the product's producers are accountable for every step of the product's life cycle, including packaging disposal, recycling, and take-back. EPR is becoming more widely acknowledged as an effective waste management strategy to enhance recycling and decrease product and material landfilling [19].

The producer, who chooses designs and marketing strategies, is primarily accountable for EPR. In general, producers offer incentives to encourage environmentally friendly product design, stop waste at its source, and support management objectives for material recycling. Benefits include waste reduction, improving environmentally friendly product design, and assisting in the accomplishment of public recycling and materials management objectives. In theory, this idea guarantees efficient end-of-life collection, environmentally friendly handling of collected products, and enhanced reuse and recycling. In actuality, the final user has some influence over the collection. EPR can be implemented as a recycling, buyback, or reuse program. The producer may alternatively opt to assign this duty to a third party, a so-called producer responsibility organization that receives payment from the producer for the management of old products [20, 21].

It integrates the following two ideas:

1. Effective product down sampling to encourage reuse, recycling, and collection for social, environmental, and commercial reasons.
2. Enhanced designs for products and product systems.

When executed efficiently, Extended Producer Responsibility (EPR) initiatives can yield several advantages and prospects, such as heightened rates of collection and recycling, decreased public expenditure on waste management, decreased total expenses associated with waste management, and environmentally friendly product design that enhances durability and reusability [21].

2.6 Community Participation and Education

To increase the community's ability to continue providing the services it has established during the project after the facilitating organizations have left, community members can be involved in the process of community participation at various points and levels of intensity during the project cycle. By involving the community at every stage of the project, from planning and execution to assessment and maintenance, communities are encouraged to maintain and carry out project operations long after the project is finished, and their priorities and needs are considered in the project's activities. In addition, community involvement can boost capacities within the community and promote cost-sharing among project participants [22].

The possible advantages from the community participation for the betterment of the project are:

- Enhancing the efficacy and design of the project. When the community participates in the project's design, it becomes easier to incorporate its requirements and limitations into the project's goals and achieve a more successful implementation.
- Improvement of the projects' sustainability and impact. Including the community in the project could boost local ownership and a sense of accountability for upkeep of the services the project provides. Both elements are necessary for projects to be long-lasting and ongoing.
- Enhancing the effectiveness of the project. The agreement and comprehension of cost sharing (financial and physical contributions) can be improved by community involvement. In addition, community involvement can be used to defuse tensions and promote mutual understanding and collaboration amongst various parties. This will minimize total expenditures and cut down on project execution delays.

Community participation in properly disposing of waste at the appropriate time and place on the street is a minimal requirement for the continued operation of any solid waste management systems. Residents have personal responsibility as users. This entails doing things like organizing waste properly in a bag or bin, separating organic or recyclable materials from other waste, placing rubbish where it belongs at the appropriate time for collection, and tidying up the environment around the residence. In addition to personal accountability, groups can exercise collective responsibility through meetings, cleanup initiatives, and awareness-raising events, among other roughly planned actions. Moreover, community involvement might take the form of material, monetary, or physical contributions to solid waste management operations, such as operating a cart or sweeper and paying garbage collection fees.

Other parties participating in solid waste management, including the municipality, microenterprises, community-based organizations (CBOs), and local leaders, are crucial to the achievement of community participation in this area. For example, the community cannot benefit from garbage separation if the municipality does not collect the waste separately. The following elements are thought to support the long-term viability of community involvement and, consequently, of services like garbage collection and separation:

1. For people in a community to have a comprehensive grasp of solid waste issues and for stakeholders to be responsive to community needs, communication methods are necessary.
2. Community-based organizations (CBOs) and representative local leaders have the power to promote community involvement and guarantee that demands are met.
3. If local government does not initially acknowledge and endorse community initiatives and CBOs, their longevity may be compromised.
4. Adherence to written agreements with stakeholders for the upkeep and management of the service system by the CBO and the local government.
5. Sufficient funding and operational feasibility to reduce reliance on outside assistance for community services.
6. After-project assistance to uphold awareness and new procedures and offer help when needed with the management and operation of newly formed organizations [23, 24].

Local authorities are crucial in encouraging community involvement. Community leaders are responsible for encouraging individuals to sign up for waste collection, ensuring that people pay the costs, encouraging waste separation, and keeping an eye on the service level's performance. They should also supervise the operations of local government agencies and private businesses, negotiate on behalf of the former, and exert pressure on the latter to provide services [25, 26].

Even though the actual tasks aren't always obvious, community members' voluntary involvement in neighborhood activities should be seen as a form of civic duty. It represents their commitment to one or more phases of a group project, such as control, awareness-raising, information sharing, promotion, and decision-making. The most successful waste collection and cleanup initiatives occur when locals actively participate in spreading awareness, keeping an eye on the service, and/or gaining true control over the content and social/sanitary scope of the initiatives [26, 27].

2.7 Barriers in Sustainable Waste Management

Sustainable waste management's (SWM) contribution to bettering sustainable development is now clear. Nevertheless, SWM is dealing with management issues that result in the depletion of resources and constrained approaches to waste management for emergencies and response maintenance. SWM is a complicated topic with several technological, social, political, and ecological issues. The challenges for the SWM, which include waste generation due to overpopulation, unplanned and rapid urbanization, social control, limited technical experience, incorrect legislation, and a lack of funding for ultimate safety handling, must be addressed by the workable system. Therefore, to address these issues as they contribute to unsustainability, sustainable solid waste management (SSWM) must reevaluate them and establish strategies and mindsets that view these issues as important portfolio items. Thus, there are still obstacles in the way of comprehending these issues, coming up with alternatives to unsustainable methods, and incorporating action plans that support waste management. The makers are the first to do it. Recycling would be much easier with a few small manufacturing-level adjustments.

A simple example is soda bottles. Different polymers are used by different companies. Automation of recycling would be simple if the beverage industry adopted a code of conduct dictating that all soda bottles were made of the same plastic. The neck rings ought to break so that they stay attached to the cap. It's simple to take off the cap. The rings are not made of the same plastic. Labels are challenging to take off. They could apply regular glue that is meant to be removed easily. Identifying these barriers is a critical first step in enhancing the effects of waste management and putting alternate solutions into place [28].

2.7.1 Improper Education and Lack of Awareness

One of the most important instruments for raising awareness among the populace, especially in developing nations, is education. Compared to younger generations, older people segregate their waste more. Therefore, in developing nations, environmental education is crucial to closing the knowledge gap between the young and the old regarding waste management, trash sustainability, and waste segregation. According to UNESCO, environmental education can raise people's knowledge of the environment and the issues that surround it. It cultivates the knowledge and abilities required to address environmental issues and encourages attitudes, drives, and commitments to make choices and act responsibly. Growing environmental knowledge increases people's awareness of environmental issues and may inspire them to take action to safeguard the environment. In most developing nations, environmental education has been part of the curriculum in recent years, although teachers' actual knowledge of the subject is still quite limited. Because of these inadequacies, most developing countries now have weaker standards for managing trash for sustainability [28, 29].

A teacher's specialization, qualifications, type of institution attended, and prior teaching experience all influence how attentive and knowledgeable they are about waste management and environmental issues. This is especially true in developing nations where individual teaching experience is highly valued. In addition to investing in technology to secure long-term advantages for the environment, sustainable development, and CP in developing nations, communities, instructors, and students can all have a beneficial influence on the cost of trash operations. Most emerging nations have witnessed environmental destruction because of a lack of practical environmental expertise. Most educators in developing nations lack environmental or SWM-related practical knowledge, which has a detrimental impact on students' attitudes and environments of knowledge. In most developing nations, the understanding and awareness of SWM among students is crucial for waste sustainability. Research has indicated that early life experiences have an impact on children's cognitive and emotional development as well as the formation of long-lasting environmental attitudes and behaviors [30].

Research has indicated that children's perceptions toward the environment begin to take shape approximately when they turn twelve. For this reason, environmental education should instill in kids a respect for the environment and its resources and help them develop a good attitude toward SWM. In developing nations, the difference in waste management practices between young and old can be minimized by adopting a favorable attitude toward environmental education. Routine tasks like sweeping, cutting the grass, providing food, and finishing classes all result in waste being produced in schools. As part of the educational process, it is anticipated that waste management activities in schools will help students acquire positive attitudes toward garbage management in developing nations.

2.7.2 Financial Factors

Sustainable waste management also requires funds or financial aspects for its efficient and effective implementation. The annual funding needed for effective SWM shows that, when the full cost pricing approach is used, the total annual payment required is frequently two or three times the current allocation (Figure 2.1). Municipalities manage a lot of other things; thus, they only devote 10% of their budget to SWM. The percentage of expenditures for various solid waste services shows that the expenditure is not appropriate and well-distributed. Areas of priority are not decided, and extra investments are made where it may not be required.

The state government must provide the municipal agency with financial support. The general budget allots funds for SWM-related initiatives. These funds are frequently insufficient because SWM is given a low priority. As a result, municipalities typically use an incremental budgeting method based on the budget from the previous year whenever additional spending is required for specific renovations or as a new capital expenditure. The projections and the SWM services should be included in the financial budget passed every year by the government for the proper funding of the program, such that sufficient and appropriate attention can be given to the SWM programs, before dividing up the budget among various facilities [31, 32].

Waste generators' taxes or surplus can cover a portion of the cost of SWM services. A family's size, the building's location and value, and the occupant's income can all be taken into consideration. The tax rate can vary according to the variation in the amount of waste generated by different

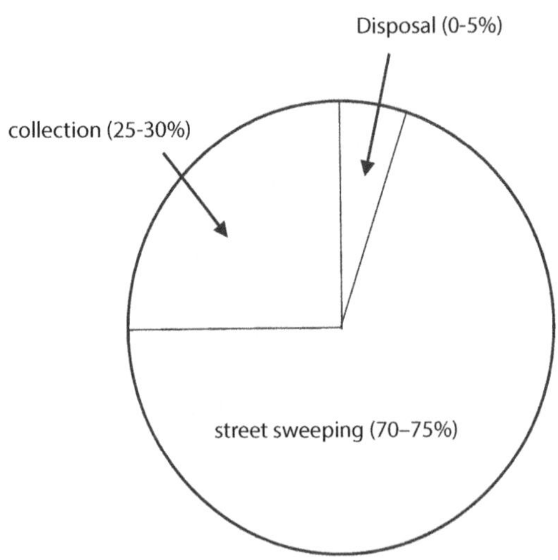

Figure 2.1 Financial distribution of the funds for various services.

commercial establishments. It is also possible to reduce SWM service costs by increasing community, nongovernmental organizations (NGOs), and business sector support.

2.7.3 Inadequate or Weak Infrastructure

Sustainable waste management can be hampered by the lack of facilities for collection of the waste, separation or segregation of different types of waste, recycling and proper disposal of the waste. The lack of suitable waste-to-energy plants, recycling hubs, composting facilities, and landfills is one example of this.

This problem mainly arises due to the improper storage or mixing of the various types of waste generated at the source. This problem mainly arises due to the lack of public awareness, motivation and education about the basic protocols about waste storage or disposal. Bad ethics and habits of the public like littering are also a main cause adding to this problem. Also, the lack of cooperation from the household and commerce adds to the problem. The lack of stringent panel provision as well as the hold on the power to levy fines is also very feeble and not significant. The problem also persists due to the resistance in the attitude of the people towards the disposal schemes and not bringing about a change.

Also, the absence of widespread coverage in print and electronic media about the methods of segregation of waste is the major barrier, as even if people wanted to dispose of their waste, they are not educated enough about the disposal processes. Despite multiple government programs and awareness campaigns to encourage garbage segregation, this problem still exists. Public education campaigns regarding the significance of waste segregation for recycling, composting, and appropriate disposal have frequently failed. Moreover, residents' motivation for appropriate waste segregation is still low because of perceived inconvenience, a lack of obvious, immediate benefits, and a general lack of awareness of how their actions affect the environment. Many people still ignore the significance of trash segregation in favor of the simplest and most convenient disposal techniques because they lack a strong feeling of personal accountability and an awareness of the long-term effects of incorrect garbage disposal. A diverse strategy is needed to address this issue, including broad education and awareness campaigns catered to various demographics, upgraded community-level facilities for disposal and segregation, and rewards and penalties to promote adherence to waste management laws. Instilling a sense of shared responsibility and inspiring residents to actively participate in sustainable waste management methods are other benefits of cultivating an environmental stewardship culture through community involvement and participation [72].

2.7.4 Technological Limitations to SWM

The most recent and affordable technologies are not used in the current SWM system. Therefore, integrating cutting-edge technology into different SWM levels, such as the utilization of Geographic Information Systems (GIS), is one example of technological factors. A conceptual framework for improving the technical features of the current SWM situation has been put forth. It includes the use of GIS for managing the entire data on trash generation, managing open dumps, and choosing disposal sites for municipal solid waste (MSW). The conceptual framework suggests that the combination of a computer model and a GIS platform can lead to the development of a user interface. Site selection, open dump management, and waste generation management are the SWM issues that are handled here for management using GIS technologies.

Tackling waste generation problem: The city's ward-by-ward waste generation data, which includes recyclable and biodegradable materials as well as inert waste and other pertinent information, can be stored in a GIS geodatabase and retrieved when making decisions about, say, the design of composting and recycling facilities or other SWM-related matters.

Management of open garbage dumpsters: The locations of different dump sites can be digitally represented using a GIS, and each dump site's attributes, including the type of waste that ends up there, its area, water quality, water table depth, geotechnical characteristics of the contaminated soil, and an analysis of different contaminants and their concentrations, can be provided. This makes it possible to identify the location with the highest risk and take appropriate action, such as landfill reclamation and engineered closure.

Selection of site for the garbage dumping grounds: It deals with the capability of eliminating environmentally inappropriate sites and reducing the number of sites that require additional evaluation using GIS-based constraint mapping. As per the site selection criteria established by CPCB, MoEF, and the Indian government, many constraints can be taken into consideration. Finding candidate sites consequently allows for the creation of multiple thematic layers of the candidate sites, which can then be overlaid using weighted overlay techniques in GIS to select the best site [27].

2.7.5 Impact of Globalization and Rapid Urbanization

Globalization refers to the swift expansion of the capitalist market, encompassing trade and private investment, across the globe. It encompasses the interchange of resources, various cultural practices, and the strengthening of ties between nations and regions. Industrialization has been a worldwide process since the colonial era, which began in the late sixteenth century, where resources, including labor, in one part of the world have been used for the advantage of those in other parts. Since the founding of the first city-states with their class structures and market economies, at least 5,000 years ago, this process has been ongoing in various ways. The current phase of this relationship, which has been increasingly linking rich areas and groups to poor ones over the past 500 years, with rich areas being characterized by overconsumption and poor areas by underconsumption, is marked by the rise of contemporary consumer society, which is focused on the use of materials produced for a mass market [33, 34].

Globalization poses a significant obstacle to the sustainable management of trash over the long run, and proper waste management is a prerequisite for sustainable globalization. There hasn't been any systematic research done on the connection between solid waste management practices on a local and global level and globalization, the dominating global trend. The past few decades have seen an increase in international trade that has improved living standards and purchasing power while reducing poverty in many emerging nations. Simultaneously, it has drastically altered the global waste management landscape. Products were primarily made close to where they were consumed, and garbage was managed close to where it was produced until a few decades ago. These days, a huge number of mass-produced consumer goods are shipped all over the world. Solid waste management is becoming more and more intertwined with resource management, resulting in a complex worldwide network of material and recyclable waste flows that impact many facets of daily life and the environment.

The number, volume, and variety of waste streams are growing, and many nations—particularly those in the developing world—are witnessing severe environmental and urban deterioration. Growing trash generation affects more than just waste management. It is a sign of insufficient production and consumption techniques used globally. Recycled materials, energy, and other valuables are lost when waste is generated, adding to the environmental risk. Furthermore, the effects of resource extraction frequently manifest themselves far from the sites of production and consumption [34].

Enhancing solid waste management is among the top three priorities for many cities, but it doesn't seem that foreign development assistance for SWM is being allocated in a way that reflects that evaluation. The comparatively low level of investment may be caused, at least in part, by earlier issues with governments' and localities' preparedness to meet loan funding requirements for SWM. Consequently, there has been a notable emphasis on grant money for technical support aimed at developing skills and capacity and addressing governance issues, which are crucial components of a sustainable and integrated waste management system.

Megacities were formerly primarily regional centers, but they are gradually evolving into global hubs that serve as interfaces between their local communities and the world's trade, population, and ideas. Megacities have more extensive waste management systems than smaller ones. The availability of commodities and resources is influenced by global trade, and how they are transformed into waste and recyclables depends on local waste management techniques. It is extremely difficult to establish, run, and maintain efficient waste management systems in megacities. It is unrealistic to expect megacities' waste management systems to be straightforward because they are intricate systems. Consequently, it is critical to comprehend the dynamics and features of megacities, as well as how they will impact and interact with the waste management system. The implementation of any solution must adhere to the integrated sustainable waste management approach, considering both the system's physical components and governance concerns. There is no "one size fits all" solution that can be used internationally. Although learning from bad experiences is far simpler, it is challenging to draw broad lessons from the strategies used in many megacities. However, a lot of developing countries' rising megacities are also centers of innovation, and their local markets feature non-industrial but promising methods of managing resources and trash [33–35].

2.7.6 Constraints Due to Cultural Practices

Waste management is significantly influenced by culture, which affects how the general population feels, makes decisions, and participates in the creation and management of garbage in metropolitan areas. The societal norms, convictions, and mindsets of urban dwellers influence their inclination towards non-biodegradable packaging and behaviors that lead to the production of solid waste. In a similar vein, First Nations communities in Canada's municipal solid waste management practices are shaped by cultural elements such as disposing of trash, conserving the environment, and feeling a connection to the land. The way that various communities respond to laws pertaining to the management of plastic garbage is also influenced by their legal cultures. In order to encourage appropriate waste management techniques, cultural practices may also be targeted for intervention. One way to do this is by emphasizing long-term behavior change in cultural designs. Creating sustainable regulations and encouraging environmentally friendly behavior require an understanding of the local waste management culture.

The collection system in certain areas is influenced by long-standing customs that forbid garbage collection from entering residential areas. In some places, where caste customs are still in place, workers for trash management and street cleaning must be recruited from specific demographic groups, and some people will not allow storage bins to be placed close to their homes. Most of the time, community social norms take precedence over what many would view as sensible fixes. Planning, designing, and operating waste management practices should thus take these local lifestyle trends into account [72].

2.7.7 Weather Patterns and Seasonal Differences

Certain areas in the far north experience extremely low temperatures for a significant portion of the year. Trenches must be excavated in the summer to conceal waste that is stored for winter usage since drifting snow and frozen ground obstruct landfill operations in cold areas. In contrast, tropical regions have abrupt fluctuations in the length of the wet and dry seasons. As a result, solid waste's moisture content varies significantly, going from less than 50% during the dry season to more than 65% during the rainy months. During the wet months, it might be difficult to collect and dispose of garbage. Solid waste disintegrates significantly more quickly in warm weather than it does in cooler areas due to high temperatures and humidity. Therefore, the frequency of garbage collection in warm, humid regions ought to be higher than in cold ones. Because of the low rainfall and limited leachate generation from sanitary landfills, there is little change in the moisture content of wastes in sub-tropical or desert climates. However, landfill sites have unique challenges from strong winds and wind-blown sand and dust. Groundwater can be impacted by landfill sites because they change the thermal characteristics of the soil, but temperature inversions can retain airborne contaminants close to the surface.

2.8 Solid Waste Management

The collection, handling, and proper disposal of garbage in an ecologically sound and safe way is known as solid waste management. To lessen the negative effects of trash on the environment and public health, waste management plans must be planned, coordinated, and put into action. It is a major global issue since it pollutes both the air and water. It demonstrates its immediate impact on economic expansion, environmental deterioration, and public health. It may result in environmental contamination and the spread of diseases carried by rodents and insects, known as vector-borne illnesses.

From municipal trash to industrial waste, solid waste is any non-liquid, non-soluble material that occasionally contains dangerous or complicated materials. A variety of waste types are included in it, such as biomedical waste, e-waste, commercial, institutional, catering, and market garbage. Many emerging cities have many tons of uncollected rubbish on their streets every day. It serves as a haven for pests that propagate illness, clog drains, and result in various problems with the infrastructure. India generates 277.1 million tons of solid garbage annually, but because of "rapid urbanization, population growth, and economic development," this amount is expected to increase to 387.8 million tons in 2030 and 543.3 million tons by 2050 [36, 37].

Solid waste originates from several sources such as commercial, industrial, institutional, and residential activities. Hazardous waste is defined as certain kinds of waste that represent an immediate risk to people or surroundings they encounter. Municipal solid waste (MSW), often known as refuse, is any nonhazardous solid waste generated in a municipality that must be collected and transported to a processing or disposal location. Rubbish and trash are examples of refuse. Rubbish is primarily dry materials like glass, paper, fabric, or wood; garbage is primarily decomposable food waste. Rubbish does not decompose or form putrescine to the same extent as garbage. Bulky things like old refrigerators, sofas, or massive tree stumps are considered trash. Garbage needs to be collected and handled specifically. Despite not being regarded as part of the MSW stream, construction and demolition (C&D) trash, also known as debris, makes up a sizeable portion of the overall amount of solid waste produced in the United States (about 20%). However, municipal sanitary landfills are typically used to dispose of C&D trash since it is nonhazardous and inert. Electronic garbage, or "e-waste," is another category of solid waste that is possibly expanding the quickest in many industrialized nations. It consists of abandoned computer equipment, televisions, phones, and other electronic gadgets. There is growing concern about this kind of waste. Among the elements of concern found in electronic equipment are lead, mercury, and cadmium. Government regulations may be necessary to control the recycling and disposal of these items [35, 38].

First and foremost, the reduction of waste is the solution. Waste recovery and energy recovery, together with waste remanufacturing and recycling into useful goods, should be the second choice in situations when waste cannot be avoided. Large resource savings result from recycling. As an illustration, 17 trees and 50% of the water may be saved for each ton of recycled paper. Recycling also produces jobs; in Brazil, China, and the United States alone, the industry employs 12 million people.

2.8.1 Organic Waste

Organic wastes are materials that come from living things, such as microbes, plants, and animals, and they can decompose into simpler organic molecules through biodegradation. Organic waste can be liquid or solid depending on how it was created in nature. Industrial goods, residential activities, and agriculture are the main producers of organic waste. Examples of biodegradable or organic wastes are green wastes, which include food wastes, food-soiled paper, non-hazardous wood waste, landscaping wastes, and pruning wastes. Even though most organic wastes enrich soil with nutrients and minerals that support plant development and soil fertility, improper disposal techniques may seriously harm the ecosystem. The notion of managing and recycling organic waste has, nonetheless, recently been presented and put into practice.

The process of managing organic waste involves recycling it or turning it into something usable using various recycling techniques. This is known as organic waste recycling. Since waste management has become a major concern in most large cities, there has been a growing demand for organic waste recycling. Most of the garbage produced in nature is organic, and because of its high moisture content, this waste has an immediate impact on urban living systems. Due to the high moisture content, waste volume increases and incinerator temperatures decrease, resulting in an overall waste disposal burden. Many approaches and procedures have been developed and put into use globally to address these problems. Enhancing soil fertility while getting rid of organic waste is also possible with the use of microbes in waste management. The organic waste recycling process involves exposing the waste to various treatments, which ultimately turn it into compost or vermicompost, which may be used as organic fertilizers.

One of the easiest and most efficient ways to manage organic waste is through biological treatments. These procedures maximize waste component recovery and recycling. The fundamental goal of recycling organic waste is to preserve a sustainable cycle in which different recycling methods are used to transform the biodegradable portion of the trash into beneficial fertilizer or manure [39, 40].

Organic waste can be disposed of through several ways and can also be converted into useful and resourceful products:

1. Composting: The organic matter that is naturally recycled into a beneficial fertilizer that may improve soil and plants is called compost. Examples of this material include food scraps and leaves. Everything that grows ultimately breaks down; composting just expedites this process by creating the perfect environment for fungus, bacteria, and other creatures that break down matter like worms, sowbugs, and nematodes. Compost is the resultant decomposed material, which frequently has the appearance of rich garden soil. Compost, which farmers lovingly refer to as "black gold," is nutrient-rich and useful in horticulture, agriculture, and gardening. Among the various methods available for processing organic waste are anaerobic digesters, community composting programs on a smaller scale, and large-scale composting facilities [41].
2. Animal feed: Supplying cattle and other animals with agricultural and food waste is one of the most popular and effective ways to recycle organic waste. One cheap, straightforward way to recycle garbage is to feed organic waste to animals. Direct feeding is a straightforward and simple approach. Donations of leftover food from kitchens and plates can be made to farmers, who will utilize them to feed the animals. A portion of the food might be fed to those who raise animals. Because they are omnivores and will eat everything, pigs perform well in this. As for feeding animals directly, many experts believe that this might result in health problems. Providing the animals with food in a regulated manner will take care of this effortlessly. If someone would like to recycle food waste for animal feed, the US Environmental Protection Agency advises them to get information from their county's agricultural extension office, public health department, or solid waste department.

There are other things to think about, including what kinds of meals to provide and how often. Nonetheless, a lot of professionals believe that feeding animals directly might cause health problems. This is relatively manageable if the procedure for feeding the animals is kept under control.

3. Incineration: The process of burning organic materials at high temperatures in controlled circumstances is known as "incineration of organic solid waste." This technique is frequently used to get rid of a variety of waste materials, such as paper, wood, yard debris, and food leftovers. While burning organic solid waste can be a useful way to manage garbage, there are benefits and drawbacks to this process. Nevertheless, burning rubbish is one of the worst methods cities may use to meet their energy and waste reduction objectives. It is costly, ineffective, and harmful to the environment. It puts municipalities on a high-carbon path by making them continue to produce a lot of waste to feed the incinerator, undermining efforts to reduce garbage output or increase recycling rates.

2.8.2 Medical Waste

By promoting health, preventing sickness, and treating it, healthcare services improve and lengthen people's lives. Still, the provision of healthcare services results in a significant volume of medical waste, 20% of which presents health concerns, including radiation and chemical exposure and infection.

There are trade-offs associated with the COVID-19 pandemic's effects on society, the economy, the environment, and people's everyday lives in many ways. The epidemic hasn't spread as much because of technological advancements like quick testing, tracking sick individuals, internet remote work and schooling, etc. However, there are also disadvantages, such as problems with waste management brought on by the rise in disposable product consumption and inequality brought on by social and digital divisions. For instance, the increased amount of plastic garbage brought on by COVID-19-related activities has serious consequences and makes maintaining a sustainable environment more difficult. Both hazardous and non-hazardous (general) trash can be categorized as medical waste. Hazardous medical waste can lead to illnesses and environmental risks, but non-hazardous medical waste does not specifically pose a risk.

1. The first stage in managing and treating medical waste is to reduce its creation as much as possible. This is known as realistic medical waste management. Usually, every treatment unit produces some medical waste, which employees can dispose of in the designated containers. Nevertheless, common garbage frequently ends up in medical waste bins despite not needing the same care as medical waste. Consequently, hospitals must launch campaigns based on assessments of what may be recycled or reused to lower the production of medical waste.
2. Given the significance of medical waste management, including proper collection and separation of the waste generated daily, all hospital staff members ought to be encouraged to take part in educational programs that emphasize the value of medical waste management and its circulation of resources. To put it another way, it is imperative to minimize the amount of medical waste generated, avoid negative environmental effects, and create an eco-friendly, green environment by recovering and recycling waste materials. Furthermore, as medical supplies and medications reach the end of their shelf life, they are either used or discarded. To prevent expensive medical goods from going to waste after their expiration dates, inventory management is crucial. Incorporating a successful inventory management program into staff education and training courses is one strategy to minimize medical waste.
3. Laws pertaining to the environment and the Medical Service Act, among other rigorous treatment guidelines, apply to medical waste disposal. Medical trash must be properly categorized during the sorting step by hospital staff in order to reduce waste creation, for example, since it is illegal to dispose of infectious medical waste as general waste [42].

2.8.3 Plastic Waste

Plastics are often considered to be a sweet poison these days, harming both people and the environment. Without a doubt, plastics play a significant role in daily life due to their affordability, resistance to decomposition, lifespan (mechanical, thermal, stability, and durability), and affordability. Many novel polymers with desired attributes have been discovered and produced. Plastic materials may be substituted for materials such as wood, metal, and glass in a variety of applications such as weaving, coatings, building, decorating, and packaging. Plastics are eventually disposed of as garbage after being utilized for a while in certain applications. The volume of plastic waste is rising at an alarming rate as a result of growing population and development [43, 44].

The plastic waste generated can be minimized using various chemical techniques which either convert the plastic polymers into other useful products or disposes of them by converting them into other neutral materials.

1. Adsorption: One of the earliest methods, adsorption, has been effectively used to remove both organic and inorganic contaminants from wastewater and water. This surface phenomenon occurs when contaminants from the gaseous or aqueous phases are carried to the surface of solid materials, which are known as adsorbents. The adsorbent is essential to this phenomenon's process of contaminants being absorbed. To remove microplastics from water, the adsorption approach has received a lot of attention lately. It has been discovered that the adsorption method works well for newly developing contaminants in air and water streams. With the use of different micro, meso, and macroporous nanomaterials, microplastics may be readily adsorbed and removed from wastewater. These methods are only applicable to plastics larger than 5 μm. With this method, removing microplastic from seas is an insurmountable task. Additionally, it is thought that adsorbent regeneration is a restriction [43, 45, 46].
2. Photocatalysis: An intriguing method of managing plastic waste, particularly micro/nano plastics, is photocatalysis. This method involves a light-mediated redox process wherein suitable light energy stimulates nanostructured semiconductors, resulting in the generation of exciton pairs. These pairs then react with the surrounding water or moisture to produce extremely reactive species, such as hydroxyl radicals and superoxides, which are capable of oxidizing organic species, including polymers. PVC, PET, PS, and PE degradation have all been studied using a variety of photo catalysts, including titanium oxide (TiO_2) and zinc oxide (ZnO) [43, 47].

3. Microorganisms: Scientists and researchers are now concentrating on creating strains that could be able to aid in the elimination of plastic trash. Numerous microbes, including fungi and bacteria, have been shown to have the ability to break down plastic waste, particularly microplastics. Plastics break down by microbial decomposition to produce biomass, methane, carbon dioxide, water, and other inorganic chemicals. Research into the intriguing science of PET's enzymatic or microbiological breakdown into its component parts is necessary. In this regard, the French business Carbios creates specially designed enzymes that can degrade PET. Together with L'Oreal and Nestle, Carbios created the first food-grade PET plastic in history through enzymatic recycling. According to the study, if the container is still unclean, these enzymes won't contaminate the product. But this technique can only work with two polyesters. Research is currently focused more on finding enzymes and microbes that can totally digest plastic waste by using polymers as their energy source.
4. Incineration and landfill techniques: A few years ago, the landfill was the method of choice for managing plastic garbage since it was affordable, simple to create, easy to dispose of, convenient, and approachable. Waste is often disposed of by communities at a location close to their homes or separately in various dustbins before being placed at certain locations (city outskirts). After decades or perhaps longer of use, plastic debris has now been released into the ocean, with landfill operations being a major factor in its movement. But some of the related hazardous substances leak into the groundwater and soil from the landfill site, seriously harming human health. Because of the loss of important materials, health and environmental risks, and other factors, they are now the least chosen management choice. In order to decrease the quantity of plastic garbage at dump sites, some nations have placed limitations on landfills and made further efforts to recover valuable materials. Researchers have attempted to leverage the incineration method in order to decrease the amount of garbage. Incineration is always used as a landfill substitute. In this method, trash is reduced in volume and energy and recovered by thermally treating it through a regulated combustion process called fast

oxidation. Although incineration is seen to be an effective way to eliminate plastic, it has the drawback of producing more harmful toxic vapors. Depending on the kind of garbage, the fumes discharge VOCs, PCBs, polyvinyl chloride, dioxins, furans, CO, and CO2 from the incinerator [43, 48].

2.9 Role of Integrated Sustainable Waste Management

Solid waste management in a city is closely connected to various aspects of urban life, including economic, social, and health factors. Inadequate handling of solid waste from households and businesses, common in many southern cities and transitioning countries, can hinder economic growth and contribute to the spread of disease and discomfort. Carefully organized and consistently carried out waste management and recycling efforts can bring a sense of satisfaction to urban residents and officials, offer employment opportunities to individuals in need, improve the accessibility of soil and water resources, and set an example for effective governance in various public services [49, 50].

Efficient management of solid waste requires a focus on technology innovation within a system-level, data-driven approach. A strong evaluation phase is vital, collecting numerical information on rates of waste production, waste composition, capacity of landfills, logistics of collection, and metrics for recycling. This evidence-based groundwork helps pinpoint systematic obstacles and directs the focus on interventions.

An all-encompassing view is necessary, considering waste management in relation to resource use, product design, and waste reduction initiatives. Life cycle analysis has the ability to show the complete material flow, which allows for the creation of closed-loop systems. Moreover, future systems will be transformed by exploring and utilizing advanced technologies in waste-to-energy conversion, materials sorting, landfill management, and smart monitoring. This will enhance resource recovery, reduce environmental impact, and enhance waste management efficiency.

An in-depth introduction to a very specific planning framework, called the Integrated Sustainable Waste Management framework or ISWM was introduced. Traditional waste management approaches in developing countries often focused on technical solutions but frequently failed due to their limited scope. In the 1990s, international agencies and NGOs recognized the need for a more holistic strategy [51]. This spurred the development of the Integrated Sustainable Waste Management (ISWM) framework. The ISWM framework, crystallized through programs like the Urban Waste Expertise Programme (UWEP), is built on bottom-up, participatory processes. It recognizes the significance of the informal sector (waste pickers and recyclers) and community initiatives in effective waste management. ISWM assessments provide a valuable tool to identify strengths and weaknesses in existing systems [52]. This comprehensive approach is critical for designing locally relevant solutions that address complex sustainability challenges in developing countries. In collaboration with organizations in Mali, Costa Rica, Peru, the Philippines, India, Bulgaria, and Egypt, the organization WASTE, Consultants on Urban Environment and Development, based in Gouda, the Netherlands, developed and implemented the UWEP program. In addition, partners from Tanzania, Kenya, Pakistan, Guatemala, and Tanzania contributed to the establishment of the knowledge base that led to the development of the ISWM framework [53].

During the UWEP Plus phase, nine cities with diverse dual objectives evaluated and planned using ISWM as the basis. These cities did so from 2001 to 2003. To promote planning and development in the cities has been the original goal. In order to improve governance and urban services in other cities, the second goal has been to learn from the trial projects in those cities and develop a number of methodological insights [54]. Program personnel were able to execute and validate the methodology for

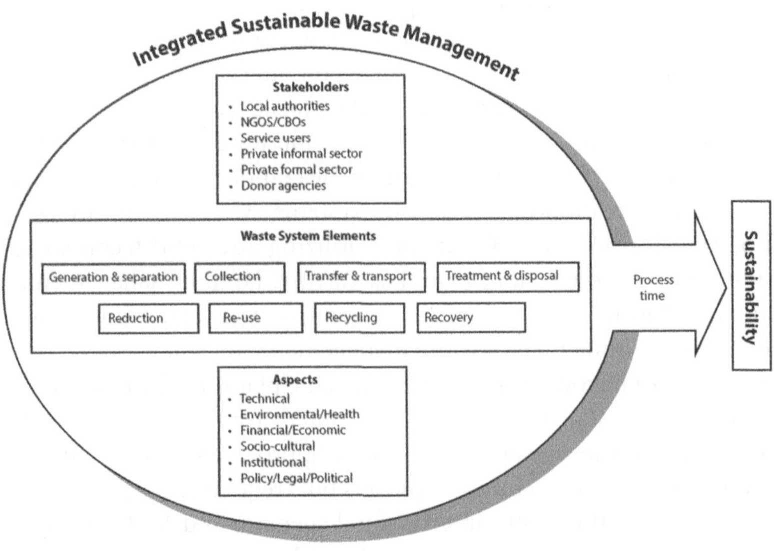

Figure 2.2 ISWM model [56].

the ISWM assessment, which forms the basis for planning, in five additional cities after the completion of the whole ISWM planning process in four cities. One secondary approach that has substantial value when used is the ISWM evaluation [55].

The concept of Integrated Sustainable Waste Management or ISWM is demonstrated in Figure 2.2, which illustrates three essential principles of waste management:

a) The stakeholders involved in and affected by waste management.
b) The (practical and technical) elements of the waste system.
c) The sustainability aspects of the local context that should be taken into account when assessing and planning a waste management system.

2.9.1 Dimensions of ISWM

a) Stakeholders
b) Water System Elements
c) ISWM Aspects

2.9.1.1 Stakeholders

The involvement of stakeholders is the primary focus of ISWM. Any individual or group with an interest in waste management, in this situation, is considered a stakeholder. Waste management usually involves the participation of residents or houses that use the system, as well as the municipality, which has overall responsibility for maintaining urban cleanliness. The success of ISWM hinges on the collaboration of diverse stakeholders within a city's unique context. These stakeholders, categorized by their shared interests in waste management, can present a significant hurdle. Their varied roles and priorities within the waste system can make achieving consensus on improvements difficult. Further complicating matters, these stakeholders are often interconnected through other city systems beyond just waste management. They share a common social and geographic environment, creating a complex web of influences [57].

2.9.1.2 Water System Elements

The physical components of a waste management system are often called its technical elements. These elements align with the stages of a material's life cycle: resource extraction, manufacturing, consumer use, processing, and finally, treatment and disposal. How solid waste is handled and its ultimate destination have significant environmental consequences. This has led many national environmental ministries to adopt a waste management hierarchy as a guiding policy framework.

2.9.1.3 ISWM Aspects

Within ISWM the third dimension consists of six sustainability aspects, or lenses, which is represented in Table 2.1. These aspects, like different lenses, help us evaluate existing systems or plan new ones. They cover a wide range of areas including:

- Rules and Regulations: This looks at the laws and policies that affect waste management.

Table 2.1 Key Aspects of integrated sustainable waste management.

Aspects	**Description**
Waste Hierarchy	Prioritizes actions in order: waste prevention, reuse, recycling/composting, energy recovery, and responsible disposal (as a last resort).
Stakeholder Participation	Involves individuals, communities, businesses, informal waste workers, governments, and NGOs in decision-making and implementation.
Systems Thinking	Views waste management as interconnected, recognizing that choices in one part of the system affect others.
Life-Cycle Approach	Considers the environmental and economic impacts of waste management across all stages: generation, collection, transport, treatment, and disposal.
Social Considerations	Addresses social equity, health, safety, and the inclusion of informal waste workers.
Policy and Legislation	Provides clear regulations, incentives for sustainable practices, standards, and enforcement mechanisms.
Economic Aspects	Balances costs with long-term benefits, explores revenue from resource recovery, and ensures fair financing.

Table 2.2 Technical aspects of integrated sustainable waste management.

Aspects	Description
Infrastructure	Includes collection bins/vehicles, transfer stations, sorting facilities, recycling plants, composting sites, waste-to-energy plants, and safe landfills.
Technologies	Employs a range of technologies from basic (manual sorting) to advanced (automated sorting, anaerobic digestion, gasification, pyrolysis).
Operations	Covers route optimization, personnel training, equipment maintenance, and monitoring system performance.
Data Management & Analysis	Tracks waste flows, evaluates treatment effectiveness, measures recycling rates, and guides decision-making.

- Social and Cultural Factors: This considers how people's habits and values influence waste generation and how waste management affects them.
- Organizations and Institutions: This examines how well waste collection and treatment facilities are run.
- Technical Performance: This focuses on how efficiently waste is processed and disposed of. The technical aspects of ISWM are represented in Table 2.2.
- Environmental and Health Impacts: This considers the effect of waste management on the environment and people's health.

Financial Considerations: This looks at the costs and potential revenue associated with waste management.

2.10 Progress, Challenges and Path Forward

Solid waste management (SWM) has undergone notable transformations in developing countries. Historically, inadequate collection coverage and reliance on open dumps posed serious threats to public health and the environment. However, recent decades reveal expanding waste collection services, with many cities nearing universal coverage. This focus on basic service provision has significantly improved the lives of urban residents who were previously without waste management options. Additionally, there's a marked shift away from open dumping towards controlled waste disposal. Engineered landfills are being constructed, replacing the hazards of uncontrolled dumpsites. This shift is often prompted by both environmental legislation and a response to high-profile disasters linked to unstable or poorly managed waste dumps, such as landslides in Manila, the Philippines, and Bandung, Indonesia [58].

Within many developing countries, informal recycling sectors flourish. Driven by the economic value of discarded materials, these informal systems often achieve recycling rates rivaling those of formal programs in developed nations. In Bamako, Mali, for example, the informal sector plays a substantial role in collecting and processing recyclables, reducing the waste management burden on the municipality [59]. These systems not only provide livelihoods for marginalized populations but also contribute to resource conservation. Despite their contributions, the informal sector is often stigmatized and faces challenges related to working conditions, child labor, and environmental pollution [60].

While SWM improvements are evident, obstacles remain. Even with increased collection coverage, low-income and informal settlements frequently lack adequate service, creating social disparities and public health risks. Adequate waste disposal facilities remain a challenge for low-income nations where financing for infrastructure, operation, and maintenance is a major barrier [61]. Furthermore, growing economies correlate with increased waste generation, placing additional strain on SWM systems in both developed and developing countries. Plastic waste, in particular, is a concern, with examples like Kampala, Uganda, where plastic bags contribute to flooding, and Mumbai, India, where a plastic bag ban was implemented after devastating floods in 2005 [62].

Successful SWM in developing countries requires solutions tailored to local needs. Collection methods must be adapted to areas with limited infrastructure, using creative solutions like handcarts or small-scale transfer stations, as seen in Kunming, China. Governance plays a critical role, with strong institutions and secure financing needed for sustainable operation [63]. Integrating the informal recycling sector, rather than displacing it, can lead to greater efficiency, environmental benefits, and the protection of livelihoods. Embracing the "3Rs" of reduce, reuse, recycle is essential to manage growing waste streams. Waste prevention, repairing and reusing products, and composting organic wastes are key strategies. International aid, innovative waste-to-energy technologies for specific contexts, and planning for climate resilient SWM systems will also play a role in the continued advancement of waste management for developing nations [64]. NGOs and women's unions play a crucial role in representing both waste generators and those working within the informal waste sector, ensuring marginalized voices contribute to the decision-making process [65, 66].

Cities like Bamako (Mali), Belo Horizonte (Brazil), Bengaluru (India), and Quezon City (the Philippines) showcase how consultation with users, platforms for participatory planning, and collaboration with informal service providers can significantly enhance SWM equity and effectiveness. These collaborations often lead to solutions that better address the specific needs and challenges of diverse communities [67–69].

Financial sustainability remains a critical concern as modern SWM is expensive. Cities must creatively secure revenue through a mix of sources to bridge the gap between limited budgets and the costs of improved services. While waste service fees are a vital component of cost recovery, cities often supplement revenue by allocating central government funds, implementing cross-subsidies (where wealthier areas support services in poorer ones), charging franchise fees, utilizing property taxes, or revenue from municipal asset sales [70].

Strong institutions and clearly defined roles within the municipality are vital for transparent contracting, combating corruption, and improving service quality. Public-private partnerships (PPPs) can increase efficiency and coverage, provided cities retain their authority and create conditions for fair competition. The "pro-poor" PPP model specifically addresses equity by intentionally serving low-income communities while empowering small businesses and the informal sector [71, 72].

2.11 Conclusion

Solid waste management (SWM) is a complex challenge for cities in developing countries, necessitating a collaborative approach for effective and equitable solutions. Successful SWM hinges on the active participation of all stakeholders, including municipal authorities, waste generators (households and businesses), service providers (public, private, and informal sectors), and external agencies within the broader enabling environment. Developing countries can borrow principles like performance-based management and increased accountability from the "new public management" model used in developed nations, adapting them to local contexts. This can help streamline efficiency and promote better waste management practices. In conclusion, while there's no single solution for SWM in developing cities, inclusivity, financial sustainability, and strong institutions with proactive policies are essential starting points towards creating equitable, effective, and accountable waste management systems.

References

1. Shekdar, A.V., Sustainable solid waste management: An integrated approach for Asian countries. *Waste Manage. (Oxford)*, 29, 4, 1438–1448, 2009, ISSN 0956-053X, https://doi.org/10.1016/j.wasman.2008.08.025 (https://www.sciencedirect.com/science/article/pii/S0956053X08003024).
2. Wong, M.H., Integrated sustainable waste management in densely populated cities: The case of Hong Kong. *Sustain. Horiz.*, 2, 100014, 2022, ISSN 2772-7378, https://doi.org/10.1016/j.horiz.2022.100014 (https://www.sciencedirect.com/science/article/pii/S2772737822000098).
3. Pappu, A., Saxena, M., Asolekar, S.R., Solid wastes generation in India and their recycling potential in building materials. 42, 6, 2311–2320, 2007, ISSN 0360-1323, https://doi.org/10.1016/j.buildenv.2006.04.015. (https://www.sciencedirect.com/science/article/pii/S0360132306001168).
4. Yau, Y., Domestic waste recycling, collective action and economic incentive: The case in Hong Kong. *Waste Manage. (Oxford)*, 30, 12, 2440–2447, 2010.
5. Xue, W., Cao, K., Li, W., Municipal solid waste collection optimization in Singapore. *Appl. Geogr.*, 62, 182–190, 2015.
6. Singh, E., Kumar, A., Mishra, R., Kumar, S., Solid waste management during COVID-19 pandemic: Recovery techniques and responses. *Chemosphere, 288*, 132451, 2022, https://doi.org/10.1016/j.chemosphere.2021.132451.
7. Adhikari, B. and Koirala, S., Gap Assessment of Solid Waste Management (SWM) Practices and Challenges in Nine Densely Populated Cities of Nepal Before and during Lockdown Due to COVID-19. Preprints 2020, 2020090304, https://doi.org/10.20944/preprints202009.0304.v1.
8. Ahuja, A., India generated over 18,000 tonnes of COVID-19 related bio-medical waste in 4 Months, experts call to reduce, reuse and segregate, 2020.
9. Energy Star, Light bulbs for consumers [Online], Retrieved from https://www.energystar.gov/index.cfm?fuseaction=find_a_product. showProductGroup&pgw_code=LB.
10. Environmental Protection Agency, Fact flash — 6: Resource Conservation and Recovery Act (RCRA) [Online], Retrieved from https://www.epa.gov/superfund/students/clas_act/haz-ed/ff06.pdf.
11. Environmental Protection Agency, Municipal solid waste in the United States: 2009 facts and figures [Online], Retrieved from https://www.epa.gov/epawaste/nonhaz/municipal/pubs/msw2009rpt.pdf, 2010.

12. Environmental Protection Agency, Municipal solid waste generation, recycling, and disposal in the United States: Tables and figures for 2010 [Online], Retrieved from https://www.epa.gov/wastes/nonhaz/municipal/pubs/2010_MSW_Tables_and_Figures_508.pdf, 2011.
13. *Procedia Environ. Sci.*, 35, 408–415, 2016, ISSN 1878-0296, https://doi.org/10.1016/j.proenv.2016.07.022.(https://www.sciencedirect.com/science/article/pii/S1878029616301116).
14. Tolba, M.K. (Ed.), *Our Fragile World: Challenges and Opportunities for Sustainable Development*, vol. II, 2001.
15. Goodroad, L.L. and Keeney, D.R., Nitrous oxide production in aerobic soils under varying pH, temperature and water content. *Soil Biol. Biochem.*, 16, 1, 39–43, 1984, ISSN 0038-0717, https://doi.org/10.1016/0038-0717(84)90123-8).
16. Wei, Y., Liang, Z., Zhang, Y., Evolution of physicochemical properties and bacterial community in aerobic composting of swine manure based on a patent compost tray. *Bioresour. Technol.*, 343, 126–136, 2022.
17. Wang, C., Liu, D., Li, Y., Wang, L., Gu, W., A multidisciplinary perspective on the evolution of municipal waste management through text-mining: A mini-review. *Waste Manage. Res. J. Sustain. Circ. Econ.*, 39, 1, 32–42, 2020.
18. Vigneswaran, S., Kandasamy, J., Johir, M.A.H., Sustainable Operation of Composting in Solid Waste Management. *Procedia Environ. Sci.*, 35, 408–415, 2016, ISSN1878-0296, https://doi.org/10.1016/j.proenv.2016.07.022.(https://www.sciencedirect.com/science/article/pii/S1878029616301116).
19. Woolridge, A. and Hoboy, S., Ch. 27 - Medical Waste, in: *Waste*, Second Ed., T.M. Letcher and D.A. Vallero (Eds.), pp. 517–530, Academic Press, 2019, ISBN 9780128150603, https://doi.org/10.1016/B978-0-12-815060-3.00027-X.(https://www.sciencedirect.com/science/article/pii/B978012815060300027X).
20. Nnorom, I.C. and Odeyingbo, O.A., Electronic waste management practices in Nigeria, Ch. 14, in: *Handbook of Electronic Waste Management*, M.N.V. Prasad, M. Vithanage, A. Borthakur (Eds.), pp. 323–354, Butterworth-Heinemann, 2020, ISBN 9780128170304, https://doi.org/10.1016/B978-0-12-817030-4.00014-0.(https://www.sciencedirect.com/science/article/pii/B9780128170304000140).
21. Biron, M., Plastics Sustainability: Drivers and Obstacles, Ch. 11, in: *Plastics Design Library, A Practical Guide to Plastics Sustainability*, M. Biron (Ed.), pp. 557–593, William Andrew Publishing, 2020, ISBN 9780128215395, https://doi.org/10.1016/B978-0-12-821539-5.00011-2. (https://www.sciencedirect.com/science/article/pii/B9780128215395000112).
22. Sewak, A., Deshpande, S., Rundle-Thiele, S., Zhao, F., Anibaldi, R., Community perspectives and engagement in sustainable solid waste management (SWM) in Fiji: A socioecological thematic analysis. *J. Environ. Manage.*, 298, 113455, 2021, ISSN 0301-4797, https://doi.org/10.1016/j.

jenvman.2021.113455. (https://www.sciencedirect.com/science/article/pii/S0301479721015176).

23. Colon, M. and Fawcett, B., Community-based Household Waste Management: Lessons Learnt from EXNORA's 'Zero Waste Management' Scheme in Two South Indian Cities. *Habitat Int.*, Elsevier Publication.
24. Moningka, L., Community Participation in Solid Waste Management Factors Favouring the Sustainability of Community Participation, A Literature Review, UWEP Occasional Paper, website: http://www.waste.nl, 2000.
25. Bulle, S., Issues and Results of Community Participation in Urban Environment Comparative Analysis of Nine Projects on Waste Management, UWEP Working Document 11, website: http://www.waste.nl, 1999.
26. Subash, A., Community participation in solid waste management. Office of Enviornmental Justice, Washington, DC, 2002.
27. Yukalang, N., Clarke, B., Ross, K., Barriers to Effective Municipal Solid Waste Management in a Rapidly Urbanizing Area in Thailand. *Int. J. Environ. Res. Public Health*, 14, 9, 1013, 2017, https://doi.org/10.3390/ijerph14091013.
28. Debnath, B., Mainul Bari, A.B.M., Ali, S.M., Ahmed, T., Ali, I., Kabir, G., Modelling the barriers to sustainable waste management in the plastic-manufacturing industry: An emerging economy perspective. *Sustainability Anal. Model.*, 3, 100017, 2023, ISSN 2667-2596, https://doi.org/10.1016/j.samod.2023.100017. (https://www.sciencedirect.com/science/article/pii/S2667259623000036).
29. Saxena, S., Srivastava, R., Samaddar, A., Sustainable Waste Management Issues in India. *IUP J. Soil Water Sci.*, 3, 1, 72–90, 2010.
30. Almarshad, S., Adopting Sustainable Behavior in Institutions of Higher Education: A Study on Intentions of Decision Makers in the Mena Region. *Eur. J. Sustain. Dev.*, 6, 2017, 10.14207/ejsd.2017.v6n2p89.
31. Lohri, C.R., Camenzind, E.J., Zurbrügg, C., Financial sustainability in municipal solid waste management – Costs and revenues in Bahir Dar, Ethiopia. *Waste Manage. (Oxford)*, 34, 2, 542–552, 2014, ISSN 0956-053X, https://doi.org/10.1016/j.wasman.2013.10.014. (https://www.sciencedirect.com/science/article/pii/S0956053X1300500X).
32. Bernstein, J. and Wright, F., Investments in Solid Waste Management, Working Paper, Infrastructure and Urban Development Dept., The World Bank, April 1990.
33. Chitotombe, J.W., Globalization of Information Communication Technology (ICT) and consumerism in developing countries: Confronting the challenges of e-waste disposal in Harare urban, Zimbabwe. *Int. J. Environ. Sci.*, 3, 6, 2172–2185, 2013.
34. Mavropoulos, A., Globalization, Megacities and Waste Management. *ISWA Conference*, Daegu, Republic of Korea, 2011.
35. Achankeng, E., Globalization, urbanization and municipal solid waste management in Africa. *Proceedings of the African Studies Association of Australasia and the Pacific 26th Annual Conference*, 2003.

36. Nathanson, J.A., Solid-waste management, Encyclopedia Britannica, 31 Oct. 2023, https://www.britannica.com/technology/solid-waste-management. Accessed 25 February 2024.
37. https://www.unep.org/explore-topics/resource-efficiency/what-we-do/cities/solid-waste-management.
38. Asnani, P.U., Solid waste management, India Infrastructure Report, 570, 2006.
39. Sun, F.-S. and Yu, G.-H., Chapter Seven - Fate of bio-contaminants in organic wastes during composting and vermicomposting processes, in: *Fate of Biological Contaminants during Recycling of Organic Wastes*, K. Huang, S.A. Bhat, G. Cui (Eds.), pp. 143–156, Elsevier, 2023, ISBN 9780323959988, https://doi.org/10.1016/B978-0-323-95998-8.00004-2. (https://www.sciencedirect.com/science/article/pii/B9780323959988000042).
40. Sharma, B., Vaish, B., Mahajan, M., Singh, U.K., Singh, P., Singh, R., Recycling of Organic Wastes in Agriculture: An Environmental Perspective. *Int. J. Environ. Res.*, 13, 2019, 10.1007/s41742-019-00175-y.
41. Keener, H., Challenges and Opportunities in Composting Organic Waste, in: *Climate Change and Food Security in South Asia*, pp. 295–324, 2011, 10.1007/978-90-481-9516-9_18.
42. Lee, S.M. and Lee, D., Effective Medical Waste Management for Sustainable Green Healthcare. *Int. J. Environ. Res. Public Health*, 19, 22, 14820, 2022 Nov 10, doi: 10.3390/ijerph192214820. PMID: 36429539; PMCID: PMC9690095.
43. Pandey, P., Dhiman, M., Kansal, A., *et al.*, Plastic waste management for sustainable environment: techniques and approaches. *Waste Dispos. Sustain. Energy*, 5, 205–222, 2023, https://doi.org/10.1007/s42768-023-00134-6.
44. Chen, G.G.-Q. (Ed.), *Plastics from Bacteria: Natural Functions and Applications*, vol. 14, Springer Science & Business Media, 2009.
45. Pandey, P. and Saini, V.K., Pillared interlayered clays for pollution remediation, in: *Green Adsorbents for Pollutant Removal*, pp. 353–376, Springer, Cham, Switzerland, 2018.
46. Pandey, P. and Saini, V.K., Pillared interlayered clays: sustainable materials for pollution abatement. *Environ. Chem. Lett.*, 17, 2, 721–727, 2019.
47. Ameta, R., Solanki, M.S., Benjamin, S., *et al.*, *Photocatalysis advanced oxidation processes for waste water treatment*, pp. 135–175, Elsevier, Amsterdam, the Netherlands, 2018.
48. Ritchie, H. and Roser, M., Plastic pollution, Our World in Data, Oxford, UK, 2018.
49. Bartone, C.R., Strategies for Improving Urban Waste Management: Lessons from a Decade of World Bank Lending, World Bank, Washington, DC, USA, 1997.
50. Crowe, M. and Carty, G., Municipal Waste Characterisation, Environmental Protection Agency, Wexford, Ireland, 1996.

51. GEF/UNDP/IMO, Integrated Waste Management Action Plan for the Batangas Bay Region, GEF/UNDP/IMO, Quezon City – the Philippines, 1996.
52. ICLEI – International Council for Local Environmental Initiatives, *The Local Agenda 21 Planning Guide – An Introduction to Sustainable Development Planning*, 1996.
53. van de Klundert, A., Muller, M., Scheinberg, A., Dulac, N., Anschütz, J., Hoffman, L., Integrated Sustainable Waste Management, A Set of Five Tools for Decision-makers, Experiences from the Urban Waste Expertise Programme (1995-2001), WASTE, Gouda, Netherlands, 2001.
54. Planact, A Framework and Guidelines for Community Participation: A Handbook for Councils and Communities, Braamfontein, South Africa, 1998.
55. UNCHS/UNEP, Sustainable Cities Programme (SCP) - Programme Approach and Implementation, UNEP, Nairobi, Kenya, 2000.
56. Mongkol, K., The critical review of new public management model and its criticisms. *Res. J. Bus. Manage.*, 2011.
57. Pontin, J.F., Waste management initiatives and innovation. Opportunities for an integrated approach to resource recovery – a way ahead. *R&D Manage.*, 1980.
58. Wilson, D.C., Development drivers for waste management. *Waste Manage. Res.*, 2007.
59. Wilson, D.C., Whiteman, A., Tormin, A., Strategic Planning Guide for Municipal Solid Waste Management, World Bank, Washington, DC, USA. See http://www.worldbank.org/urban/solid_wm/erm/start_up.pdf, 2001.
60. Wilson, D.C., Araba, A.O., Chinwah, K., Cheeseman, C.R., Building recycling rates through the informal sector. *Waste Manage. (Oxford)*, 2009.
61. Wilson, D.C., Blakey, N.C., Hansen, J.A.A., Waste prevention: its time has come. *Waste Manage. Res.*, 2010a.
62. Wilson, D.C., Rodic, L., Scheinberg, A., Alabaster, G., Comparative analysis of solid waste management in cities around the world, in: *Proceedings of Conference Waste 2010: Waste and Resource Management – Putting Strategy into Practice, Stratford-upon-Avon, UK*, Townshend, M. (Ed.), 2010b, See http://www.warrr.org/901/.
63. Wilson, D.C., Rodic, L., Scheinberg, A., *et al.*, Comparative analysis of solid waste management in 20 cities. *Waste Manage. Res.*, 2012.
64. World Bank, Urban Development – Urban Solid Waste Management – Key Obstacles. See http://web.worldbank.org/wbsite/external/topics/exturbandevelopment/extuswm/0,contentMDK: 20239149~menuPK:497508~pagePK: 210058~piPK:210062~theSitePK:463841,00.html, 2012.
65. Xi, B.D., Su, J., Huang, G.H., *et al.*, An integrated optimization approach and multi-criteria decision analysis for supporting the waste-management system of the City of Beijing, China. *Eng. Appl. Artif. Intell.*, 2010.

66. Zhao, Y., Xing, W., Lu, W., *et al.*, Environmental impact assessment of the incineration of municipal solid waste with auxiliary coal in China. *Waste Manage. (Oxford)*, 2012, http://dx.doi.org/10.1016/j.wasman.
67. Zotos, G., Karagiannidis, A., Zampetoglou, S., *et al.*, Developing a holistic strategy for integrated waste management within municipal planning: challenges, policies, solutions and perspectives for Hellenic municipalities in the zero-waste, low-cost direction. *Waste Manage. (Oxford)*, 2009.
68. Zurbrügg, C. and Rothenberger, S., Determinants of resilience in community-led waste management. *Proc. Inst. Civ. Eng.–Waste Resour. Manage.*, 2013.
69. Zurbrügg, C., Gfrerer, M., Ashadi, H., *et al.*, Determinants of sustainability in solid waste management – the Gianyar waste recovery project in Indonesia. *Waste Manage. (Oxford)*, 2012, https://doi.org/10.1016/j.wasman.2012.01.011.
70. Stehlik, P., Chapter 5, in: *Up-to-Date Waste-to-Energy Approach: From Idea to Industrial Application*, Springer, Czech Republic, 2016.
71. Tisi, Y.S.A.B., Matos, F.A., Carneiro, M.L.N., Development of waste-to-energy through integrated sustainable waste management: the case of ABREN WtERT Brazil towards changing status quo in Brazil. *Waste Disposal Sustain. Energy*, 5, 3, 295–308, 2023.
72. Wilson, D.C., Velis, C.A., Rodic, L., Integrated sustainable waste management in developing countries. *Proc. Inst. Civ. Eng.-Waste Resour. Manage.*, 166, 2, 52–68, 2013, May, ICE Publishing.

3

An Insight into the Management of the Biomedical Waste Produced During a Pandemic Like Covid-19

Deeksha Ranjan[1], Roopali Sharma[2], Anjali Bhati[3] and Choudhary R.[3]*

[1]Department of Applied Sciences and Humanities, Faculty of Engineering and Technology, Rama University, Mandhana, Kanpur, U.P., India
[2]School of Basic Sciences, Galgotias University, Greater Noida, U.P., India
[3]School of Biological and Life Sciences, Galgotias University, Greater Noida, U.P., India

Abstract

The COVID-19 pandemic posed unprecedented obstructions in managing biomedical waste (BMW), necessitating rapid and adaptable methods to limit environmental, public health, and occupational challenges. Moreover, due to the widespread utilization of disposable and personal protective equipment (PPE), treatment centers and hospitals generated considerable amounts of biomedical waste (BMW) from used gloves, surgical masks, disposable aprons, syringes, and other PPE items. The initial wave of this pandemic generated biomedical waste of 126 metric tons per day. The unexpected and unusual biomedical waste generated from the COVID-19 pandemic has raised concerns among environmentalists and the worldwide waste management sector, aggravating the already difficult problem of environmental pollution. To alleviate the panic, the Central Pollution Control Board (CPCB) issued updated recommendations for the proper treatment and disposal of biomedical waste. This study proposes a thorough understanding of the biomedical waste management process during the global health crisis, illustrating the difficulties encountered, creative solutions used, and lessons learned. The analysis emphasizes that specific protocols and guidelines are needed to assure the safe handling, disposal, and separation of biomedical waste. This study explores various approaches to manage BMW during a pandemic, including legislative frameworks, legislative actions for COVID-19-based BMW, treatment and

**Corresponding author*: richa.choudhary@galgotiasuniversity.edu.in

Priyanka Singh, Pooja Agarwal and V. Vivekanand (eds.) Solid Waste Management: Challenges, Sustainability and Advancements, (93–116) © 2026 Scrivener Publishing LLC

disposal facilities' capacity to meet challenges in India. This study also emphasizes the sustainable and environmentally friendly methods for managing BMW.

***Keywords*:** Biomedical waste (BMW), central pollution control board (CPCB), personal protective equipment, segregation, guidelines, COVID-19

3.1 Introduction

The COVID-19 disease outbreak significantly impacts public health, socioeconomic welfare, and environmental sustainability. One of the less discussed but equally critical issues that emerged during this period is the management of biomedical waste (BMW) that is produced by health care facilities and households [1]. The biomedical waste includes used personal protective equipment (PPE), masks, diagnostic kits, syringes, vials, and other medical supplies (Figure 3.1), all of which pose a potential risk of infection if not managed properly [2]. In general, BMW accounts for only 15% of healthcare waste [3]. It has been well documented that India is also the most severely impacted country and stands second after the USA with 45,035,393 COVID-19 infections till then [4]. Due to insufficient biomedical waste management systems and a lack of resources during the

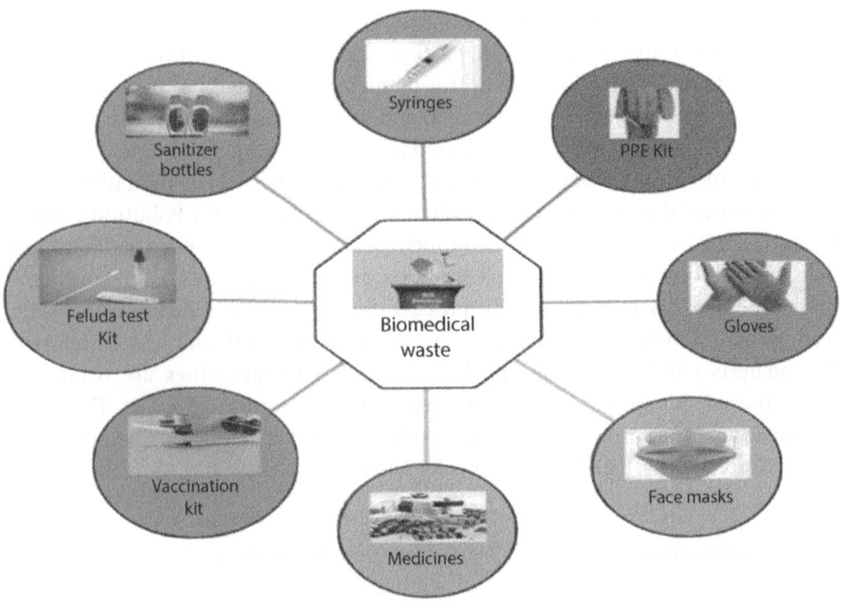

Figure 3.1 Various types of biomedical waste produced during the pandemic.

COVID-19 pandemic, India is suffering from major consequences. As per the report released by the Central Pollution Control Board, the total BMW generated in India was approximately 5238.45 tons/day in August 2020, 5490 tons/day in September 2020 and 164 tons/day in June 2021 [5] (Figure 3.2). The detailed distribution of biomedical waste production in different states of India is represented by Figure 3.3. The improper disposal of this huge amount of infectious waste can lead to the spread of the virus, posing a threat to healthcare workers, the community and the environment at large. It is now documented that COVID-19 can persist on inert surfaces for durations that can range from three hours to nine days and is one of the major contributory factors for their easier spread [6]. Kitajima and Kumar (2020) detected the existence of SARS-CoV genetic material in the sample of wastewater procured from India and another region of the world, suggesting the strong need for proper guidelines and implementation of rules to manage biomedical waste [7, 8]. Moreover, Núñez-Delgado (2020) also reported traces of the COVID-19 virus in soil [9]. These data represent that mishandled biomedical wastes somehow mix with water and soil and hence become the source of recurrent infection. Therefore, stringent protocols and innovative solutions are required to handle the increased volume of biomedical waste safely and efficiently.

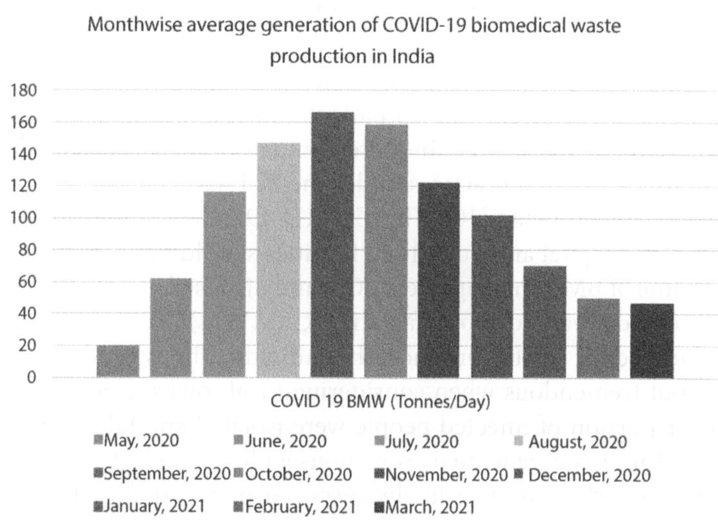

Figure 3.2 Bar graph displaying the average data of biological waste generated in India from May 2020 to March 2021 during the COVID-19 pandemic.

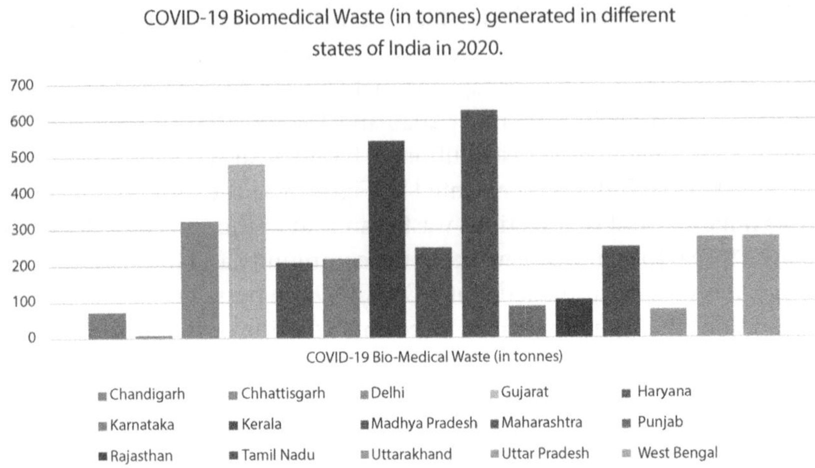

Figure 3.3 Bar graph representing the COVID-19-associated BMW produced in different states of India in 2020.

For biomedical waste management, India established the Biomedical Waste Management (BMWM) rules in 1998 under the Central Pollution Control Board (CPCB), which were later expanded into a more comprehensive framework with the BMWM rules of 2016, and further amended in 2018 and 2019. When COVID-19 was declared a pandemic in March 2020, India responded with stringent guidelines issued by the CPCB. On March 18, 2020, the CPCB released specific protocols for the handling, treatment, and disposal of waste generated during the treatment, diagnosis, and quarantine of COVID-19 patients, with a fourth revision issued on July 17, 2020. These guidelines are mandatory for all stakeholders, in addition to the existing practices under the BMWM rules of 2016 and its subsequent amendments [10]. The COVID-19–related biomedical waste generated by hospital and household facilities should be segregated as per the guideline of BMW management 2016 and disposed of properly in common biomedical waste disposal facilities (CBMWFs).

The biomedical wastes produced by individual households are less in amount but tremendous when considering local towns, cities, and states. The major portion of infected people were isolated and taking treatment at home. The waste generated from households were not segregated or collected properly and were finally disposed of by landfilling and open dumping. No pretreatment prior to disposal exhibits a strong tendency to contaminate water and soil, which may adversely impact the health of the public.

On the other hand, public awareness also plays a key role in managing biomedical waste management. A recent study in Karnataka revealed that only 43% of the participants properly knew the categorization and disposal of BMW in color-coded bags in a hospital setup. Moreover, a significant portion (54%) have not undergone any training related to BMW [11]. It is also well documented that biomedical waste management is practiced globally and recently grabbed attention due to COVID-19 [12, 13]. The requirement of comprehensive training on BMW management, covering aspects such as segregation, in-house transportation, waste storage in color-coded bins, and final disposal for treatment is strongly advised. Additionally, strict enforcement of biomedical waste management is also essential.

3.2 Biomedical Waste During the Pandemic

The COVID-19 pandemic has significantly impacted biomedical waste management, altering the kind and volume of biomedical waste produced. Municipal solid waste (MSW) includes various types of solid waste generated by residents in both urban and rural communities, where direct physical contact between individuals is possible. When individuals infected with or suspected of having COVID-19 are treated or cared for at home, household waste management becomes a potential vector for transmission. Contaminated surfaces and objects, as well as aerosols generated during waste handling, can facilitate the spread of infectious diseases such as COVID-19 through household waste. The COVID-19 virus can live from 3 hours to 9 days depending upon the type of surfaces [14–16].

COVID-19 primarily disseminates through touching the virus-infected area, person, and object, with droplets of saliva/fluids from sneezing and coughing being key transmission sources. To mitigate the spread of the virus, the World Health Organization (WHO) recommended wearing masks in public (WHO, 2020). Consequently, the face mask market surged from a value of US$0.79 billion in 2019 to approximately US$166 billion in 2020 [16]. Before January 2020, India did not produce PPE kits or N-95 face masks domestically, relying on imports of 275,000 units. However, by the end of May 2020, India had become the second-largest producer of PPE kits, manufacturing approximately 450,000 kits and over 1.04 million N-95 masks daily [15].

India's waste disposal system, already under strain, has faced additional challenges because of this global pandemic. COVID-19's highly infectious nature has caused a spike in requirement for the PPE kits along with various different disposable medical equipment, greatly increasing the output of BMW falling in the yellow category (Y-BMW) [17]. Y-BMW includes scrapped or perished medicines, anatomical and animal organ waste, clinical waste from labs, soiled waste, blood/body fluid contaminated bedding, dressing, etc. The daily disposal of disposable face masks by each person could produce 66 kT of Y-BMW annually [18, 19]. To control the spread of contagious COVID-19 virus, numerous safety measures were taken in hospitals and medical centers utilizing advanced medical kits/equipment which also contributed to the substantial rise in biomedical waste generation.

During 2017 India generated 557 tonnes of BMW daily, with 517 tonnes per day remaining untreated, as per the report of the CPCB of India. The country has 198 authorized biomedical waste treatment facilities, with 28 more under construction. These BWTFs handled 710 tonnes per day of BMW through May 2020, out of which 101 tonnes per day waste was generated from COVID-positive people [20]. The annual report from CPCB (2019) stated that 233 incinerators were functional across various states of India [21]. The daily production of various types of COVID-related biomedical waste in the pandemic period of March to August 2020 underscores the significant effect of the global crisis situation of highly infectious COVID-19 on the waste management system [22].

After May 2020, India saw a sharp rise in COVID-19 cases, which resulted in a surge in the production of BMW of yellow category (T-Y-BMW). By around mid-July 2020, it became evident that India's CBWTFs incineration capacity—both with and without captive incineration—had reached its limit. This capacity was represented by two dotted horizontal lines: one for captive incineration (854 tonnes/day) and one for non-captive incineration (782 tonnes/day). The volume of C-Y-BMW alone surpassed the total Y-BMW incineration capacity in India. Additionally, the production rate of C-Y-BMW remained constant until August 2020, underscoring the ongoing pressure on the waste management system throughout the pandemic [23–26].

The sudden surge in use of PPE and medical equipment, etc., due to the COVID-19 outbreak has negatively impacted the global as well as India's waste management system [27, 28]. In 2020, the pandemic caused an increase of 850 tonnes per day in BMW. There was an increase of about 46% in BMW in India by 2021 through COVID-19 waste produced across nationwide hospitals, authorized quarantine centers, testing centers, and collection sites. Delhi, Kerala, Gujarat, Maharashtra, and Andhra Pradesh were amongst the top in producing COVID-19 waste [23–26].

In Delhi, the increased biomedical waste generation rate can be attributed largely to the cotton swabs used in COVID-19 assays of RT-PCR. The pandemic caused a dramatic increase in waste generation, reaching around 3.4 kg/person/day. In Delhi, clinics produced BMW around 0.5 kg to 4 kg/bed/day [29].

3.3 Steps Involved in Bio-Medical Waste Management

BMW management requires a high order of organization, as even a single mistake can result in harm to the person in charge [30]. Biomedical waste management involves the first five steps for its proper disposal according to BMWM Rule 2016 (Figure 3.4). The five crucial steps, i.e., segregation, collection, internal transportation, pre-treatment and storage are the responsibility of healthcare facilities before handing over to the Common Bio-medical Waste Treatment Facility (CBWTF) operator. If CBMWFs are not available, the generated garbage can alternatively be taken to a nearby hospital that has been authorized for incineration; however, it must be disposed of following the BMW 2016 regulations [5]. To achieve effective waste management in healthcare facilities, it is essential that all involved personnel understand and implement the technical guidelines for waste handling outlined in the BMW management Rules 2016.

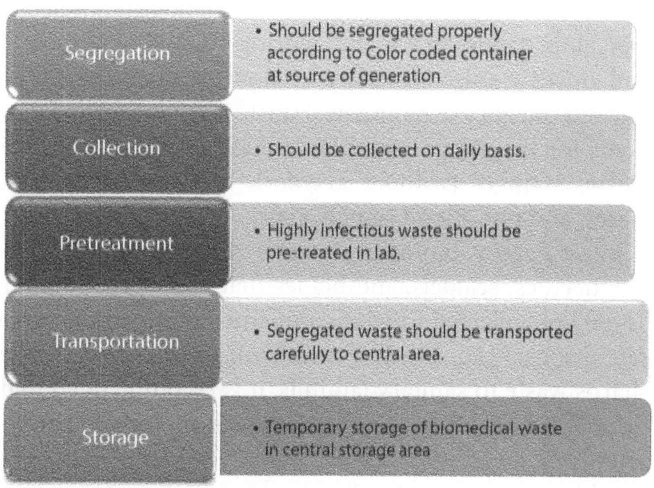

Figure 3.4 The first five steps—segregation, collection, pretreatment, transportation and storage—of BMW at medical facilities.

3.4 Segregation of BMW

Biomedical waste, especially throughout the COVID-19 outbreak, necessitates special control and disposal protocols to minimize the transmission of illnesses. The inside management of biomedical waste is the sole responsibility of the healthcare facilities (HCFs) that generated them. The biomedical waste generated at medical facilities should be segregated according to color coding on a daily basis. Color coding is commonly used to differentiate between various forms of biomedical waste and ensure correct segregation and disposal as per the regulation of the Bio Medical Waste Management Rules, 2016. Segregation is the crucial step for the separation of various types of biomedical waste in well-labeled colored bins according to their category. Proper segregation reduces the risk of secondary infection. Biomedical waste should not be stored in hospital care facilities for more than 8-10 hours [31]. Biomedical wastes should be divided into four categories [30]. There are four types of colored bins: yellow, red, blue, and white bags (Table 3.1).

- The yellow category indicates several different types of waste. It consists of human and animal anatomical waste contaminated with blood or other bodily fluids, such as bandages, dressings, and swabs [32]. The yellow category is also known as incinerable waste or infectious waste. It contains hazardous waste, which can be microbiological, biotechnological, or laboratory waste. During COVID-19, used PPE and gloves are also considered yellow category waste [33].
- The red category of biomedical waste is contaminated plastic items for recycling. Examples include IV tubing, catheters, plastic syringes (without needles), ICT test cards, ELISA plates, and non-blood sample vials.
- White-category biomedical waste refers to sharp objects like needles, scalpels, and blades that can cause punctures. These are collected in white, puncture-proof containers for safe disposal [32, 34]. If it is difficult to transport sterilized sharps waste to sanitary landfills, it is recommended to dispose of it in a concrete pit or recycle it in a steel furnace or foundry.
- The blue category comprises two types of waste: glassware and metallic body implants. To recycle glass bottles or

Table 3.1 Biomedical waste segregation and the methods used to dispose of them.

Category	Segregation	Treatment and disposal
Yellow category waste		
Waste should be collected in yellow coloured non-chlorinated plastic bags or container.	Human and animal Anatomical Waste. Tissues, organs, body parts of human and animal, animal carcasses, dead fetuses, placenta and extracted tooth.	Disinfection by autoclaving/ microwaving/ hydroclaving and can be handed over to CBWTF for incineration.
	Soiled Waste Cotton swabs, dressings as well as plaster casts, bags containing residual contamination with body fluids/blood or disposed blood and blood elements.	
	Disposed mattresses, linen, beddings, routine mask and gown. Microbial and Other Biomedical Laboratory Waste.	
	Waste blood bags, laboratory cultures, waste vaccines, human cell cultures residual toxins, petri plates, tubes and devices, etc., utilized during cultures.	
	Expired and Discarded Medicine. Pharmaceutical waste like medicines, antibiotics, cytotoxic drugs.	All the expired and discarded medicines should be returned to the manufacturer or given to the CBWTF to be removed through incineration.

(*Continued*)

Table 3.1 Biomedical waste segregation and the methods used to dispose of them. (*Continued*)

Category	Segregation	Treatment and disposal
Yellow category waste		
Separate collection system leading to effluent treatment system.	Chemical Waste. Solid or liquid residual chemicals.	
Red Category		
This category waste should be taken in red colored non-chlorinated plastic bag and store in red coloured bins.	Plastic items such as catheters, waste pipette tips, intravenous tubes, syringes (without needles), tubing, bottles, and sets, urine bags, drains, gloves, plastic pipette, vacutainers, Eppendorf and oxygen mask.	Vials/vacutainers containing samples of blood should be pre-treated and disinfected. Such biomedical waste needs to be handed over to CBWTF for final treatment and disposal.
White Category		
This category waste should be collected in white translucent, puncture proof, leak-proof, tamper-proof container.	Scalpels, syringes having fixed needles, needles, needles from needle burners/syringe destroyers, blades, or any types of contaminated sharp objects.	Autoclaving/Dry Heat Sterilization followed by shredding/mutilation or encapsulation in metal container/cement concrete and handed over for final disposal through landfilling/ selected concrete waste sharp pit/iron foundries.
Blue Category		
This category waste should be collected puncture-proof, leak- proof boxes or containers with blue coloured marking.	Glassware and Metallic Body Implants Infected or broken glass and body metallic implants.	Blue category wastes are decontaminated by cleaning with detergent and Sodium Hypochlorite; after that *via* autoclaving/ microwaving/ hydroclaving followed by recycling.

broken glass, they must be sterilized or disinfected using methods such as autoclaving or microwaving.

Additionally, in order to provide enough strength without leakage, multiple-layered bags must be utilized as a precaution when collecting waste from a COVID-19 isolation ward. It is also important to store COVID-19 waste in a specified collecting container marked "COVID-19" and separate it in a temporary storage facility before transferring it to CBWTF-authorized workers. Cleanliness and disinfection of vessels, bins, vehicles used to collect COVID-19 waste are also required from time to time with 1% NaClO solution [35, 36] as COVID-19 virus. If highly infectious material is suspected, then it should be pretreated in the lab, then transported and stored in a common area for final proper disposal at CBWTF.

After segregating BMW in different color-coded bags, it is treated and finally disposed of with the help of various methods (Table 3.1).

3.5 Treatment, Disinfection and Disposal Methods of Covid-19 BMW

Biomedical waste treatment is the procedure employed to eliminate the harmful effect of BMW. Various treatment methods are available to maintain the highest degree of safety while disposing of biomedical waste [1]. Various methods of treatment and disposal of COVID-19 BMW are as follows:

3.5.1 Autoclaving

It is a low heat thermal process operating between 93 - 177 °C and it uses steam for disinfection of waste. These are of two types, viz., gravity flow autoclave and vacuum autoclave, depending on the method they use for removing air pockets.

3.5.2 Microwaving

It is also a low heat thermal process operating between 177 to 540 °C. Microbes are destroyed by the moist heat and steam created by microwave radiation (having wavelengths of 1 mm to 1 m in the frequency of hundreds of megahertz to 3000 MHz). This method shall not be used for cytotoxic, hazardous or radioactive wastes, contaminated animal carcasses, body parts and large metal items.

3.5.3 Incineration

It employs combustion to destroy the BMW. It completely kills microbes and burns 90% organic waste. Wastes to be incinerated shall not be chemically treated with any chlorinated disinfectants. Also, the toxic metals should be within the regulatory limits. It involves high capital and operating cost. The process of incineration produces dioxins and furans [37, 38].

3.5.4 Pyrolysis

This technique operates between 540–830 °C and encompasses various processes such as pyrolysis-oxidation, plasma pyrolysis, induction-based pyrolysis, and laser-based pyrolysis. In pyrolysis-oxidation, air is supplied to the primary combustion chamber at a level below the theoretical chemical reaction requirement. In this chamber, organic solid and liquid waste is vaporized at around 600 °C under air turbulence, leaving behind residual ash, glass, and metallic fragments. In the second combustion stage, the flammable gaseous vapor is burned in a chamber at higher temperatures ranging from 982 °C to 1093 °C, ensuring the complete destruction of toxic substances like dioxins and releasing clean exhaust steam. Given the rapid spread potential of SARS-CoV-2, using plasma energy for the quick decomposition of COVID-19 waste is recommended over traditional laser or gaseous combustion methods [38]. This technique has been observed to have a low emission rate, produce inert residuals, reduce volume by up to 95%, and reduce mass by up to 90% [37, 38].

3.5.5 Chemical Disinfection

This method is majorly used to pre-treat COVID-19 waste, often combined with prior mechanical shredding. To prevent aerosol formation during shredding, the exhausted air is filtered through a high-efficiency particulate air (HEPA) filter. The shredded waste is then mixed with chemical disinfectants and kept in a closed system or under negative pressure for a specified period. This process decomposes organic substances and inactivates or kills infectious microorganisms. The main advantages of chemical disinfectants include their low effective concentration, stable performance, rapid action, broad sterilization spectrum, and lack of residual hazards. These properties allow the disinfectants to effectively kill microorganisms and inactivate bacterial spores. Chemical treatment of COVID-19 waste can be divided into chlorine-based and non-chlorine-based systems. In chlorine-based treatment, sodium hypochlorite (NaOCl) or chlorine dioxide (ClO2) is used as the disinfectant. The electronegativity of chlorine oxidizes peptide links and denatures proteins, allowing it to penetrate cell layers even at neutral pH. NaOCl, one of the first chemical disinfectants, releases haloacetic acids, dioxins, and chlorinated aromatic compounds. ClO2, a strong biocide, became more widely used due to its effective disinfection properties, although it is unstable and must be used on-site. It decomposes into salt and less-toxic products that are non-reactive to alcohol and ammonia. In non-chlorine-based systems, hydrogen peroxide (H2O2) is commonly used. It oxidizes and denatures proteins and lipids, causing membrane disorganization by swelling saturated H+-ions. This system is advantageous due to its high reactivity and lack of toxicity compared to chlorinated systems. Other chemical solutions like povidone-iodine (>0.23%), formaldehyde (>0.7%), isopropanol (>70%), and ethyl alcohol (>75%) can also inactivate SARS-CoV-2 [38, 39].

3.5.6 Mechanical Processes

Shredding and compaction are key mechanical methods used to modify the physical properties or shape of waste to facilitate easier management. Compaction reduces the volume of waste, making it more manageable. Shredding destroys paper and plastic waste to prevent its reuse. It's important to note that shredders should only be used on waste that has already been disinfected [40].

Once treated, BMW become non-hazardous and non-dangerous. It can be given for final disposal through landfilling, even after combining with other solid waste. However, landfilling is not ideal for disposing of biomedical waste or any other solid waste due to its adverse environmental and health impacts. Landfills generate leachates and release toxic gas into groundwater and soil and thus contaminate them. For safe disposing of BMW, the following should be kept in mind:

- Frequent covering of BMW
- BMW should be buried for at least three months beneath municipal solid waste,
- Bottom of landfills should be water-resistant
- Layer of BMW should be at least 2 meters above the level of water
- Chemical dumping should be prohibited.

3.6 Economical and Sustainable Methods of BMW Management

With the disadvantages and limitations associated with the aforementioned treatment and disposal methods, which adversely affects not only various forms of life but also pollutes the environment, priority is being given to various low-cost and environment-friendly methods of BMW management. These methods which are in use nowadays are as follows:

3.6.1 Bioremediation

In this method, living organisms or their by-products are employed for the treatment of toxins, present in medical waste, to their less toxic or non-toxic forms. Recombinant enzymes, e.g., Horseradish Peroxidase, is used to catalyze the breakdown of organic pollutants in BMW like phenols and polycyclic aromatic hydrocarbons (PAHs). Another copper-containing recombinant enzyme, Laccase can oxidize lignin, found in some medical waste streams. Enzymes like lipases, proteases, and cellulases, are reported for degradation of proteins, cellulose and lipids. It has also been reported that a fungus found in cow dung *Periconiella sp.*, is effective in decomposing biomedical waste, such as dressings, soiled cotton pieces, and surgically removed materials, within approximately 40 days [41, 42].

Phytoremediation is also a type of bioremediation, where plants and microbes associated with it are used to reduce toxic contaminants. It has been reported that Vetiver grass (*Chrysopogon zizanioides*) effectively treats BMW containing various pathogens, organic and inorganic pollutants. A combination of tobacco and neem extracts was reported to be very cost-effective for treating BMW. Willow (*Salix spp.*) has also been reported to treat biomedical wastes [41, 43, 44].

Another eco-friendly and biological approach to treat BMW is "Bio-Converter" 9 Biomedical Disposal, Inc. [37, 45]. This device utilizes a cocktail of enzymes to disinfect the biomedical waste and produced sludge is sent to an extruder to extract water for sewage disposal and solid waste for landfilling.

3.6.2 Composting

It is a biological process that helps recycle the biodegradable organic components, present in BMW, by converting it into nutrient-rich soil fertilizer or conditioner with the help of microbes, which break down complex organic substances into simpler forms like organic acids, CO2 and H2O vapors. It is used to treat organic compounds like expired or discarded hospital food and medicines. Microbes like *Bacillus subtilis* and *Aspergillus niger*. Recombinant enzymes like cellulase, lipase, and protease are also employed to enhance the composting process [41, 44].

3.6.3 Vermicomposting

A research study investigated the potential of using earthworms to break down infectious medical waste. Researchers tested various earthworm species, both individually and together, on medical waste that had been initially sterilized with bleach and then pre-treated with cow dung. The results showed that the earthworm species *Eisenia fetida* was most effective at decomposing the medical waste compared to other species tested. Interestingly, a combination of different earthworm species worked just as well as using *Eisenia fetida* alone. These findings suggest that vermicomposting could be a safe and environmentally friendly method for managing medical waste [43].

3.6.4 Innovative and Sustainable Way to Remodulate Biomedical Waste

These biomedical wastes can be demodulated and reused in a sustainable and innovative way rather than disposing of them through incineration. The exponential increase of PPE during COVID-19 results in the generation of massive biomedical waste. The main drawback of use of PPE kit is associated with its single usage policy. Mohan *et al.* (2021) reutilized the PPE kit to prepare constructive material like eco-friendly bricks, wall panels, construction blocks and interlocking blocks [46].

3.7 Challenges During Management of COVID-19 Biomedical Waste Management

With the sudden uncontrollable rise in COVID-19 cases, several loopholes in the existing BMW management of the country came to the forefront due to the panic situation witnessed among ordinary people. An individual COVID-19 patient generated ~4.11 kg/day of BMW in Delhi, India [47]. Lack of BMW management training and awareness for the generated COVID-19 waste for the technical and/or non-technical staff and ordinary people led to a number of challenges faced in managing the infectious BMW. The inadequacy can be traced to the unsatisfactory operation of institutional and financial systems for supervision, monitoring, remuneration, and fines. The major challenges during management of COVID-19 BMW can be summed up as follows (Figure 3.5):

- The initiatives and non-compliance for BMW management systems were much impacted due to high demand of medical facilities required; to cater to this, various unauthorized local healthcare establishments operated. Due to these unaccounted operations, significant discrepancies were observed which led to the challenges faced in COVID-19 BMW collection and treatment.
- The unaccounted BMW generated at quarantine centers, home-care facilities, and at domestic levels also resulted in manifold increase in the contagious COVID-19 waste unexpectedly. The waste generated here also faced challenges of segregation due to lack of awareness, facilities, and funds.
- The high cost of segregation, disposal, and treatment of the hazardous lethal BMW generated during all the different

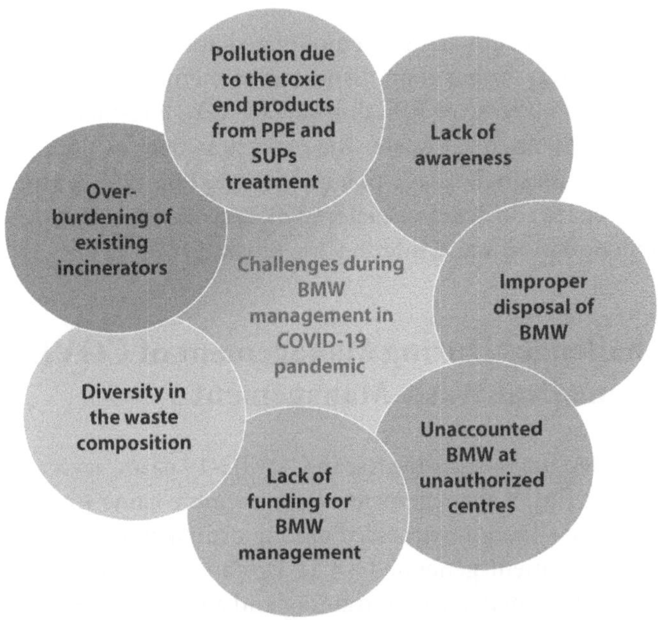

Figure 3.5 Challenges faced by biomedical waste management during COVID-19.

levels of COVID-19 care and management centers, such as PPE kits, face masks, hand gloves, face shields, plastic shoe covers, syringe needles, cotton swabs, tissues, etc., remained as the prime issue in implementing BMW management efficiently. The cost of safe BMWM during pandemic situations was reported to be ten times higher than the waste generated during non-pandemic situations [48]; which in India came into the picture as the chief problem in executing the COVID-19 biomedical waste management in an effective way [47]. This cost might have grown further considerably in India, especially for managing the contagious and deadly COVID-19 related biomedical waste.

- Common disposal activities came into practice for various categories of waste generated during COVID-19 due to inadequacy of awareness, manpower, and formal training at the quarantine centers and local unauthorized healthcare. People threw PPE kits, masks, gloves, food, and disposable cutlery into the same double-layered yellow bins which were sent for the common treatment like incineration later on [49].
- The shortage of engineered landfills for the PPE and infected single-use plastics (SUPs) disposal in India resulted in their common disposal practice along with municipal solid waste (MSW) at the local dumping sites. As PPEs can easily degrade into microplastics and SUPs can spread over the nearby land and water bodies, increasing the plastic pollution also, both may pose a serious threat to the environment *via* land and water contamination in the upcoming years.
- Due to the unexpected rise in COVID-19 cases, the surge in the generation of BMW resulted in crossing the treatment capacity of the authorized incineration plants in most of the states [24]. Despite this, the treatment plants were compelled to feed in more COVID-19 biomedical waste like PPE and SUPs, due to which the incineration efficiency was significantly reduced, resulting in incomplete combustion. During the process of incineration, the PPEs and SUPs used for COVID-19 management, generate hazardous organic pollutants like furans, polychlorinated biphenyls, dioxins, polychlorinated biphenyl, etc. [33]. Additionally, the ashes released from their incineration contained microplastics,

further increasing the long-term environmental concerns [50]. Apart from the abovementioned issues, some of the other concerns observed were lack of attention on the modification of action plans for sewage water treatment, which can be a strong source of COVID-19 infection for the involved workers and community, unorganized door-to-door collection system for COVID-19 BMW generated from home/quarantine centers, and transportation of BMW in open vehicles along with manual handling of these infectious wastes by the workers [51].

3.8 Conclusion and Future Perspectives

This chapter provides an overall view of the management and the related challenges of biomedical waste generated from the COVID-19 pandemic. The steps involved in the biomedical waste management at the local healthcare sites prior to their final disinfection, disposal, and treatment have been discussed in detail. This chapter also summarizes the economical and sustainable methods opted for COVID-19 waste generated in situations of global health crisis. The analysis of the overall scenario of biomedical waste management during COVID-19 emphasizes the designated measures and guidelines required for the conscious handling, segregation, disposal, and treatment of the highly infectious biomedical waste. Various challenges encountered while handling biomedical waste during the COVID-19 pandemic period have been discussed in detail, including the lack of awareness, funds, skilled workers, treatment plants, specifically designed transportation vehicles, etc. The chapter thoroughly discusses the legislative framework and actions, the treatment capacity of the biomedical waste management plants, and the future perspective and management strategies for the challenges faced during COVID-19 in India.

As a result of the experience of handling and managing the biomedical waste during the worldwide pandemic, people and authorities alike have learned to develop well-equipped biomedical waste management facilities and to educate people for better preparedness for any disaster situations in the future. Some of the steps that can be taken in the future to combat the pandemic situations are discussed below.

- Better funding amount should be released to develop biomedical waste segregation and disposal at medical centers and for conducting training for healthcare professionals and sanitation workers to handle the pandemic waste. Funding should also be reserved to set up mobile equipment for the BMW management to overcome the overburdening of stationary waste management stations during a pandemic. This will also help in managing the state-wise variable biomedical waste amount; for instance, in India, the number of active COVID-19 cases were reportedly more in southern states than in northern states, generating similar trends in biomedical waste generation. So, the mobile incineration units can be shifted as per the requisite.

- A universal management plan should be set up instead of making localized policies. The local waste management authorities should provide specific color-coded containers and bags to the quarantine centers or home isolation units to segregate the infected PPEs, gloves, face masks, shoe covers, etc., to improve on-site COVID-19 waste management. This will help in managing later stages of biomedical waste degradation effectively.
- The designs of the existing incinerators can be improved by incorporating appropriate monitoring devices for controlling air pollutants released during the incineration process, for instance, organic contaminants like dioxins, PCPs, furan, etc., can be removed by utilizing filters based upon activated carbon. The new incinerators shall be designed with these activated carbon filters along with larger capacity to tackle the issue of overburdening as well with environmentally benign products.
- With advancement in IoT (Internet of Things), better technologically sound automated BWM management systems can be developed involving the least amount of manpower. The existing facilities can also be improved using artificial intelligence for the steps involved in biomedical waste management like segregation, treatment, processing, etc., to reduce involvement of workers.
- The mobile apps can be developed for managing the door-to-door collection and treatment and maintaining area-wise databases by the waste management authorities, especially to handle any such pandemic-generated infectious biomedical waste in future. The database generated will also help in modifying action plans and preparing protocols well in advance during such global crisis situations.

References

1. Jindal, M.K. and Sar, S.K., Medical waste management during COVID-19 situation in India: Perspective towards safe environment. *Waste Manage. Bull.*, *1*, 1, 1–3, 2023, https://doi.org/10.1016/j.wmb.2023.03.002.
2. Chand, S. and Hiremath, S., Updates on biomedical waste management during COVID-19: The Indian scenario. *Clin. Epidemiol. Global Health*, *11*, 100715, 2021, https://doi.org/10.1016/j.cegh.2021.100715.
3. Kumar, S.R., Abinaya, N.V., Venkatesan, A., Natrajan, M., Bio-medical waste disposal in India: From paper to practice, what has been effected. *Indian J. Health Sci. Biomed. Res. (KLEU)*, *12*, 3, 202, 2019, https://doi.org/10.4103/kleuhsj.kleuhsj_112_19.
4. Worldometer, Coronavirus Toll update: Cases & Deaths Country, Worldometer, 2024, April 13, https://www.worldometers.info/coronavirus/.
5. Central Pollution Control Board, Cpcb.nic.in, 2019, https://cpcb.nic.in/covid-waste-management/.
6. Das, A.K., Islam, M.N., Billah, M.M., Sarker, A., COVID-19 and municipal solid waste (MSW) management: a review. *Environ. Sci. Pollut. Res. Int.*, 28, 23, 28993–29008, 2021, doi: 10.1007/s11356-021-13914-6.
7. SARS-CoV-2 in wastewater: State of the knowledge and research needs. *Sci. Total Environ.*, *739*, 139076, 2020, https://doi.org/10.1016/j.scitotenv.2020.139076.
8. Kumar, M., Patel, A.K., Shah, A.V., Raval, J., Rajpara, N., Joshi, M., Joshi, C.G., First proof of the capability of wastewater surveillance for COVID-19 in India through detection of genetic material of SARS-CoV-2. *Sci. Total Environ.*, *746*, 141326, 2020, https://doi.org/10.1016/j.scitotenv.2020.141326.
9. Núñez-Delgado, A., SARS-CoV-2 in soils. *Environ. Res.*, *190*, 110045, 2020, https://doi.org/10.1016/j.envres.2020.110045.
10. Central Pollution Control Board, Cpcb.nic.in. Retrieved December 25, 2022, from https://cpcb.nic.in/bio-medical-waste-rules/.
11. Golandaj, J.A. and Kallihal, K.G., Awareness, attitude and practises of biomedical waste management amongst public health-care staff in Karnataka, India. *J. Humanit. Appl. Social Sci.*, 3, 1, 49–63, 2021, https://doi.org/10.1108/JHASS-08-2019-0041.
12. Hosny, G., Samir, S., El-Sharkawy, R., An intervention significantly improves medical waste handling and management: a consequence of raising knowledge and practical skills of health care workers. *Int. J. Health Sci.*, 12, 4, 56–66, 2018.
13. Joseph, L., Paul, H., Premkumar, J., Rabindranath, Paul, R., Michael, J.S., Biomedical waste management: study on the awareness and practice among healthcare workers in a tertiary teaching hospital. *Indian J. Med. Microbiol.*, 33, 1, 129–131, 2015.
14. Yousefi, M., Oskoel, V., Jafari, A.J., Farzadkia, M., Firooz, M.H., Abdollahinejad, B., *et al.*, Municipal solid waste management during

COVID-19 pandemic: effects and repercussions. *Environ. Sci. Pollut. Res. Int.*, 28, 25, 32200–32209, 2021, doi: 10.1007/s11356-021-14214-9.
15. Khosla, R., Jha, A., Dua, S., Varmani, S.G., Rajput, N., Pani, B., Upsurge in biomedical waste due to COVID-19 in India: A statistical correlation, challenges and recommendations. *Front. Environ. Sci.*, 10, 2022, doi: 10.3389/fenvs.2022.1022098.
16. World Health Organization (WHO), Rational use of personal protective equipment for COVID-19 and considerations during severe shortages: interim guidance, 23 December 2020, [cited Mar. 19, 2023]. Available from: https://iris.who.int/handle/10665/338033?show=full.
17. Das, A.K., Islam, M.N., Billah, M.M., Sarker, A., COVID-19 pandemic and healthcare solid waste management strategy - A mini-review. *Sci. Total Environ.*, *778*, 146220, 2021, https://doi.org/10.1016/j.scitotenv.2021.146220.
18. Selvaranjan, K., Navaratnam, S., Rajeev, P., Ravintherakumaran, N., Environmental challenges induced by extensive use of face masks during COVID-19: A review and potential solutions. *Environ. Challenges*, *3*, 100039, 2021, https://doi.org/10.1016/j.envc.2021.100039.
19. Fadare, O.O. and Okoffo, E.D., Covid-19 face masks: a potential source of microplastic fibers in the environment. *Sci. Total Environ.*, 737, 2020.
20. Central Pollution Control Board, Annual Report on Biomedical Waste Management, Retrieved August 6, 2024, from https://cpcb.nic.in/uploads/Projects/Bio-Medical-Waste/AR_BMWM_2022.pdf.
21. Bhawan, P. and Nagar, A., Revised Guidelines for Common Bio-medical Waste Treatment and Disposal Facilities, Central Pollution Control Board, (Ministry of Environment, Forest and Climate Change). Retrieved August 6, 2024, from https://cpcb.nic.in/uploads/Projects/Bio-Medical-Waste/Common_Bio_Medical_Waste_treatment_facilities.pdf.
22. Savino, A. and de Titto, E., The COVID-19 pandemic challenges the future. *Waste Manage. Res.*, 2020, 0734242X2094298, https://doi.org/10.1177/0734242x20942989.
23. Thind, P.S., Sareen, A., Singh, D.D., Singh, S., John, S., Compromising situation of India's bio-medical waste incineration units during pandemic outbreak of COVID-19: Associated environmental-health impacts and mitigation measures. *Environ. Pollut.*, 276, 116621, 2021, doi: 10.1016/j.envpol.2021.116621.
24. Dehal, A., Vaidya, A.N., Kumar, A.R., Biomedical waste generation and management during COVID-19 pandemic in India: challenges and possible management strategies. *Environ. Sci. Pollut. Res. Int.*, 29, 10, 14830–14845, 2022, doi: 10.1007/s11356-021-16736-8.
25. Bhar, A., Biswas, R.K., Choudhury, A.K., The influence of COVID-19 pandemic on biomedical waste management, the impact beyond infection. *Proc. Indian Sci. Acad.*, 88, 117–128, 2022, doi: 10.1007/s43538-022-00070-9.

26. Saxena, P., Pradhan, I.P., Kumar, D., Redefining bio-medical waste management during COVID-19 in India: A way forward. *Mater. Today Proc.*, 60, 849–858, 2022, doi: 10.1016/j.matpr.2021.09.507.
27. Emanuel, E.J., Persad, G., Upshur, R., Thome, B., Parker, M., Glickman A., Phillips J.P., Fair Allocation of Scarce Medical Resources in the Time of Covid-19. *N. Engl. J. Med.*, 382, 21, 2049–2055, 2020, doi: 10.1056/NEJMsb2005114.
28. World Health Organization, *Global analysis of health care waste in the context of COVID-19*, 2022, February, who.int.https://www.who.int/publications/i/item/9789240039612.
29. Singh, M., Karimi, N., Wai Ng, K.T., Mensah, D., Stilling, D., Adusei, K., Hospital waste generation during the first wave of COVID-19 pandemic: a case study in Delhi. *Environ. Sci. Pollut. Res. Int.*, 29, 33, 50780–50789, 2022, doi: 10.1007/s11356-022-19487-2.
30. Singhal, L., Tuli, A.K., Gautam, V., Biomedical Waste Management Guidelines 2016: What's done and what needs to be done. *Indian J. Med. Microbiol.*, 35, 2, 194–198, 2017, doi: 10.4103/ijmm.ijmm_17_105.
31. Bansod, H.S. and Deshmukh, P., Biomedical Waste Management and Its Importance: A Systematic Review. *Cureus*, 15, 2, 2023, 10.7759/cureus.34589.
32. Central Pollution Control Board, Revision 4 Guidelines for Handling, Treatment and Disposal of Waste Generated during Treatment/Diagnosis/Quarantine of COVID-19 Patients 17, 2020, https://cpcb.nic.in/uploads/Projects/Bio-Medical-Waste/BMW-GUIDELINES-COVID_1.pdf.
33. Singh, N., Tang, Y., Ogunseitan, O.A., Environmentally Sustainable Management of Used Personal Protective Equipment. *Environ. Sci. Technol.*, *54*, 14, 2020, https://doi.org/10.1021/acs.est.0c03022.
34. Central Pollution Control Board, Revision 5 Guidelines for Handling, Treatment and Disposal of Waste Generated during Treatment/Diagnosis/Quarantine of COVID-19 Patients 26, 2022, https://cpcb.nic.in/uploads/Projects/Bio-Medical-Waste/BMW-GUIDELINES-COVID_2.pdf.
35. Ilyas, S., Srivastava, R.R., Kim, H., Disinfection technology and strategies for COVID-19 hospital and bio-medical waste management. *Sci. Total Environ.*, *749*, 141652, 2020, https://doi.org/10.1016/j.scitotenv.2020.141652.
36. Nzediegwu, C. and Chang, S.X., Improper Solid Waste Management Increases Potential for COVID-19 Spread in Developing Countries. *Resour. Conserv. Recycl.*, *161*, 104947, 2020, https://doi.org/10.1016/j.resconrec.2020.104947.
37. Datta, P., Mohi, G.K., Chander, J., Biomedical waste management in India: critical appraisal. *J. Lab. Physicians*, 10, 6–14, 2018, doi: 10.4103/JLP.JLP_89_17.
38. Wang, J., Shen, J., Ye, D., Yan, X., Zhang, Y., Yang, W., Li, X., Wang, J., Zhang, L., Pan, L., Disinfection technology of hospital wastes and wastewater: suggestions for disinfection strategy during coronavirus disease 2019 (COVID-19) pandemic in China. *Environ. Pollut.*, 262, 2020, doi: 10.1016/j.envpol.2020.114665.

39. Duarte, P. and Santana, V.T., Disinfection measures and control of SARS-COV-2 transmission. *Global Biosecur.*, 1, 3, 2020.
40. Hirani, D.P., Villaitramani, K.R., Kumbhar, S.J., Biomedical waste: An introduction to its Management. *Int. J. Innov. Res. Adv. Eng.*, 1, 8, 82–87, 2014.
41. Chandra, V., Arpita, K., Yadav, P., Raghuvanshi, V., Yadav, A., Prajapati, S., Environmental biotechnology for medical waste management: A review of current practices and future directions. *Ann. Biomed. Sci.*, 6, 2, 1–9, 2023, doi: 10.23880/aabsc-16000214.
42. Rajan, R., Robin, D.T., Vandanarani, M., Biomedical waste management in Ayurveda hospitals-current practices and future prospective. *J. Ayurveda Integr. Med.*, 10, 3, 214–221, 2019.
43. Dinesh, M.S., Geetha, K.S., Vaishnavi, V., Kale, R.D., Murthy, V.K., Ecofriendly treatment of biomedical wastes using Epigeic earthworms. *J. Indian Soc. Hosp. Waste Manage.*, 9, 1, 5–20, 2010.
44. Patil, P.M. and Bohara, R.A., Nanoparticles impact in biomedical waste management. *Waste Manage. Res.*, 38, 11, 1189–1203, 2020, doi: 10.1177/0734242X20936761.
45. Emmanuel, J., Non-Incineration Medical Waste Treatment Technologies, Health Care without Harm, Washington, DC, 2001.
46. Mohan, H.T., Jayanarayanan, K., Mini, K.M., A sustainable approach for the utilization of PPE biomedical waste in the construction sector. *Eng. Sci. Technol. Int. J.*, 2021, https://doi.org/10.1016/j.jestch.2021.09.006.
47. Goswami, M., Goswami, P.J., Nautiyal, S., Prakash, S., Challenges and actions to the environmental management of bio-medical waste during COVID-19 pandemic in India. *Heliyon*, *7*, 3, e06313, 2021, doi: 10.1016/j.heliyon.2021.e06313.
48. World Health Organization, Health-care Waste Fact sheet Updated February 2018, http://www.who.int/mediacentre/factsheets/fs253/en/, 2018.
49. Basavaraj, T.J., Shashibhushan, B.L., Sreedevi, A., To assess the knowledge, attitude and practices in biomedical waste management among health care workers in a dedicated COVID hospital in Bangalore. *Egypt J. Intern. Med.*, 33, 1, 37, 2021, doi: 10.1186/s43162-021-00066-9.
50. Shen, M., Hu, T., Huang, W., Song, B., Qin, M., Yi, H., Zeng, G., Zhang, Y., Can incineration completely eliminate plastic wastes? An investigation of microplastics and heavy metals in the bottom ash and fly ash from an incineration plant. *Sci. Total Environ.*, 779, 146528, 2021, https://doi.org/10.1016/j.scitotenv.2021.146528.
51. Mondal, R., Mishra, S., Pillai, J.S.K., Sahoo, M.C., COVID 19 Pandemic and biomedical waste management practices in healthcare system. *J. Fam. Med. Primary Care*, 11, 2, 439–446, February 2022, DOI: 10.4103/jfmpc.jfmpc_1139_21.

4

Role of Artificial Intelligence in Waste Management

Diksha Srivastava

Department of Biotechnology and Chemical Engineering, Manipal University Jaipur, Jaipur, Rajasthan

Abstract

Waste disposal and management is one of the major challenges faced today in attaining sustainable development goals. Due to uncontrolled waste generation either by the domestic or industrial sectors, waste management is the need of the hour. Deteriorating environmental and public health conditions due to improper waste management systems pose a great threat to the current situation. An integrated waste management system combined with artificial intelligence (AI) provides a great solution to improve the current strategies and policies adapted for waste management in most countries. Artificial intelligence not only speeds up the rate of waste management but also improves the quality of management by better identification, segregation and accurate recycling of the waste generated. AI helps not only in waste recycling but also in developing and innovating new technologies as per the situation demand, consumer behavior pattern and waste generation pattern analysis. This chapter will provide a detailed overview of the use of artificial intelligence for efficient waste management for encouraging sustainable development.

***Keywords*:** Artificial intelligence, smart waste management, machine learning, image processing, recycling, convolution neural network

4.1 Introduction

Exponential population growth and advancements in urbanization have tremendously increased the waste generation rate in the world. According to

Email: diksha.srivastava@jaipur.manipal.edu

Priyanka Singh, Pooja Agarwal and V. Vivekanand (eds.) Solid Waste Management: Challenges, Sustainability and Advancements, (117–138) © 2026 Scrivener Publishing LLC

the latest statistics, 2.01 billion tonnes of municipal solid waste was generated globally in 2016. This figure is expected to increase to 3.4 billion tonnes by 2050 [1]. The waste is generated from all the sectors of services such as production, processing, commercial, agricultural, etc. These wastes are different from each other in one or many ways; some may differ in their heat resistance, some in size, some are organic, etc. All these types of waste combined constitute different categories of waste such as domestic, construction, processing, medicinal, etc. The most basic type of waste classification is according to its origin or source, which is depicted in Figure 4.1.

As it is rightly said, the 3 Rs—Reduce, Reuse and Recycle—are the heart of any of the waste management policies. These not only contribute to waste reduction and management but also strongly support sustainable development. Improper waste disposal is not only a potent environmental threat, but it drastically effects the health and mortality rate of the surroundings [2]. In the present context, where developed countries are more focused towards handling and reducing the waste generation rate, developing countries are still greatly in need of waste disposal and storage. The major lacuna in both sets of countries is the age-old practices for waste management, which involves traditional practices. These traditional methods of waste management include more of human intervention, which not only increases the time required for the whole process to be completed but also makes the humans involved in it more prone to infection and poses

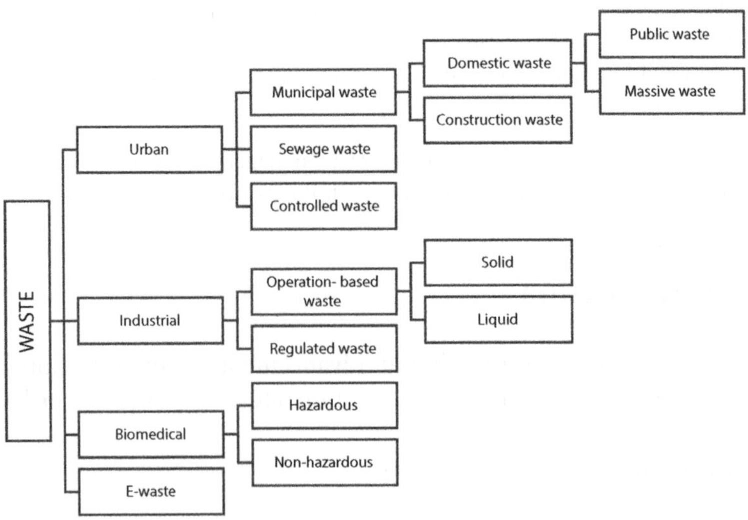

Figure 4.1 Waste classification according to source or origin.

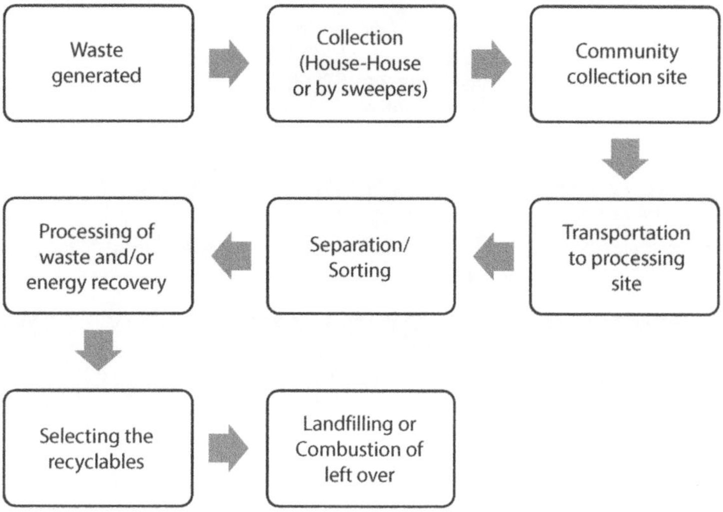

Figure 4.2 Waste treatment practices.

detrimental health hazards. These waste management practices include the steps as shown in Figure 4.2.

Irrespective of the development status of the country, waste management is a serious problem which needs to be addressed with priority. The nature or type of waste generated has also differed over time. As a result of technological advancement, the rise in electronic, medical, and biomedical waste has posed a serious problem. Not only waste generation but waste sorting is also an issue which needs to be addressed due to different routes of waste management. The unexpected fluctuations in waste composition and quantity require a dynamic response from policymakers [3], which will lead to an increase in the demand for new products and services that can balance resources and waste management with health and safety concerns [4]. Even though a huge amount has been invested in improving waste management, no remarkable achievements have resulted so far. It has been noticed that these investments are mainly focused on waste collection and transportation rather than waste segregation or treatment [7]. The following block diagram, Figure 4.3, illustrates the issues which need to be addressed to resolve the issues raised by improper waste management.

Therefore, a new technology is needed which can not only give advancements in collection, sorting and managing the waste generated but can also keep a check on the rate of composition of waste generated, long-term operating costs of waste management and the geographical location

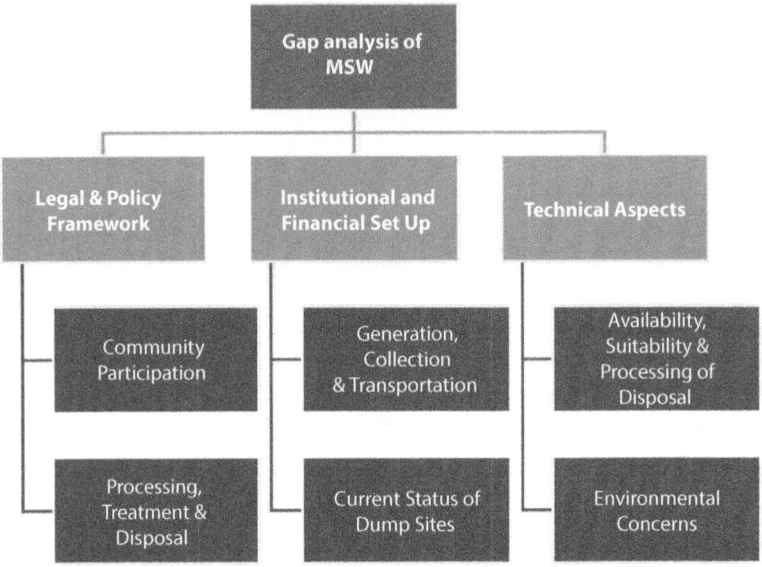

Figure 4.3 Issues that need to be addressed for proper waste management. (Adapted from: M D Meena *et al.*, [11]).

of waste processing sites. These studies would help to define the updated management and recycling policies which can be adapted to reduce and reuse the waste. This chapter elucidates the intervention of artificial intelligence in waste management which has been established and implemented for attaining sustainable development.

4.2 Artificial Intelligence as a Revolutionary Innovation in Waste Management

Artificial intelligence can be explained as a process or tool in which machines mimic human intelligence by acquiring cognitive abilities like that of a brain. The basic idea behind the development of artificial intelligence is to make a tool which can perceive and interpret information just like a brain but in a much faster rate with more accuracy. The implementation of artificial intelligence in the environmental engineering field is gaining attention nowadays mainly in reducing the pollution load and carbon mitigation. With the use of satellite data new policies regarding reducing the energy load have also been proposed for waste management. This technology initially uses historic or previously collected data, and using the supervised learning algorithm it trains and learns to solve the problem statement. In simple terms the working of artificial intelligence can be depicted as shown in the following workflow as Figure 4.4:

With the increasing complexities in waste generated and waste management, using traditional techniques is a perplexing task. But with the involvement of artificial intelligence, this task could be done with more accuracy without harming biological lives. Not only this, these artificial intelligence–based models can also predict and propose the modifications needed in the future for waste management [8].

In the present context, the main problem faced today is the waste collection and segregation according to the treatment types. By using artificial intelligence, a huge amount of waste can be handled at a time and can be easily collected, sorted and predicted for treatment [9, 58]. Artificial intelligence uses various prediction models to predict and estimate the waste generated and also suggest the options for waste treatment. Some of the innovations and advancements in this field are waste characterizing, waste compression, process and its output prediction, vehicle routing according to waste-bin fullness level, etc. Combining technologies like Robotics, Geographical Integration Systems (GIS), Smart-chip-based Sensor technology, Data access networks and Combinatorial optimization this could be done more accurately. Even this waste could be used to make many carbon neutral and environmentally friendly value-added products like solar panels, etc. Therefore, it will be appropriate to state that artificial intelligence plays a critical role in transforming today's world into a "zero

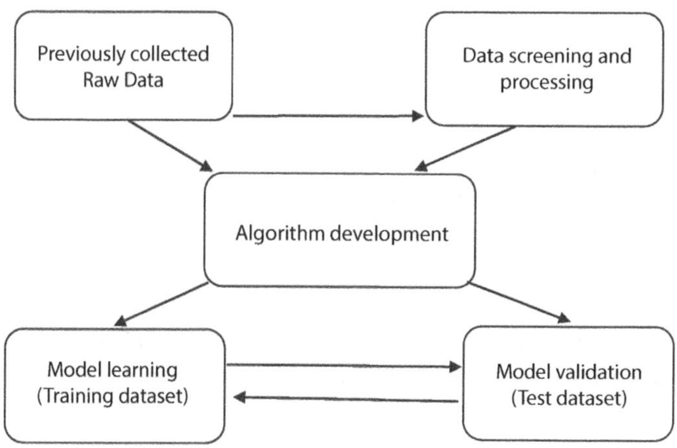

Figure 4.4 Working of artificial intelligence.

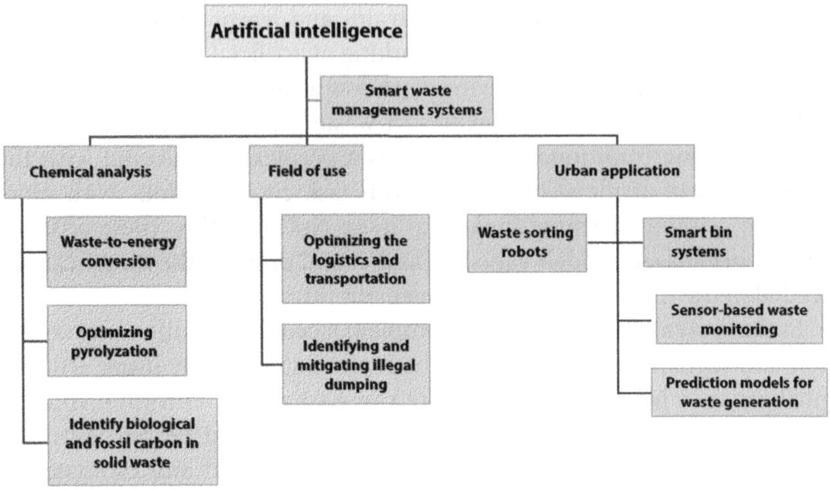

Figure 4.5 Application of artificial intelligence in waste management. (Adopted from: Fang, B. *et al.*, [12]).

waste circular economy" which will not only improve the social and environmental factors but also the economic factors [10].

The waste management system consists mainly of five different stages: collection, transportation, analysis and processing, recycling, and disposal. Figure 4.5 illustrates the role of artificial intelligence in a smart waste management system with respect to these stages.

4.3 Smart Bin Systems

Several methods and technologies have been applied to waste management systems yet one of the major problems faced is waste collection. This not only increases the manpower required to collect but also poses severe health and sanitation threats. Waste collection is one of the most time-consuming and labor-intensive steps in which the major share of the waste management expenses is utilized. With the integration of Internet of Things (IoT) and data access networks this issue can be solved and managed easily.

Waste bins, or as popularly known "smart bins," are IoT-based sensor chip–enabled bins which are integrated with data access networks using satellites. These bins can sense, collect and send the waste volume collected in respective bins. Once filled or on reaching the threshold level these bins generate real-time signals using graph theory optimization algorithms so that the bins could be emptied. These signals are then sent using satellites or Geographic Information System (GIS) to the waste collection centers for emptying these bins.

The principal behind this technology is echolocation combined with microelectronics. The sensor chip embedded in the smart bins generates high-level acoustic signals whose echoes are perceived by ranging modules and microcontrollers located in the signal receptor towers. Various types of sensors like temperature sensors and weight sensors, along with metal detectors, are mostly used for detecting and sorting the biodegradable, recyclable and reusable items from the trash cans. The time taken from the signal transmission to the signal reception is calculated and is used to locate the position of the smart bin, whereas the intensity of signals received by different types of sensors is used for the quantification analysis of the filled status of smart bin. Ardino Uno microcontrollers are mostly used to receive the signals transmitted. These are one of the most used and available microcontrollers which can work with high accuracy over a range of operating voltage. Usually at the time of microcontroller selection for these smart bins, parameters like minimal power consumption, memory, processing and cost are considered. The data collected by these

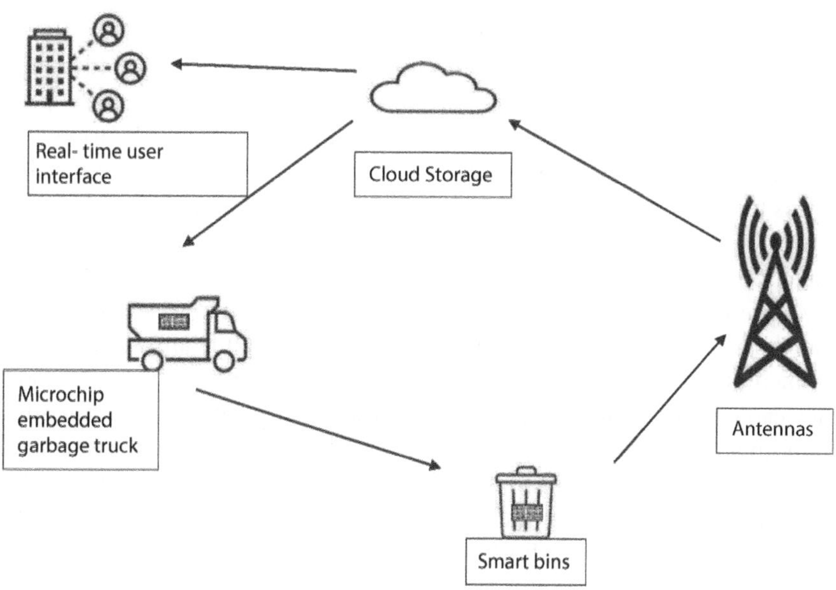

Figure 4.6 Smart bins in waste management.

microcontrollers is then sent to the cloud where it is stored and retrieved at the time of need. The path of signal transmission and signal movement along with its angel should be unique for each sensor to avoid the mixing or loss of signals. This phenomenon is called "crosstalk" in technological terms.

This technology is being used in many developed cities like Copenhagen, Amsterdam, New York, or San Francisco. Using this technology, the waste bins can be emptied before the waste can exceed the capacity of the bin along with reducing the manpower required for surveillance in the waste collection. Figure 4.6, IoT-based sensor chip–enabled smart bin systems for efficient waste collection.

This technology can cut approximately 50% of the total waste management costs by saving the fuel and manpower required. This technology has been applied not only on bins but also on the garbage trucks to ensure their on-time emptying at the waste collection or processing sites. Nowadays these bins are even connected with cellular Wi-Fi so that they can be accessed easily and can share their location to the surrounding people. Integrating cameras, sensors, Wi-Fi or radio-frequency identification (RFID) tracking systems are just a few advancements made in this technology for better access to bins, surveillance of the bin filling level, and asset and manpower management in waste management.

The use of radio-frequency identification (RFID) tracking systems has recently gained the interest of many researchers. These radio technology–inspired devices are combinations of antennas and embedded microchips. The microchips have their own unique ID. These devices use radio tags which work by transmitting radio frequency signals, which are then perceived by antennas. The data is stored in the radio tags which stores the information in the form of numbers. The added advantage of these radio tags is that they have in them an in-built battery which activates when the source battery is not available. But their main limitation is the low range detection.

At present the focus of researchers is on automation in waste collection and the bin monitoring. As these smart bins can significantly reduce the problems caused by waste littering and spread of diseases through waste, they can potentially improve the health and sanitary conditions of the surroundings. However, as promising as these bins look, the cost associated with their manufacturing and implementing are not economically feasible. Also, environmental factors such as humidity, rainfall or temperature considerably impacts the working and monitoring of them. Regular check-ups and maintenance of these bins is also a crucial point which must be considered before implementing them on a commercial level.

The data stored in the cloud can also be used for various statistical analysis for determining the rate of waste generation or can act as a dataset for prediction models used for waste generation prediction. Also, this data will help us to identify the sites based on the quantity and type of waste generated, and using prediction models it will help to identify the routes of processing required. It will also help us to identify the routes of waste processing that should be more focused on.

4.4 Waste Monitoring and Prediction

This is a recent technology which helps in the identification and quantification of the waste generated with the use of Wireless Sensor–based technology. This technology, when integrated with the smart bins, can help in the real-time assessment of environmental parameters like humidity, temperature, pollution load, etc., in a Zigbee network structure [19]. In this structure terminal nodes are installed in the smart bins which help us to detect the level of bin fullness. But the drawback of this suggested model was that it was battery operated, which needs charging or replacement. In such cases, resources are needed, which can increase the complexity as well as the need of proper monitoring. Therefore, a solar power–based node network was suggested to overcome these drawbacks and increase the lifespan of developed systems [20, 21]. This system was further improved by installing a progressive bar which uses graphical user interface to show the type of waste and bin-filling status in a smart bin [23].

Machine Learning, a new advanced method which works on a defined dataset, statistics and complex algorithms, can also be used to solve this complex problem of waste prediction. This technology uses supervised learning algorithms and can complete tedious tasks with great precision and perfection. The dataset used to train these programs are past historical data and usually use regression for analysis. The algorithms include Artificial Neural Networks, Support Vector Machines, Naïve Bayes, K-nearest neighbor, decision trees, random forest and adaptive network fuzzy inference system (ANFIS). All these algorithms have their own benefits and limitations and can be used according to the type of dataset.

Statistical prediction based on Machine Learning and Fuzzy models about the quantity and composition of waste generated have also gained the attention of researchers. The models based on Artificial Neural Networks are the most dependable in the current scenario. Their predictability is more than accurate, which helps in the new policy building and altering the existing policies of waste management as per the demands of current population waste generation trends. The past records of the waste generated, and their composition can be used as a dataset for better predictability. Use of Big Data technology can help a lot in better predicting by handling huge data sets at a time with precision.

Models for waste generation prediction are mainly reliant on various factors such as demographic data, environmental conditions, economic growth and individual behavior, etc. According to the literature reviewed, Artificial Neural Network is the most commonly used method but its performance is affected by data over-fitting, local minima and poor generalization.

These models are still in the development stage and more focus and innovations are needed for improvising them to the applicable form. Many researchers are now focused on these prediction models based on artificial intelligence using support vector regression, linear regression, decision trees and genetic algorithm. The use of real-time geo-localized data can enhance the precision and efficiency of these models with proper information. The major problem in the development of such models is the variation in the environmental and societal differences which creates a huge number of small datasets with minor differences.

4.5 Waste-Sorting Robots

Mechatronics is a branch of multidisciplinary engineering which combines technologies from mechanics, electronics, electrical engineering and computer science. This technology when integrated with artificial intelligence or machine learning produces very potent applications. Robots can be said to be the mechatronic substitute of humans. These substitutes, when imparted with the intelligence of humans through artificial intelligence, can create wonders by reducing human intervention in the given tasks without compromising the efficiency of the task in hand.

In the context of waste management, waste segregation is considered as one of the most tedious and hazardous tasks with the maximum human intervention needed. Most of the rag pickers or waste segregators are the part of the population associated with waste management who are more prone to diseases or sanitation problems. The major reason behind this is the lack of public awareness and education for waste segregation at the site of waste generation only. Source segregation of waste can be a boon for waste management as it not only reduces the difficulties of waste treatment but also creates a vast possibility for recycling of waste. The science where robots are used for sorting the waste by characterizing and identifying it into recyclable and non- recyclable is known as Data Annotation. This technique is more reliable and accurate as compared to human efficiency in waste segregation.

In 2011, Zen Robotics came out with a practical solution to this problem by creating robots with a combination of machine learning and artificial intelligence. These robots can segregate waste with the help of metal sensors, 3-D laser cameras, Machine Learning and Artificial Neural Networks (ANN) into recyclable, degradable and non-degradable and non-reusable materials. These robots are used just before the waste materials are sent for processing or treatment. Artificial Neural Networks and Hyperspectral imaging techniques are used for segregating plastics and electronic wastes, whereas Multilayer Perception–based models are used for segregating recyclable and non- recyclable waste materials. Using Deep Learning technologies like simultaneous localization and mapping technology and instance segmentation, robots can segregate construction and demolition waste materials. The five most used deep learning architectures are DenseNet121, DenseNet169, InceptionResnetV2, MobileNet, and Xception. The overall efficiency of these robots depends on the quantity of waste segregated along with the efficiency by which it has been segregated. Using Adaptive and Interactive Modelling Systems (AIMS) a high throughput and efficient system could be devised. These systems give us the real-time acquisition data with a very high speed. AIMS is a meta-learning algorithm as it can improve itself and decide its parameters according to the situation.

Use of optical sensors in such robots has recently been focused on for better waste segregation. Segregation based on the differences in size, shape, dimensions, color, etc., using different cameras and sensors will help in better segregation. With the advancement of technologies integrated in these robots, waste segregation can be significantly simplified without compromising efficiency. But the cost of their development, installation and maintenance is again a major constraint in their application. Most of the research nowadays is focused on reducing the overall cost of the production of these robots and on improving their efficiency.

Mobile robotic arms with Pi cameras are also available for on-site waste segregation. These robotic systems are much smaller in size than the conventional robots and can easily segregate waste. The image is first captured by a Pi camera and then processed and detected by various algorithms. On detection, these arm-like robotic systems pick the waste and segregate it into its type. Various bins are allocated for different types of waste generated. These arms, based on supervised learning, segregate the waste and drop it in its type of waste bin. For object detection these robotic arms use Haar Cascade files, which enables waste detection in picture or video form. Using such files, waste detection and segregation is done easily and with accuracy. Sensors are also placed on these robotic arms for providing precision and accuracy to the whole process. The different types of sensors mostly used are infrared sensors for waste bin filling level detection, copper clads for moisture detection, ultrasonic sensors for biodegradable and non-biodegradable waste detection, inductance-based electronic sensors for metal detection, and so on.

4.6 Chemical Analysis and Waste Treatment

Waste to energy is the new paradigm that has been evolving these days. Generating energy and energy resources from waste is the most popularly researched area and novel techniques are developed every day. These techniques not only help to manage the waste and reduce the pollution load but also address the issue of energy management. There are many ways by which waste can be converted into energy resources, but the major technologies are either thermo-chemical processes or biochemical processes. Both these technologies are serving as better alternatives than traditional practices of waste management like landfilling or combustion. The use of machine learning in waste analysis is gaining great interest from researchers.

Among various thermo-chemical processes specifically pyrolysis and gasification are the two technologies in which technological advancements are significant. In both technologies the biomass is converted into biochar, gas, or oil. All these products reduce the carbon footprint and offer a wide range of valuable products like biodiesel, biochar, or the producer gas. Raw material traits which are included in these algorithms for determining the waste characteristics are composition of different types of waste. These are mostly found based on the dataset provided by history of the location, physical parameters like moisture or texture analysis, and particle size (mm). These models will also provide suggestions on the various catalyst or reagents that can be used for the treatment adopted and their products formed. Additionally, these models can be utilized to optimize hetero-catalyst loading during hydrothermal gasification and replicate the sodium hydroxide-catalyzed hydrothermal gasification of waste biomass to investigate the environmental impact of the overall process adopted.

Researchers from all over the world are now exploring the new paradigm of waste characterization using artificial neural network and various imaging techniques like infrared thermal imaging, etc. Infrared thermography is an imaging technique which helps to predict the thermo-chemical changes occurring in the reaction. These integrated technologies not only help in analyzing the nature of the waste product but also in determining the chemical reaction involved in its remediation with high precision and speed. Some of such developed models have been adopted at the industrial level to reduce the time required for research and process development. Certain miniatured bioreactors are also designed to analyze and remediate the waste in less than a minute with the use of microquantities of reactants. These bioreactors have drastically reduced not only the overall cost of the waste management process but also the problem associated with the space acquired by the waste management plants.

Estimation about the carbon emission is also one of the major deciding factors when it comes to the feasibility of the process adopted. Machine learning algorithms such as random forests and support vector machines have become powerful tools for identifying hidden relationships that can predict the characteristics of different solid waste groups using established datasets and literature data. However, there are three main reasons why these methods may not be widely adopted: limited data size and quality can hamper performance [35], poor interpretability can make it challenging to extract relevant information [36] and heavy computation requirements can result in extended processing times [37]. Machine learning techniques, such as random forests and support vector machines, can uncover latent connections and forecast the properties of solid refuse groups. The carbon content of biological sources and fossils can be calculated in terms of mass using infrared spectroscopy and machine learning, allowing researchers to evaluate the effect of solid refuse burning on decreasing carbon emissions and saving substantial labor and reagents. In biological processing, machine learning algorithms can be used to separate impurities from raw materials, compost, and solid digests. This can help reduce possible environmental risks and improve the profitability of compost and anaerobic digestion products [38–40].

The most popular biochemical technology adopted for waste treatment is composting. All the organic waste can be used as a raw material for composting. It is a method in which microorganisms degrade or decompose the organic matter into a nutrient-enriched soil like material, called compost. The temperature and environmental conditions have to be strictly maintained during composting so that the microorganisms can grow and flourish by using the organic matter. As stated in many research papers and articles, the waste generated has the maximum share of organic material in it. Therefore, by using various segregating methods we can segregate the organic component of the waste and that can be used for composting. After composting we not only have got rid of foul-smelling waste but also have created a value-added source for soil enrichment. Artificial intelligence in many ways can help us to improve and check the progress of the composting process by controlling the environmental parameters for composting. Using machine learning the maturity of compost and the real-time progress in composting can be tracked.

The composting process is defined by various characteristic changes occurring in the waste or organic matter. The color, texture, odor, moisture and temperature are the prime parameters which change during the composting process. A few studies have attempted to use traditional Artificial Neural Network and Convolutional Neural Network to learn the morphology, texture, and color characteristics of compost images and establish compost maturity prediction models [56, 57]. These models are still in the development stage and have many research gaps which need to be taken care of.

4.7 Recycling of Waste Using AI

Improper waste disposal and management not only creates environmental and health hazards but often many resources are lost due to lack of knowledge about recycling. Recycling not only helps in resource conservation but also, if appropriate recycling approaches are adapted, the issue associated with waste management can be minimized. Use of artificial intelligence in waste sorting can aid in recycling but if artificial intelligence can be used in waste identification based on its recyclability, then it can be an outstanding approach. Various models have been proposed and some are still in their research phase where waste can be recognized by various levels of screening for its ability to be recycled. However, the algorithm has limitations, such as only identifying one waste type in mixed waste and low accuracy in classifying paper, tissue, and fabric.

Deep learning methods with convolutional neural networks are one of the major advancements in waste identification. Convolutional neural networks are one of the most significant methods of deep learning. In this method, algorithms based on material types, shape or nature are used for waste identification using image recognition. These models are associated with great efficiency and speed. The major limitation associated with such models is inaccuracy, due to a variety of waste generated and location-based variation in the type of waste generated. Application of transfer learning in such cases has been proved to be of great importance in terms of increasing the efficiency of the model. But still these models are in the development stage and need more focus. Many models are reported to provide better efficiency but either the cost is very high, or the technology has some glitches in it. Some models that have overcome both the problems can recycle only a particular type of waste material, either textile, plastic or organic, etc.

But recently, Thumiki and Khandelwal [42] have developed a real-time mobile application based on integration of image recognition and convolutional neural network which can classify waste by the type of material used in it to classify it as recyclable. In summary, artificial intelligence, particularly deep learning models based on convolutional neural networks, has been shown to be effective in classifying waste according to its material and shape. These models can accurately identify and classify different types of waste, which can help with waste management and recycling efforts.

4.8 Limitation and Prospects

Artificial intelligence is proving to be a boon in every field where human interference is limited or should be limited. But like any other technology it also has some limitations, which leaves some gaps for further improvement. Like in waste collection and transportation, we have suggested the use of sensor chips, but the limitation associated with such tracking devices is that the speed of the vehicle should be fixed. This parameter is hard to achieve in real-time situations. The drivers need to change the speed according to the route and the traffic. Likewise, a convolutional neural network–based model, which is proposed to identify waste dumped near the shores of rivers, is working with great accuracy. But it cannot be implemented in urban areas due to complexities in the waste generated. Also, detecting the images through drones is also a troublesome task. The algorithms used in identifying drone images are incapable of identifying waste generated in canopy regions. Some models have also been developed which help us to identify illegal dumping sites by real-time pictures captured in high resolution. But still these pictures suffer from errors in identification or have blurred aerial view. When artificial intelligence is used for waste identification and sorting, the waste images used to train transfer learning and convolutional neural network models for waste classification are much less complex than real-world waste, leading to reduced accuracy rates in practical applications [45]. Additionally, researchers have explored using image data augmentation and transfer learning to identify and classify construction waste. The major limitation identified in such models is the low quality of images captured, which further reduces the efficiency of the model in addition to increasing the cost.

The major gap in this area clearly depicts the lack of knowledge and provides many research gaps to the researchers. There are a variety of algorithms developed, each with its own unique feature. Still there are no clear guidelines about which algorithm will be best suited for different processes of waste treatment. Also, integrated technologies, where artificial intelligence is combined with microelectronics, robotics, Big Data or IoT, etc., are offering a great field for multidisciplinary research to solve this problem of waste management. In summary, the limitations of artificial intelligence in waste management are primarily related to its difficulty in practical operation compared to theory, the limited scope of application, and lower efficiency in practical use. The potential for future progress lies in comprehensively understanding the various mechanisms behind artificial intelligence models, combining artificial intelligence with other technologies, and using multiple models, such as convolutional neural network, residual network model, and gradient enhancement regression models to provide a full-fledged solution to waste management.

4.9 Conclusion

Waste management is a serious issue in today's era and if not addressed with proper strategy it can lead to not only environmental issues but also could produce serious health-related concerns. This issue is faced worldwide and to provide a solution to it various artificial intelligence–based models are proposed. These models can improve the waste management strategies in various steps and can highly reduce the human intervention, thus reducing the environmental degradation and health- and sanitation-related issues. This chapter focuses on such proposed or some small-scale implemented models. It also highlights various technologies implemented by artificial intelligence to improve their efficiency. But due to lack of technological knowledge and the high cost of implementation, these models are still not in the full-fledged working mode.

On a concluding note, this chapter gives us an insight to implementation of artificial intelligence in waste management systems and highlights the various research gaps which need to be addressed. It also gives us a clear vision for the future solutions which can not only be used in waste management but also can help us attain sustainable development by addressing socioeconomic issues and creating an efficient and intelligent waste management system.

Bibliography

1. Kaza, S., Yao, L., Bhada-Tata, P., Van Woerden, F., *What a waste 2.0: a global snapshot of solid waste management to 2050*, 2018, Available at: http://datatopics.worldbank.org/what-a-waste/.
2. Triassi, M., Alfano, R., Illario, M., Nardone, A., Caporale, O., Montuori, P., Environmental pollution from illegal waste disposal and health efects: a review on the "Triangle of Death." *Int. J. Environ. Res. Public Health*, 12, 1216–1236, 2015, https://doi.org/10.3390/ijerph120201216.
3. Sharma, H.B., Vanapalli, K.R., Cheela, V.R.S., Ranjan, V.P., Jaglan, A.K., Dubey, B., Goel, S., Bhattacharya, J., Challenges, opportunities, and innovations for effective solid waste management during and post COVID-19 pandemic. *Resour. Conserv. Recyl.*, 162, 105052, 2020, https://doi.org/10.1016/j.resconrec.2020.105052.
4. Neumeyer, X., Ashton, W.S., Dentchev, N., Addressing resource and waste management challenges imposed by COVID19: an entrepreneurship perspective. *Resour. Conserv. Recyl.*, 162, 105058, 2020, https://doi.org/10.1016/j.resconrec.2020.105058.
5. Priyadarshi, H., Priya, S., Jain, A., Khursheed, S., A Literature Review on Solid Waste Management: Characteristics, Techniques, Environmental Impacts and Health Effects in Aligarh City", Uttar Pradesh, India, in: *Recent Thoughts in Geoenvironmental Engineering. GeoMEast 2019. Sustainable Civil Infrastructures*, H. Ameen, M. Jamiolkowski, M. Manassero, H. Shehata, (Eds.), Cham, Springer, 2020, https://doi.org/10.1007/978-3-030-34199-2_6.
6. Kumar, A. and Agrawal, A., Recent trends in solid waste management status, challenges, and potential for the future Indian cities – A review. *Curr. Res. Environ. Sustain.*, 2, 100011, 2020, ISSN 2666-0490, https://doi.org/10.1016/j.crsust.2020.100011.
7. Rana, R., Ganguly, R., Gupta, A.K., An assessment of solid waste management system in Chandigarh City, India. *Electron. J. Geotech. Eng.*, 20, 1547–1572, 2015, https://www.researchgate.net/publication/276275442_An_Assessment_of_Solid_Waste_Management_System_in_Chandigarh_City_India.
8. Abdallah, M., Abu Talib, M., Feroz, S., Nasir, Q., Abdalla, H., Mahfood, B., Artificial intelligence applications in solid waste management: a systematic research review. *Waste Manage. (Oxford)*, 109, 231–246, 2020, https://doi.org/10.1016/j.wasman.2020.04.057.
9. Andeobu, L., Wibowo, S., Grandhi, S., Artificial intelligence applications for sustainable solid waste management practices in Australia: a systematic review. *Sci. Total Environ.*, 834, 155389, 2022, https://doi.org/10.1016/j.scitotenv.2022.155389.
10. Osman, A.I., Chen, L., Yang, M., Msigwa, G., Farghali, M., Fawzy, S., Rooney, D.W., Yap, P.-S., Cost, environmental impact, and resilience of renewable

energy under a changing climate: a review. *Environ. Chem. Lett.*, 2022, https://doi.org/10.1007/s10311-022-01532-8.

11. Meena, M.D., Dotaniya, M.L., Meena, V.S., *et al.*, Municipal solid waste: Opportunities, challenges and management policies in India: A review. *Waste Manage. Bull.*, 1, 1, 4–18, 2023, https://doi.org/10.1016/j.wmb.2023.04.001.
12. Fang, B., Yu, J., Chen, Z., *et al.*, Artificial intelligence for waste management in smart cities: a review. *Environ. Chem. Lett.*, 21, 1959–1989, 2023, https://doi.org/10.1007/s10311-023-01604-3.
13. Xiao, W., Yang, J., Fang, H., Zhuang, J., Ku, Y., Zhang, X., Development of an automatic sorting robot for construction and demolition waste. *Clean Technol. Environ. Policy*, 22, 1829–1841, 2020, https://doi.org/10.1007/s10098-020-01922-y.
14. Koskinopoulou, M., Raptopoulos, F., Papadopoulos, G., Mavrakis, N., Maniadakis, M., Robotic waste sorting technology: toward a vision-based categorization system for the industrial robotic separation of recyclable waste. *Rob. Autom. Mag.*, 28, 50–60, 2021, https://doi.org/10.1109/MRA.2021.3066040.
15. Wang, Z., Li, H., Yang, X., Vision-based robotic system for on-site construction and demolition waste sorting and recycling. *J. Build. Eng.*, 32, 101769, 2020, https://doi.org/10.1016/j.jobe.2020.101769.
16. Chen, X., Huang, H., Liu, Y., Li, J., Liu, M., Robot for automatic waste sorting on construction sites. *Autom. Constr.*, 141, 104387, 2022b, https://doi.org/10.1016/j.autcon.2022.104387.
17. Leveziel, M., Laurent, G.J., Haouas, W., Gauthier, M., Dahmouche, R., A 4-DoF parallel robot with a built-in gripper for waste sorting. *Rob. Autom. Lett.*, 7, 9834–9841, 2022, https://doi.org/10.1109/LRA.2022.3192582.
18. Mao, W.-L., Chen, W.-C., Fathurrahman, H.I.K., Lin, Y.-H., Deep learning networks for real-time regional domestic waste detection. *J. Clean. Prod.*, 344, 131096, 2022, https://doi.org/10.1016/j.jclepro.2022.131096.
19. Karthikeyan, S., Rani, G.S., Sridevi, M., Bhuvaneswari, P.T.V., IoT enabled waste management system using ZigBee network, in: *2017 2nd IEEE International Conference on Recent Trends in Electronics, Information & Communication Technology (RTEICT)*, pp. 2182–2187, 2017, https://doi.org/10.1109/RTEICT.2017.8256987.
20. Raaju, V.A., Meeran, J.M., Sasidharan, M., Premkumar, K., IoT based smart garbage monitoring system using ZigBee, in: *2019 IEEE International Conference on System, Computation, Automation and Networking (ICSCAN)*, pp. 1–7, 2019, https://doi.org/10.1109/ICSCAN.2019.8878742.
21. Jino Ramson, S., Jackuline Moni, D., Alfred Kirubaraj, A., Senith, S., Self-powered wireless sensor network framework to monitor bin level. *J. Solid Waste Technol. Manag.*, 43, 295–304, 2017, https://doi.org/10.5276/JSWTM.2017.295.
22. Ramson, S.J., Moni, D.J., Vishnu, S., Anagnostopoulos, T., Kirubaraj, A.A., Fan, X., An IoT-based bin level monitoring system for solid waste

management. *J. Mater. Cycles Waste Manag.*, 23, 516–525, 2021, https://doi.org/10.1007/s10163-020-01137-9.
23. Joshi, L.M., Bharti, R.K., Singh, R., Malik, P.K., Real time monitoring of solid waste with customized hardware and Internet of Things. *Comput. Electr. Eng.*, 102, 108262, 2022, https://doi.org/10.1016/j.compeleceng.2022.108262.
24. Wu, F., Niu, D., Dai, S., Wu, B., New insights into regional differences of the predictions of municipal solid waste generation rates using artificial neural networks. *Waste Manage. (Oxford)*, 107, 182–190, 2020, https://doi.org/10.1016/j.wasman.2020.04.015.
25. Kolekar, K.A., Hazra, T., Chakrabarty, S.N., A review on prediction of municipal solid waste generation models. *Procedia Environ. Sci.*, 35, 238–244, 2016, https://doi.org/10.1016/j.proenv.2016.07.087.
26. Coskuner, G., Jassim, M.S., Zontul, M., Karateke, S., Application of artificial intelligence neural network modeling to predict the generation of domestic, commercial and construction wastes. *Waste Manage. Res.*, 39, 499–507, 2020, https://doi.org/10.1177/0734242X20935181.
27. Cha, G.-W., Moon, H.J., Kim, Y.-M., Hong, W.-H., Hwang, J.-H., Park, W.-J., Kim, Y.-C., Development of a prediction model for demolition waste generation using a random forest algorithm based on small datasets. *Int. J. Environ. Res. Public Health*, 17, 6997, 2020, https://doi.org/10.3390/ijerph17196997.
28. Cha, G.-W., Moon, H.J., Kim, Y.-C., A hybrid machine-learning model for predicting the waste generation rate of building demolition projects. *J. Clean. Prod.*, 375, 134096, 2022, https://doi.org/10.1016/j.jclepro.2022.134096.
29. Cha, G.-W., Choi, S.-H., Hong, W.-H., Park, C.-W., Developing a prediction model of demolition-waste generation-rate *via* principal component analysis. *Int. J. Environ. Res. Public Health*, 20, 3159, 2023, https://doi.org/10.3390/ijerph20043159.
30. Ponis, S., Plakas, G., Aretoulaki, E., Tzanetou, D., Maroutas, T.N., LoRaWAN for tracking inland routes of plastic waste: introducing the smart TRACKPLAST bottle. *Clean Waste Syst.*, 4, 100068, 2023, https://doi.org/10.1016/j.clwas.2022.100068.
31. Liu, W., Xu, Y., Fan, D., Li, Y., Shao, X.-F., Zheng, J., Alleviating corporate environmental pollution threats toward public health and safety: the role of smart city and artificial intelligence. *Saf. Sci.*, 143, 105433, 2021b, https://doi.org/10.1016/j.ssci.2021.105433.
32. Tcheng, D.K., Lewis, M.R., Newell, T.A., Applications of machine learning to sorting of recyclable containers, in: *2015 Conference on Solid Waste Research in the Midwest*, Chicago, Illinois, Oct 1990.
33. Lewis, M.R. and Newell, T.A., Development of an automated clear/color sorting system for recycling containers, in: *Conference on Solid Waste Research in the Midwest*, Chicago, Illinois, Oct 1990.
34. Ahmad, I., Khan, M.I., Khan, H., Ishaq, M., Tariq, R., Gul, K., Ahmad, W., Pyrolysis study of polypropylene and polyethylene into premium oil

products. *Int. J. Green Energy*, 12, 663–671, 2015, https://doi.org/10.1080/15435075.2014.880146.

35. Li, J., Suvarna, M., Pan, L., Zhao, Y., Wang, X., A hybrid data-driven and mechanistic modelling approach for hydrothermal gasification. *Appl. Energy*, 304, 117674, 2021b, https://doi.org/10.1016/j.apenergy.2021.117674.
36. Visser, H., Evers, N., Bontsema, A., Rost, J., de Niet, A., Vethman, P., Mylius, S., van der Linden, A., van den Roovaart, J., van Gaalen, F., Knoben, R., de Lange, H.J., What drives the ecological quality of surface waters? A review of 11 predictive modeling tools. *Water Res.*, 208, 117851, 2022, https://doi.org/10.1016/j.watres.2021.117851.
37. Yan, J., Yan, X., Hu, S., Zhu, H., Yan, B., Comprehensive interrogation on acetylcholinesterase inhibition by ionic liquids using machine learning and molecular modeling. *Environ. Sci. Technol.*, 55, 14720–14731, 2021, https://doi.org/10.1021/acs.est.1c02960.
38. Guo, H.-N., Wu, S.-B., Tian, Y.-J., Zhang, J., Liu, H.-T., Application of machine learning methods for the prediction of organic solid waste treatment and recycling processes: a review. *Biores. Technol.*, 319, 124114, 2021, https://doi.org/10.1016/j.biortech.20.
39. Schwarzböck, T., Aschenbrenner, P., Spacek, S., Szidat, S., Rechberger, H., Fellner, J., An alternative method to determine the share of fossil carbon in solid refuse-derived fuels—Validation and comparison with three standardized methods. *Fuel*, 220, 916–930, 2018, https://doi.org/10.1016/j.fuel.2017.12.076.
40. Wang, D., Tang, Y.-T., He, J., Yang, F., Robinson, D., Generalized models to predict the lower heating value (LHV) of municipal solid waste (MSW). *Energy*, 216, 119279, 2021, https://doi.org/10.1016/j. energy.2020.119279.
41. Sousa, J., Rebelo, A., Cardoso, J.S., Automation of waste sorting with deep learning, in: *2019 XV Workshop de Visão Computational (WVC)*, pp. 43–48, 2019, https://doi.org/10.1109/WVC.2019.88769 24.
42. Thumiki, M. and Khandelwal, A., Real-time mobile application for classifying solid waste material into recyclable and non-recyclable using Image recognition and convolutional neural network. *2022 IEEE International Students' Conference on Electrical, Electronics and Computer Science (SCEECS)*, pp. 1–6, 2022, https://doi. org/10.1109/SCEECS54111.2022.9740863.
43. Takahashi, Y., Fujii, J., Amakata, M., Training a robust UAV river patrol AI for different river and an analysis of the training dataset, in: *Proceedings of the National Congress of the Society for Artificial Knowledge*, JSAI2022, 3S3IS2e01–3S3IS2e01, 2022, https:// doi.org/10.11517/pjsai.JSAI2022.0_3S3IS2e01.
44. Youme, O., Bayet, T., Dembele, J.M., Cambier, C., Deep learning and remote sensing: detection of dumping waste using uav. *Procedia Comput. Sci.*, 185, 361–369, 2021, https://doi.org/10.1016/j.procs.2021. 05.037.
45. Zhang, Q., Yang, Q., Zhang, X., Bao, Q., Su, J., Liu, X., Waste image classification based on transfer learning and convolutional neural network.

Waste Manage. (Oxford), 135, 150–157, 2021a, https://doi.org/10.1016/j.wasman.2021.08.038.

46. Suvarnamma, A. and Pradeepkiran, J.A., SmartBin system with waste tracking and sorting mechanism using IoT. *Clean. Eng. Technol.*, 5, 100348, 2021, ISSN 2666-7908, https://doi.org/10.1016/j.clet.2021.100348.
47. Kumar, S., Kumaran, T., Kumar, A., Mathapati, M., Smart garbage monitoring and clearance system using internet of things. *2017 IEEE International Conference on Smart Technologies and Management for Computing, Communication, Controls, Energy and Materials (ICSTM)*, pp. 184–189, 2017.
48. Xia, W., Jiang, Y., Chen, X., Zhao, R., Application of machine learning algorithms in municipal solid waste management: A mini review. *Waste Manage. Res.*, 40, 6, 609–624, 2022, doi: 10.1177/0734242X211033716.
49. Abbasi, M. and El Hanandeh, A., Forecasting municipal solid waste generation using artificial intelligence modelling approaches. *Waste Manage. (Oxford)*, 56, 13–22, 2016.
50. Abu Qdais, H. and Shatnawi, N., Assessing and predicting landfill surface temperature using remote sensing and an artificial neural network. *Int. J. Remote Sens.*, 40, 9556–9571, 2019.
51. Al-Refaie, A., Al-Hawadi, A., Fraij, S., Optimization models for clustering of solid waste collection process. *Eng. Optim.*, 2020, Epub ahead of print 23 November 2020. DOI: 10.1080/0305215X.2020.1843165.
52. Antanasijevic, D., Pocajt, V., Popovic, I., *et al.*, The forecasting of municipal waste generation using artificial neural networks and sustainability indicators. *Sustain. Sci.*, 8, 37–46, 2013.
53. Arebey, M., Hannan, M.A., Begum, R.A., *et al.*, Solid waste bin level detection using gray level co-occurrence matrix feature extraction approach. *J. Environ. Manage.*, 104, 9–18, 2012.
54. Ayeleru, O.O., Fajimi, L.I., Oboirien, B.O., *et al.*, Forecasting municipal solid waste quantity using artificial neural network and supported vector machine techniques: A case study of Johannesburg, South Africa. *J. Clean. Prod.*, 289, 125671, 2021.
55. Bircanoğlu, C., Atay, M., Beser, F., *et al.*, RecycleNet: Intelligent waste sorting using deep neural networks, in: *2018 Innovations in intelligent systems and applications (INISTA)*, Thessaloniki, Greece, 3–5 July 2018, IEEE, New York, NY, pp. 1–7, 2018.
56. Kujawa, S., Mazurkiewicz, J., Czekala, W., Using convolutional neural networks to classify the maturity of compost based on sewage sludge and rapeseed straw. *J. Clean. Prod.*, 258, 120814, 2020.
57. Xue, W., Hu, X., Wei, Z., *et al.*, A fast and easy method for predicting agricultural waste compost maturity by image-based deep learning. *Bioresour. Technol.*, 290, 121761, 2019.

58. Wilts, H., Garcia, B.R., Garlito, R.G., Gómez, L.S., Prieto, E.G., Artificial intelligence in the sorting of municipal waste as an enabler of the circular economy. *Resources*, 10, 28, 2021. https://doi.org/10.3390/ resources10040028.
59. Xiangzhou, Y., Manu, S., Sean, L., Pavani, D.D., Ki Bong, L., Jie, L., Xiaonan, W., Yong Sik, Ok., Applied Machine Learning for Prediction of CO2 Adsorption on Biomass Waste-Derived Porous Carbons. *Environ. Sci. Technol.*, 55, 17, 11925–11936, 2021. DOI: 10.1021/acs.est.1c01849.

5

Application of Remote Sensing and GIS in Waste Management

Tishar Chander and Divya Bajpai Tripathy*

Division of Chemistry, School of Basic Science, Galgotias University, Greater Noida, India

Abstract
The role of Artificial Intelligence (AI) in environmental remediation is expanding rapidly. AI is being used to monitor and analyse environmental data, to model and predict the behaviour of pollutants, and to develop new technologies for remediation. While there are some challenges and limitations to using AI in this context, the benefits are significant. AI can help by improving the accuracy and efficiency of environmental monitoring and in assessment, and it can also help to develop more effective and efficient remediation strategies. As AI continues to evolve continuously, it is likely to play an increasingly important role in the effort to protect and preserve our environment.

Keywords: Environmental remediation, artificial intelligence, machine learning, data analysis, image analysis, remote sensing, geographic information systems

5.1 Introduction

Environmental remediation refers to the processes that revert or stop the harmful impacts of human activities on the environment. The environment can be understood as the entirety of the air, water and land and the other components associated with it or may affect it, like temperature, humidity, etc. Environmental remediation may include the cleaning of the contaminants from soil, air and water bodies; restoration of natural habitats; conservation of natural species of animal and plants which are

**Corresponding author*: divyabaj@gmail.com

Priyanka Singh, Pooja Agarwal and V. Vivekanand (eds.) Solid Waste Management: Challenges, Sustainability and Advancements, (139–182) © 2026 Scrivener Publishing LLC

continuously decreasing, endangered or already at the point of extinction; and to prevent further damage to the environment. Use of the five R's is the current basis of environmental remediation; these are "Refuse, Reduce, Reuse, Repurpose and Recycle". The circular economy is another advance initiative that promotes the extraction or development of new products from the waste products instead of scrapping or dumping them. The ultimate goal of environment remediation is the improvement in its health conditions through sustainable developments in order to assure the health of the living creatures, especially humans, that depend on it [1, 2].

Artificial intelligence (AI) is an approach or more specifically the learning capability or intelligence of machines to perceive, synthesize and infer the information. AI itself is an approach that is directly or indirectly involved in the environment remediation as it can produce the virtual representation of many chemical and environmental processes without involving or exploiting the natural resources and predict the further consequences even without the involvement of actual natural species or resources [3].

The use of AI straightaway in environmental remediation is still relatively new but is rapidly evolving. AI can be used for monitoring and analyzing the environmental data, to model and to predict changing behaviour of the pollutants in the environment, and for developing new technologies for remediation. AI can also help for optimizing the exploitation of resources, for making the process more cost effective and to improve the overall impact of the remediation efforts [4].

Environmental monitoring and assessment are the two fundamental approaches that confirm their involvement in environmental remediation. AI helps to collect environment-related data such as water quality, air quality index (AQI) and soil contamination. Collection of data can be done through a variety of methods, like remote sensing, satellite imaging, or through sensors. In the interpretation or assessment of environmental condition, AI allows scientists and policymakers to clearly recognize the state of the environment and identify areas in need of remediation [5].

AI can also be exploited to model and to predict the behaviour of the pollutants in the environment through developing complex algorithms and models for simulation of the movement and interaction of pollutants in the air, water, and soil. This may help scientists and engineers develop more effective remediation strategies [6].

Development of new technologies for environmental remediation is another important area where AI works and helps to develop more efficient and effective methods for cleaning up contaminated soil, air and water.

AI also has the ability to optimize the use of resources, such as energy and water, in the remediation process [7].

AI is a tool with numerous advantages for environmental remediation. It can help to reduce costs and improve the efficiency of remediation efforts. It has potential for optimizing the use of resources and predicting the behaviour of pollutants, as well as to streamline the remediation process and ensure that resources are used effectively. It helps to improve the accuracy of environmental monitoring and assessment, to analyze large amounts of data, and to identify patterns and trends that might be overlooked or not be visible to human analysts. AI can help to identify many areas in need of remediation more quickly and accurately, and can also help to identify emerging environmental threats before they become more serious. Although AI has countless benefits, there are some challenges and limitations to using AI in environmental remediation, such as a high-quality database, reliable data in order to produce accurate data and to avoid the creation of algorithms that may produce inaccurate or misleading results. Additionally, there is a need for greater collaboration between scientists, policymakers, and other stakeholders so as to ensure that AI is being used effectively and ethically. Moreover, there is a requirement of equilibrium between the human intelligence and artificial intelligence in order to ensure its best optimization to get maximum benefits for novel cause. This review paper shows how effectively AI can be exploited in the burning field of environment remediation. It describes the special benefits of using AI in different fields of environment remediation with authenticated case studies and validated research. Challenges and the future scope of role of AI in environmental remediation has also been covered [8].

5.2 AI in Environmental Management and Assessment (Chronology and Advancements)

The introduction of environmental assessment as a tool meant to evaluate the state of the environment qualitatively and quantitatively was reported late in the 1960s. Currently, it has been acknowledged as an assessment procedure that repeatedly overevaluated or underevaluated the ecological influences [9]. Systemic complications associated with the earliest assessment procedures related to ecological status and susceptibility accentuated some specific and independent research areas [10]. In the recent past, several procedures have been projected, like the comprehensive assessment procedure, the fuzzy assessment process, and the gray assessment approach, as well as the artificial neural network assessment process, the osculation value technique, and the landscape assessment procedure [11].

Environmental Management (EM) can be understood as a combination of scientific researches, strategies and socioeconomic operations to assess the consequences of the ecological procedures [12]. The objective of environmental management is to recognize the areas to increase the productivity with savings or more specifically the "sustainable development". Environmental management accentuates exploring the best solutions to the complications created by living beings' collaboration with the environment. EM can be explained as a set of procedures that offer a strong foundation to diminish its environmental consequence and to augment its functioning competence. Environmental management was also claimed to be purely humancentric context [13] as the human brain better understands that all the natural components, living or non-living, are required to establish the balanced ecosystem that is required for the survival of humans and all natural components are interconnected as an integral part of the ecosystem. The technical approach of EM can be stated as "a scaffold and catalogue that integrates the procedures and progression for training and supervision, evaluation and reporting to stakeholders of a firm of specialised ecological performance info." EM systems encompass the acknowledgement and formation of environmental aspects and planning [14]. Styles and his co-workers have disclosed the eight BEMPs, i.e., "Best Environmental Management Practices hierarchy," intended to systematically emphasize the retail supply chains which were the most destructive for the ecosystems. However, this BEMP received lower priority because of the well-documented restrictions on customer options of environmental products [15].

Environmental assessment is a process through which it can be checked whether individuals, societies, startups, companies or organizations are taking appropriate measures so that their involvement is not leaving, in the short term or the long term, an immediate or delayed negative impact on the environment. For instance, this involves the formation and appraisal of regulatory bodies that assure the sustainable environmental practices of an organization. Environmental assessment can alert a company or organization on their waste management strategies [16, 17]. Over the last few decades, environmental assessment has emerged as one of the foremost concerned ecological public policies [18, 19] that involve noteworthy scientific researches and operative information, even though the environmental assessment couldn't reach the expected or near to expected level of environmental protection for sustainable growth.

Hence, there is a need of a well-organized and accurate environmental management and assessment system that can augment the performance rate over conventional evaluation approaches to economic accountability. AI may be the best solution for these purposes. A few scientific researches, specifically done in the 2020s, have already defended the role of AI in environmental management and assessment. Nazarova and his co-scientists [24] established a prototype that supports the audit and accounting and was found very appropriate to be used by forest industries. The USP of this innovation is the Forestry Accountability and Auditing (FAA) framework that is the source of quantifiable, artifacts, procedures, and research topics linked with the two features: accounting and forest principal financial and ecological audit. In the same year, Aslam *et al.* [98] discussed the significance of an external audit of the ecosystem in connection to Energy and Resource Management Practices. This study emphasizes general lesser quadratic regression with the random consequences to examine the link between variables. Empirical outcomes revealed that trades with the EMP's internal framework are more realistic so as to acquire superior ecological efficacy through exploiting an external audit tool. This work endorsed the importance of efficacy of internal EMP tools to create improved ecological procedures *via* internal environmental assessment and management practices [20].

In a technical research work done by Castka *et al.* [2], Financial Statements Evaluation Technique showed competence to disclose the data on latent ecological factors assessment. This method deliberates economic reporting aiming to interrelate all monetary and commercial procedures in the audited association. Magazzino *et al.* endorsed such procedures

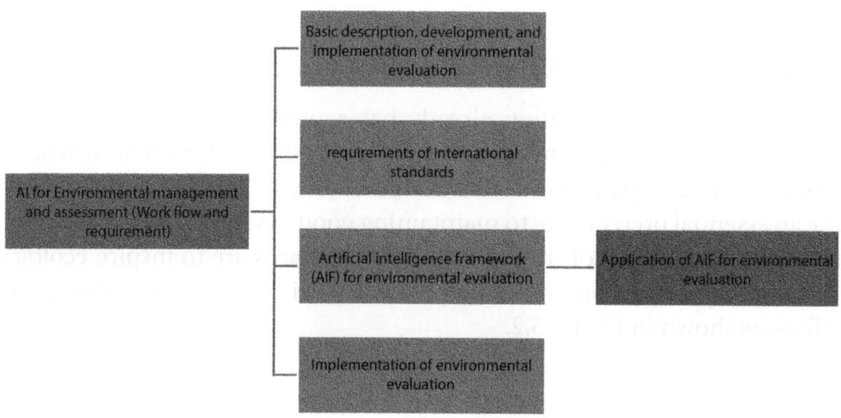

Figure 5.1 Work flow of AI in environment management and assessment.

in which the ecological concerns should be addressed in terms of their financial statements [21, 22]. Marwa *et al.* studied the relationship of environmental audit along with environment quality measured *via* voluntary revelation [23].

An AI framework and its involvement in environmental management and assessment can be understood by the flowchart shown in Figure 5.1 above.

5.3 Environmental Evaluation: Explanation, Expansion, and Employment

In order to evaluate the theories and applications involved in environmental explanation, expansion, and employment, three worldwide approaches have been emerged:

1. Environmental assessment: To identify the extent to which the business procedures are in compliance with environmental protocols.
2. Environmental evaluation: For designing a method that can provide ranking of the efficacy of impactful environmental management.
3. Environmental assessors: To judge the environmental impact on the organization.

For an effective AI system for environmental evaluation, all of the above three points should be completely and seriously taken into account.

5.3.1 Need of International Standards

Environmental interpretation already has a predefined, well-established and highly unified international standards designed for environmental management and protection. Constant amendments of ecological attitudes are an essential prerequisite to maintaining good environmental standards, and the eventual aim of the international standards are to inspire ecological security, reductions in emissions with sociological and commercial effects as shown in Figure 5.2.

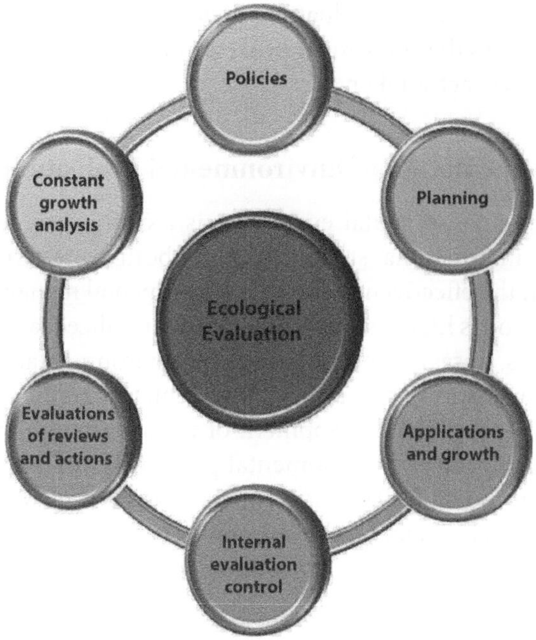

Figure 5.2 Aim of the international standards.

5.3.2 Artificial Intelligence Framework (AIF) for Environmental Evaluation

There is a need of an all-inclusive environmental assessment system in order to achieve the precise research findings. This is a dynamic procedure, so that the investigators may create a more scientific and sensible assessment model to explained the assessment procedure. In addition, the quantitative assessment system can be exploited to provide the maximum sustainable commercial advantage. Along with the criteria of environmental supply competence and key ecological approach, the commerce benefit estimation is also provisionally required, as shown in Figure 5.3.

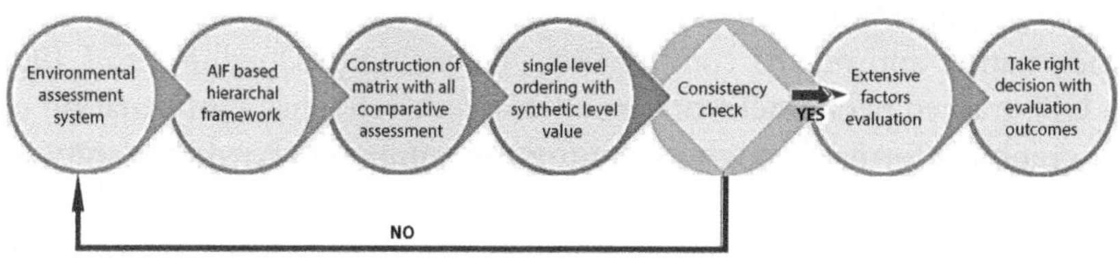

Figure 5.3 Quantitative assessment system of AI for environment.

AIF for environmental evaluation divides the environmental assessment index, defines the assessment process and the level of the dependable protocol for every factor taken.

5.3.3 Implementation of Environmental Evaluation

The process of environmental evaluation is a continuous feedback loop, as outlined by international standards. It has both protective and modifying impacts on the effectiveness of the environmental management framework. The method is highly standardized and formalized, and derives from universal norms. Research indicates that the environmental manager plays a crucial role in ensuring successful implementation of the environmental management system. The development of the system enhances the company's comprehension of environmental protection and assessment criteria. Thus, environmental evaluation entails consistent consideration of the company's external impact on the environment resulting from its internal activities, in line with environmental policies.

5.4 Different Approaches of AI in Environmental Remediations

5.4.1 Remote Sensing (RS) and Geographical Information System

The principle on which the RS works is based on the reflection theory of light. According to this hypothesis, water of a specific constituent reflects a specific amount of energy in distinct wavelength bands of the electromagnetic radiations fall on them. This is because of the property of pollutants such as their physical, chemical and structural features that alter the reflectivity of water [24]. Sensors boarded on different RS satellites and aerial vehicles have potential to record the energy of the emitted energy/reflected light and process that into an image which depicts the important statistics on the level and concentration of pollutants in aquifers [25].

Geographic information systems (GISs) are another strong tool used to study nature and related components. According to Dhedhey *et al.* [99], in GIS, trained personnel work on all available referenced information on geographical forms through exploiting the computer's hardware/software system and geographical database. Moreover, topographical features are integrated with physical parameters, enabling decision making and extensive monitoring of alterations and changes in flood zone mapping, population density, land use–land cover and housing. GIS databases use a variety of thematic maps or layers that contain diverse types of information, including but not limited to soil type, elevation, population density and land use. This information can be joint and modified into new layers for more investigate purposes. The two primary data types used in GIS are raster data and pixel data. Raster data stores geographical information in the form of pixels whereas vector data, storing it in point form, line form, and area form. The three-dimensional resolution of raster is determined pixel size, and is well-suited for points that demonstrate significant spatial variation [26].

Geographic information systems are an essential tool for decision-making and rely on structural data retrieved through remote sensing and field sampling. One of the key applications of GIS in water quality administration and remediation examination is the recording of field locations by means of handheld global positioning system (GPS) units, which can then be entered into the GIS database. Additionally, GIS can incorporate data from various sources such as aerial photographs, satellite images, topographic maps, soil maps and local survey maps. This information can be exploited to explore the structural correlation of contaminant concentrations and factors like land use–land cover, soil type and topography. Hazard zone mapping due to the polluted aquifers and adjacent agricultural areas can also be created using GIS. Furthermore, GIS allows the monitoring of alterations that have occurred over a period within a watershed or area, allowing for the linkage of alterations in water quality with changes [27–30].

Space technologies, like integrated remote sensing (IRS), geographical information system (GIS) and modeling procedures are the known prevailing tools exploited for local ecological evaluation and estimation [31]. Such technologies have potential to offer a good foundation to favour many-level and ordered integrated investigation of resources and ecological features, as well as to integrate the acquired information in a relative hypothetical ecosystem examination. Over the last two decades, RS has showed tremendous performance in the sector of resource management and environment assessment. Additionally, the GIS has evolved as a strong tool to assess and manage the resources and ecosystem. Furthermore, GIS technique is indispensable in the advancements of automated procedures to quantify the structural inconsistency ecological components and is extensively exploited to support structural modeling and investigation [32].

In an investigation done by Rahman *et al.* [27] an effort was made to examine the local ecological conditions through exploiting IRS, GIS and spatial multiple criteria evaluation (SMCE) approach analysis and to venture a measuring technique appropriate to recognize the level of the local environment of the area under examination. The SMCE model was exploited to assess an integrated REQI of that local regional area. Findings were computed and ranking of the conditions was made on the scale of 5 from very good, down to good, moderate, poor and very poor. In addition, IRS and GIS methods were used to developed the required factors for the SMCE method. Results of this investigation disclosed that human activities, soil erosion, vegetation condition, climate, soil conditions and topography have great impact on the local ecological conditions of any area [33].

5.4.2 Remote Sensing and Image Analysis in Water Remediation

Sagan *et al.* [30] endorsed the exploitation of remote sensing techniques in order to monitor the quality of inland water and also discussed the latent and limits of bio-optical simulations, spectral indices, cloud computing and machine learning. In this work, more than 200 sets of water quality data of two years, from 2016 to 2018, were analyzed that included chlorophyll-a, dissolved oxygen, blue-green algae phycocyanin, specific conductivity, turbidity, pollution-sediments and fluorescent dissolved organic matter. The water quality data for eight different rivers and lakes of the U.S. Midwest were obtained through *in situ*, real-time monitoring sensors and lab analysis of grab samples. This data was then combined with tripod-mounted hyperspectral imagery, synchronic proximal spectra and satellite data. The findings indicated that the last two were too accurate in the approximation of optically active compounds, whereas remote sensing was found suitable for the indirect estimation of non-optically active parameters with few challenges. The study suggested that deep learning performed well over other conventional procedures and has potential to provide a promising opportunity for using RS in water quality management, assessment and decision-making [34].

Ramadas & Samantaray [28] illustrated the status of GIS and RS for water quality assessment, administration and water resources remediation and drew conclusions based on the exhaustive review on RS and GIS exploitation for water quality monitoring, assessment and remediation. Remote sensing offers several benefits over field measurements and sampling for assessing water quality parameters. It provides synoptic-scale spatial coverage and real-time monitoring capabilities, allowing for comprehensive and up-to-date data collection. It is a cost-effective method that outperforms traditional sampling approaches. Remote sensing has proven to be a valuable tool for monitoring water quality, both in near real-time and when analyzing past data. This is particularly useful for research aimed at testing the hypothesis that changes in land use and land cover patterns can affect water quality. Although remote sensing is a powerful tool for monitoring water quality parameters, atmospheric effects can interfere with accurate measurements. Therefore, researchers are continuously developing new methods for atmospheric correction in order to obtain reliable information. However, the effectiveness of these methods depends on the specific ecosystem being studied, and the algorithms used for atmospheric correction can be quite complex. GIS and geo-statistics are closely related and often used together for analysing water quality parameters. Various spatial interpolation procedures like universal kriging, ordinary kriging and inverse distance weighting can be employed to achieve accurate spatial patterns of water quality parameters. The selection of an appropriate technique depends on the desired level of accuracy in the resulting spatial patterns. Surface water and groundwater susceptibility mapping and assessments can be accomplished by overlapping contaminant concentrations with terrestrial use maps. This method involves using high-resolution RS satellite images and a user-friendly GIS environment that supports complex spatial operations to process the data. In addition, the usage of models like DRASTIC, in combination with GIS, has made it possible to investigate the susceptibility to groundwater pollutants on a local scale. This has enabled the planning of remediation efforts for moderate to large aquifers like seas, long shoreline waters and even lakes. Upcoming satellite missions are aiming to monitor and manage water pollution and water quality on a synoptic scale. In the future, it may be possible to incorporate the 4th dimension of duration into GIS to point to the structural modeling of water quality with reference to the land use along with land cover changes, natural climate changeability, and anthropogenic climate change [25].

Rummell *et al.* [29] exploited a high-resolution remote sensing method for the quantification of the impact of wetland restoration. Restoration of coastal wetlands is a phenomenon to regenerate the lost ecosystem amenities. Having precise and regular measurements of the extent and distribution of coastal wetlands is crucial to assess the efficacy of restoration efforts. Traditionally, obtaining such data requires time-consuming and costly field surveys or conventional remote sensing techniques that may yield inaccurate results [35, 36].

In recent years, remote sensing technologies have advanced rapidly, with the development of high-resolution sensors with a resolution of less than 2 meters, shallow learning classifiers and object-based image analysis. However, despite their potential, these techniques have not been widely used in the context of ecosystem restoration. To address this gap, a study was conducted to evaluate changes in wetland societies at a 200-hectare restored littoral wetland in eastern Australia exploiting the object-based image analysis, worldview-2 imagery and random forest categorization. The study utilized physical raster, including numeral elevation, canopy height models, and a multi-temporal method to differentiate among land cover types that are spectrally similar. The land cover maps generated from the analysis had a high accuracy, ranges from 91% and 95%, which enabled the fast recognition of upsurges in the zone of key ecosystems, encompassing mixed paperbark and she-oak (10 ha), saltmarsh (4.31 ha), and mangroves (0.91 ha) over a 5-year assessment duration [37].

5.4.3 Remote Sensing and Image Analysis in Air Remediation

One more study suggests integrating Spectral Mixture Analysis and Endmember Remote Sensing Indices to obtain land surface temperature (LST) data, identifying urban heat islands (UHIs), and examining their associations with land use/land cover (LULC) and air contamination in Tabriz, a city in Iran. LST values are obtained from ASTER remotely sensed thermal images, and a statistical approach is employed to evaluate the intensity of UHIs. Results of this research revealed that the land surface temperature (LST) is significantly affected by land use/land cover (LULC) and urban heat islands are strongly related to both LST and LULC. LST is found to be sensitive to the humidity and the vegetation, hence low temperatures are observed in vegetated areas and aquifers. In contrast, elevated temperatures are associated with industrial sites and construction zones at the city's center. Although the results establish a strong association of UHIs and of air pollution, structural analysis revealed a dissimilar pattern while correlating recognized UHI zones with air pollution [38].

In approximately the last half-decade, satellite data has become increasingly accessible to monitor or assess the urban-scale air quality combined with the ground-based data. This integration adds synoptical and geospatial info to air quality data and modelling. To assess the effectiveness of this combination, a study was conducted including several short case studies. The study found potential for air quality applications of U.S. satellites. This work described several of these sensors as illustrations. However, usage of satellite data for air quality applications had been limited. It is due to a lack of collaboration between air quality and satellite scientists, difficulty in accessing and limited resources, understanding new data, and agency priorities focused on developing new techniques [39].

Urbanization is occurring rapidly worldwide, with developing countries such as India experiencing unplanned and uncontrolled growth that can lead to ecological imbalances and collapse. One of the most significant environmental hazards resulting from this urbanization is air pollution. Deteriorating air quality has become a critical concern for regulating the environment in which we live. In India's large cities, air quality has suffered due to industrialization, uncontrolled population growth, and increased reliance on automobiles. According to estimates, the Particulate Matter ambient pollution ($PM_{2.5}$ and PM_{10}) is responsible for approximately three million deaths per year. While ground monitoring networks provide primary data on exposure of population to air pollution, monitoring is limited in some areas. To address this, a GIS model has been developed that uses remotely sensed data in order to provide air quality information. The model, which incorporates spatially varying relationships between ground measurements and other factors that estimate air quality, is based on multiple linear regression modeling. The Landsat 8 OLI-TIRS satellite imagery has been used to demonstrate the potential of GIS as an aiding tool for monitoring air quality. The study found that using different Landsat 8 bands, accurate estimation of particulate matter ($PM_{2.5}$ and PM_{10}) can be achieved, but the model is not suitable for gaseous pollutants such as NOx and CO. Validation of the model was performed using 2017 ground truth data [40].

Over the past two decades, China has been working towards better air governance with some successes, although the results are not ideal as pollution still affects people's daily lives and their physical and mental health. Zhou *et al.* [100] introduced a double star cooperative AOT inversion algorithm to overcome the difficulty of AOT inversion in urban areas; it uses double satellite remote sensing data and the atmospheric radiation transmission model. This method, by using satellite remote sensing image processing, allowed the inverse reflection of AOT parameters in urban areas. This research work exploited the remote sensing image processing technology to study the assessment of mountain air pollution and the construction of tourism [41].

5.4.4 Remote Sensing and Image Analysis in Soil Remediation

Soil degradation refers to a range of processes, such as soil erosion and contamination, that decrease the ability of soil to serve as a foundation for plant roots. It is important to develop methods for accurately quantifying soil degradation due to contamination on a large scale. Jensen *et al.* [16] developed a concept of environmental sensitivity index (ESI) mapping for oil spills. The aim of this study was to support the spill-response controllers to evaluate the latent effect of oil through shoreline and to allocate the resources during an oil spill and after an oil spill. This study also illustrated some amendments to the ESI approach [42]. For instance:

- A base-map with a planimetric projection was created using SPOT multispectral and/or panchromatic data that had been resampled to a structural resolution of 10 × 10 m with an accuracy of ± 5 m root mean square error.
- The data for a shoreline Environmental Sensitivity Index (ESI) can be obtained through a combination of digital and visual image processing methods used for remotely sensed information.
- One approach for quickly accessing and analysing information about wildlife that is affected by oil is to use a geographic information system (GIS). By storing this information in GIS format, it can be easily retrieved and queried for rapid analysis. To ensure quick access and retrieval of all access and protection details, the GIS can be utilized and integrated with a relational database for easy querying.

These augmentations to the ESI mapping approach make it additionally efficient to develop eventuality tactics earlier in case of an oil spill and to make administration decisions just after such oil spill.

The issue of heavy metal soil pollution is a pressing global ecological concern, and accurately mapping its distribution is crucial for policy-makers and environmental managers. In a 2018 review, Shi *et al.* explored various proximal and remote sensing spectroscopy techniques to obtain soil reflectance spectroscopy and ecological covariates. It can be exploited to create maps of heavy metal soil pollution. The authors emphasized the importance of mapping the spatial distributions of different metal species because of their varying levels of bioavailability and toxicity. They also suggested that combining multiple proximal and remote-sensed sensors could facilitate both vertical and horizontal mapping of heavy metals in soil. Additionally, the integration of satellite and unmanned aerial vehicle-based hyperspectral imaging systems could lead to the progress of a space-aeronautic incorporation hyperspectral observation technology capable of accurately and rapidly monitoring soil environments at a large scale [43, 44].

Kaplan *et al.* studied using Remote Sensing for oil-polluted soil modeling and remediation monitoring in arid zones. In this study, advanced RS techniques and ML methods were utilized to explore the impact of oil spills during the first Gulf Conflict (1990-91). By employing unsubstantiated categorization on pre- and post-oil spill data, identification could be made for oil polluted and non-contaminated areas in Kuwait. Spectral signatures with varying wavelengths and multiple spectral indices were developed to detect oil contamination and assess the pre- and post-war conditions of the study area. The collected data were used to train and model various levels of oil-contaminated soil. Additionally, three modeling procedures, namely Random Tree (RT), Random Forest (RF) and Support Vector Machine (SVM) were employed on two distinct datasets. The outcomes demonstrate that the proposed method is efficient in detecting oil-contaminated soil and determining the zone and level of contamination. This analysis provides valuable support to future remediation efforts [18].

5.5 Data Analysis and Machine Learning

Antibiotics are increasingly becoming a major environmental pollutant due to their persistence and toxicity in the environment. Carbon-based materials (CBM) like the activated carbon and biochar are considered a promising technology for the treatment of wastewater containing antibiotics. However, the adsorption capacity of CBMs for different antibiotics under various environmental conditions and is difficult to predict using traditional isotherm models. To address this issue, this investigation utilized machine learning (ML) approaches to provide progress to the case of generic prediction models for the adsorption of sulfamethoxazole (SMX) and tetracycline (TC) antibiotics on CBMs. The obtained outcomes showed that random forest models outpaced gradient boosting trees and artificial neural networks for both SMX and TC adsorption. The developed ML models somewhat accurately predicted the adsorption efficacy of antibiotics on CBMs exploiting the material characteristic and adsorption circumstances as inputs. The ML models displayed better simplification aptitude over conventional isotherm models, allowing for accurate predictions under varying environmental situations and various types of adsorbents. ML models based partial dependence plots and relative importance analysis were performed well in comparison to SMX and TC adsorption on CBMs. The results showed that the specific surface area plays a critical role in both SMX (45%) and TC (24%) adsorption, although other material features, like H/C, (O + N)/C, and pHpzc, showed variable influences due to dissimilarities in molecular structures, functional groups, and pKa values of SMX and TC.

The progress made in the developing of computer models for simulating heavy metals (HMs) over the last twenty years has been significant. During this time, numerous machine learning (ML) models have been created to model HMs, resulting in impressive advancements. While there have been several investigations into various ML models, it is crucial to have a comprehensive understanding of the progress made in these computer-aided models. Yaseen reviewed the simulation of heavy metals in polluted soil, aquifers, and their elimination from aqueous solutions. The paper provides a detailed analysis of various methodological and conceptual aspects of heavy metal modeling, such as the limitations of conventional analytical approaches, heavy metal datasets types, the need for new ML models, selection of HM input parameters, performance metrics selection, internal parameter tuning, and the HM models types used. The review highlighted the importance of addressing these modeling aspects for ML developers and environmental scientists to achieve credibility and scientific consistency in environmental science research and concluded by suggesting future research directions that could enhance scientists' understanding of heavy metal simulation [45].

Environmental pollution is a growing issue globally, with urbanization and industrial developments being major contributors. In particular, air pollution in metropolitan areas worldwide has emerged as a significant issue concerning public health, and adverse effects of air pollution on the health of the city's residents have been a long-standing concern. The pollutants PM10 and PM2.5 are major contributors to air pollution. The presence of an elevated amount of fine particulate matter with a diameter of less than 2.5 μm (PM2.5) has been linked with the development of lung cancer, respiratory disease, cardiovascular disease, and metabolic disease. Accurately predicting PM 2.5 concentrations could aid government bodies in alerting individuals at high risk, thereby reducing complications.

In an effort to address this issue, Delavar *et al.* [5] developed the prediction models for air pollution based on these two pollutants. This study involved the utilization of machine learning techniques, including geographically weighted regression, regression support vector machine, auto-regressive nonlinear neural network and artificial neural network with external input, using data connected to day of the week, month of the year, meteorology, topography and pollutant rate of two nearest neighbours as input parameters. The proposed prediction model was found to significantly improve the accuracy of the prediction methods, with error reduction rates of 57%, 47%, 47%, and 94%, respectively. The most reliable algorithm for predicting air pollution was the autoregressive nonlinear neural network with external input, which achieved a one-day prediction error of 1.79 μg/m3. Using a genetic algorithm, the study identified the most effective parameters for predicting air pollution for the day of the week, month of the year, wind direction, maximum temperature, topography and pollutant rate of the two nearest neighbours [46].

Zamani studied the featured importance to predict the PM2.5 concentration in Tehran's urban area, through the implementation of extreme gradient boosting, random forest, and deep learning machine learning (ML) procedures. In this study, 23 features were examined encompassing the ground-measured PM2.5, satellite and climatological data and geographical data, in the modeling. The best model approach attained was MAE = 9.93 μg/m3, R2 = 0.81 (R = 0.9) and RMSE = 13.58 $\mu g/m^3$ exploiting XGBoost method, integrating eradication of insignificant topographies. Although all the three ML approaches performed equally, R2 changed from 0.63 to 0.67 when Aerosol Optical Depth at 3 km resolution was involved, and 0.77 to 0.81 when AOD at 3 km resolution was omitted. Differing from the PM2.5 lag data, satellite-derived AODs was not proven to improve model performance [47].

During the COVID era, Magazzino and his co-workers attempted to assess the correlation of COVID-19-linked deaths, PM_{10}, $PM_{2.5}$, and NO_2 concentrations and economic growth in New York state through city-level daily data *via* two Machine Learning researches. The study claimed that NO2 and PM2.5 were the most hazardous air pollutants accountable to accelerate the COVID-19 attributed causalities. However, it was also disclosed that only six out of various investigated causal inferences to be noteworthy and factual within the AUPRC examination. The study's causal analysis revealed that $PM_{2.5}$ and NO_2 had a direct causal effect on COVID-19 deaths, while economic growth was found to have a causal effect on both pollutants. These findings were consistent with previous research indicating that air pollution can exacerbate the severity of COVID-19. Additionally, the study found that unsustainable economic growth was associated with higher levels of air pollutants, further underscoring the need for sustainable development strategies. This shows the unsustainable economic growth could increase environmental pollution by drastically escalating emissions of pollutant agents (PM2.5 and NO2) in New York [48].

5.6 Geo Spatial Analysis

Spatial analysis involves a series of interconnected steps aimed at selecting an inferential model that takes into account the spatial connections within a phenomenon [49]. Geographical Information Systems (GIS) play a crucial role in addressing socioeconomic and environmental challenges. They integrate hardware, software, data, and human expertise within an institutional context. GIS focuses on spatial entities and relationships, enabling powerful data integration by geography. The evolution of GIS involves three stages: data collection, analytical operations, and decision support systems. While many systems remain in the inventory phase, only a few have advanced to spatial analysis due to technical challenges [50].

5.6.1 GIS-Based Multi-Criteria Decision Analysis

GIS along with multi-criteria decision analysis can be simplified as it involves allocating land based on specific objectives and a range of attributes that the chosen areas should exhibit. However, it is important to note that the approaches used in vector and raster systems often yield divergent solutions. Additionally, standardizing, and aggregating criteria to arrive at a final decision for land allocation can be ambiguous [51].

Multi-Criteria Decision Analysis (MCDA) is a powerful methodology for the structured decision-making. It aims to determine the best alternative by considering multiple criteria during the selection process. MCDA has evolved over time, resulting in various methods and software tools to address specific problems. To apply MCDA, one has to define the problem, alternatives available, and relevant criteria such as costs, social indicators, environmental impact and energy efficiency. When faced with numerous alternatives, MCDA helps identify the most suitable option. Indeed, there exists a diverse array of methods for solving problems, each with its unique characteristics and calculation approaches. When employing Multi-Criteria Decision Analysis (MCDA), it's essential to recognize that different methods may yield distinct outcomes even when provided with the same input data. The choice of method is often based on the type of result or objective desired.

A study by Zlaugotne, Beate *et al.*, "Multi-criteria decision analysis methods comparison," used five different MCDA models for renewable energy models in Latvia. Their findings showed VIKOR, TOPSIS and PROMETHEE-GAIA had similar priority selection and the highest ranking was for hydropower plants. COPRAS and MULTIMOORA results were beneficial to Solar Photovoltaic. The Hydroelectric Power Plant (HPP) and Solar Photovoltaic (Solar PV) consistently emerge as top contenders. For Latvia's renewable energy technology, the HPP's superior installed electrical capacity (27%) sets it apart. While results from different MCDA methods exhibit similarities, they are not identical. Employing a consistent methodology for both the main problem and subproblems enhances reliability and meaningful comparisons [52]. There are various types of data used in Geographical Information Analysis. The use of geo-referenced data is widespread across various fields, including environmental sciences, city planning, earth observation and real-time processing. Managing geospatial data efficiently has become a central task. While big data platforms and containers are employed, very little benchmarking has already been done to compare their efficiency, especially concerning spatial functionalities. Additionally, geospatial data management involves integrating data from diverse sources [54, 55].

To further understand it, the review [53] on Geospatial Data management describes in detail the significant progress made in the last decade and highlights the convergence of Geographic Information Systems (GIS) and Building Information Modeling (BIM) is transforming how we manage spatial data. These two approaches—each with its own philosophy of representing space—are coming together. Meanwhile, GIS uses composition models like the triangle networks for surfaces and tetrahedron networks for solids. On the other hand, BIM employs constructive solid geometry (CSG) and sweep models. In geospatial databases, both types of spatial data models must be considered.

Additionally, point clouds from terrestrial laser scanning plays a critical role in BIM. The point clouds need to be effectively managed within geospatial databases. Meanwhile, the concept of 3D virtual city spaces has gained prominence in smart sustainable cities. Cities worldwide are releasing 3D open data to the public, and then geospatial communities actively collect 3D data using laser point clouds and imaging. Simultaneously, construction communities provide rich 3D geometric and semantic information about building interiors through BIM [56].

However, integrating BIM and GIS presents challenges due to their different geometries, semantics, and data models. Despite these differences, the need for cross-domain information integration has grown. Researchers have explored various methods to bridge the gap between BIM and GIS, from data exchange standards to application-level integration. The goal is to unlock the potential for informed decision-making and knowledge discovery in smart cities and the built environment including individual buildings [57]. Efforts to achieve interoperability between BIM and GIS data formats have made significant progress. These include geometric and semantic integration methods. Geometric integration tackles issues like reference systems, 3D geometry, and level of detail. However, most studies focus on individual building structures rather than the broader context of Geo- BIM at the city scale. Semantic methods involve schema extension, simplification, and ontology-based approaches. Semantic Web Technology shows promise but faces challenges in ontology generation. Future research opportunities lie in using the Web Ontology Language (OWL) for improved BIM-GIS integration [58, 59].

Topology plays a crucial role in Geographical Information Systems (GIS) and Building Information Modeling (BIM) by modeling spatial relationships between entities. It encompasses concepts like adjacency graphs, Egenhofer's 9-intersection matrices, and simplicial complexes, enabling connectivity queries and pathfinding [60]. While GIS/BIM topologies deal with partial orders, they extend beyond spatial relationships to include non-spatial entities. Topological structures like simplicial complexes and combinatorial maps facilitate efficient representation and storage of spatial relationships, vital for applications such as emergency response and indoor network modeling. In 3D/4D applications, topology, geometry, and semantics converge in enhancing planning scenarios and geo-scientific analysis. Overall, topology serves as a foundational concept in managing geospatial data, facilitating advanced spatial analysis, and modeling [53].

This relationship between GIS in the age of machine learning and big data is highlighting [61] the potential to integrate geospatial data and its methodologies with machine learning for urban planning and livability enhancement. It emphasizes the interdisciplinary nature of urban systems analysis, acknowledging the importance of collaboration between urban planners, scientists, GIS experts, and policymakers. While exploring key aspects like data reliability and ethical concerns, it underscores the need for deeper research on data ethics and privacy. The authors advocate for a holistic approach to urban analysis, recognizing the importance of qualitative aspects alongside data-driven methodologies to truly enhance urban livability.

5.7 GIS Optimization and Various Uses

5.7.1 Geospatial Information Analysis

According to [62], Google Earth Engine is a very useful cloud-based tool for large-scale geospatial analysis, harnessing Google's extensive computing power. It addresses critical societal challenges like deforestation, drought, disaster response, and environmental conservation. Unlike traditional methods requiring technical expertise or access to supercomputers, it offers an integrated platform accessible to a broader audience. This includes not only remote sensing scientists but also individuals without the technical know-how to use large-scale computing resources.

A diverse array of IoT devices has been utilized in various research endeavors, ranging from RFID tags and barcodes to open hardware platforms like Arduino. Mobile phones, equipped with sensors like GPS, cameras, and microphones, have been particularly prevalent, especially in participatory sensing and tourism studies. Additionally, sensors for environmental monitoring, such as pollution sensors, ultrasonic water sensors, and temperature/humidity sensors, have been employed extensively. Asset management has been facilitated by combining Global Navigation Satellite System (GNSS) receivers with RFID tags [64]. Transportation systems benefit from vehicular sensors for traffic management and road safety, while cultural heritage areas have seen the use of smart glasses and wearable devices for enhanced experiences. Satellite imaging sensors have played a crucial role in biodiversity studies, agriculture, forest fire risk assessment, disaster relief, disease monitoring, and environmental impact assessments. Meteorological stations and radar data contribute to climatological modeling, disease control, and flood forecasting. Long-term sensor deployment is common in water monitoring, air quality monitoring, landslide monitoring, vehicular sensor networks, weather stations, and wildlife monitoring, particularly in cases where sensor replacement is costly or challenging. Sensor discovery is explored in IoT-based wireless sensor networks, with platforms like Pachube [65] and CityPulse [66] facilitating registration and monitoring. Additionally, text data mining and natural language processing aid in content discovery, while APIs like Flickr and Panoramio support exploration and analysis of attractive areas [63].

5.7.2 GIS in Earthquake and Disaster Management

GIS and MCDA have been used in other fields which ensure the availability of diverse types of data and methods that can be used in multidisciplinary or interdisciplinary studies. A few studies are highlighted for information about the types of data that can be used along with the methods and tools to consider while focusing on the waste disposal.

A case study by Choukolaei *et al.* to understand efficiency of relief centers in a disaster or epidemic using GIS and MCDA provided some crucial insight for preventing and mitigating natural disasters like earthquakes, emphasizing preparedness and capacity identification.

The Tehran Disaster Management Organization has designated shelters for victims during such events. However, the COVID-19 pandemic had strained resources and significantly highlighted the need for proactive crisis planning to minimize losses. Evaluating relief center efficiency in crises, this study assessed Tehran Region 1's centers across three scenarios: natural disasters, epidemics, and combined disaster-epidemic conditions. Expert evaluation and GIS analysis prioritized alternatives using PROMETHEE, with sports complexes and universities emerging as top choices due to safety, space availability, and proximity to essential services. The model's comprehensiveness enables its adaptation to various scenarios beyond earthquakes and epidemics. Despite limitations like restricted access to some centers and online evaluations due to the pandemic, the proposed model provides a robust framework for future assessments. Further research should explore relief center operations and efficiency under varying disaster scenarios, incorporating factors like seismic activity and fault types. Simulation and modeling can optimize relief resource allocation and warehouse placement for enhanced disaster response [67].

One such use case of GIS analysis has been done in Nepal which frequently encounters natural hazards like floods, landslides, and earthquakes, including the 2015 Gorkha earthquake. This study evaluates risks in Dharan using geospatial and social data from sources like the 2011 census, socioeconomic information, WorldView-3 imagery, and PERSIANN rainfall data. Twelve physical variables were modeled individually and combined into an integrated hazard map, with social vulnerability assessed using demographic and economic data. The model faced challenges due to varied data formats, outdated geological data, and limited historical records. The study's methods aid urban planning and risk management, and GIS-based models hold promise for data-scarce regions like Nepal and other Himalayan countries. Future work should evaluate operation and efficiency in different scenarios [68].

Another similar earthquake risk assessment study [69] proposes the use of Deep learning–based earthquake risk-mapping model for Northeast India, integrating convolutional neural networks (CNN), analytical hierarchy process (AHP), and spatial analysis using GIS. The CNN model achieves high accuracy in earthquake probability assessment. Results show significant risk zones, aiding disaster management planning. Challenges include data availability and model limitations, suggesting future improvements like high-resolution DEM data integration. Despite limitations, the model provides an effective framework for earthquake risk assessment, applicable to other disaster scenarios with GIS-based modifications.

5.8 Application of GIS in Optimizing Solid Waste Collection

5.8.1 Multi-Purpose Bibliometric Analysis

Waste management is a complex and cross-disciplinary field, significantly benefiting from multi-criteria decision-making (MCDM) techniques. The highlighted bibliometric analysis of 216 publications (1992-2022) reveals trends and advancements, highlighting significant contributions from major journals. The key findings include the dominance of Asian countries in publications, with China leading in citations. Collaborative research dominates, suggesting continued focus by academics, environmentalists, and policymakers [70]. Sustainable waste management is crucial as urbanization increases waste production. Effective strategies reduce environmental harm and health risks and contribute to a circular economy. Proper management, considering socioeconomic factors, is essential for achieving sustainable development goals and minimizing costs [71].

Multi-Criteria Decision Making (MCDM) methodologies address complex decision-making processes by evaluating multiple criteria simultaneously, making them suitable for waste management issues. MCDM's main strength is its adaptability to decision-makers' judgments. allowing for performance weights and scores distribution across both quantitative and qualitative criteria. These approaches can identify preferred alternatives, rank options, or separate acceptable from undesirable ones using a performance matrix. Common MCDM methods include TOPSIS, Analytical Hierarchy Process (AHP), ELECTRE, and PROMETHEE. Their ability to compare alternatives across various criteria makes them valuable for evaluating waste management strategies [72]. The most common MCDM approaches are AHP and Fuzzy-AHP. Limitations include reliance on Scopus and quantitative analysis, excluding qualitative insights. Future research could explore MCDM in the circular economy, urging investment in Industry 4.0 approaches for efficient resource utilization and sustainable waste management [70].

5.8.2 Trace Gas Emissions from Solid Landfill

This paper reviews literature from 1983-2019 on trace gas emissions present due to municipal solid waste (MSW) landfills, discussing their origins, transportation, and environmental impacts. The paper discusses sampling, analytical techniques with quantitative measurement methods for trace gas emissions from the waste degradation. It highlights challenges in measurement due to the temporal and spatial variations present. Direct techniques involve field sampling, while indirect methods rely on modeling. Inverse dispersion modeling estimates emissions using ambient concentrations downwind of landfills, emphasizing the importance of accurate measurements and GIS integration for precise analysis [73].

Another paper [74] provides an overview of unmanned aerial photo imaging technologies in waste disposal site management, outlining current practices and future prospects. It aims to bridge the gap in understanding UAV capabilities, envisioning their integration similar to satellite imagery and precision positioning methods in waste management. UAV usage in waste management lags behind other industries like construction and mining. Scientific literature on UAV applications in waste management is limited, indicating a gap between industrial practice and academic analysis. Despite growing reports, scientific understanding of UAV capabilities for landfill monitoring remains incomplete. Integration of GIS technologies are very crucial in order to harness the full potential of UAV use in landfill monitoring and environmental protection.

A major study [75] from China highlights the significant methane emissions–detected municipal solid waste (MSW) landfills, with implications for global warming. It underscores the importance of understanding factors influencing methane production, such as waste composition, landfill age, and cover system design. Spatial and temporal variations in emissions are crucial, necessitating comprehensive monitoring and management strategies. Direct measurement methods, including portable laser detectors, offer valuable insights into methane distribution and hot spot identification, aiding in landfill cover design and emission control. This paper examines methane emissions from a landfill with high kitchen waste content using direct measurements *via* a portable laser measurement detector. The study visualizes seasonal methane spatial distribution, identifies hot spots, and evaluates factors influencing emissions. It suggests that Geographic Information Systems (GIS) do play a crucial role in understanding spatial and temporal variations in methane emissions resulting from municipal solid waste (MSW) landfills. GIS helps visualize seasonal variations in methane spatial distribution, and screen and provide data for localized methane hot spots to optimize landfill cover design. Additionally, GIS aids in assessing the validity of numerical models for emission of methane from landfill covers and optimizing landfill gas hotspots for recovery system.

5.8.3 Analysis of Municipal Waste Disposal Site Use of GIS in Mexico

Poorly managed landfills can pose environmental and health risks, converting them into open dumps. A study done in Chiapas, Mexico, employed the Geographic Information Technology (GIT) tools and spatial analysis techniques to evaluate emissions, temperatures, land use changes, and regulatory compliance, offering valuable insights for decision-making and future research in the area. The current operation of the site and its location were found to comply partially with the regulations. Gaseous dispersion results showed concentrations for landfill gas ranges from 100 to 8725 $\mu g/m^3$ and for H_2S ranges 0.01 to 0.35 $\mu g/m^3$, posing a potential odor issue. Liquid emissions moved eastward, posing a risk to water supply wells. Thermal emissions from the site created microclimatic zones, with temperature differences between 8.37°C in summer season and 2.49°C in winter, which it attributed to the decomposition of waste and human activities in that region. Land use changes around the site had increased steadily [76].

The OBIA (Object-Based Image Analysis) method started by segmenting images using Segment Mean Shift. Then, image classification was performed using ArcGIS Training Sample Manager with four classes, namely water, urban/anthropized areas, tree coverage, and shrub/herbaceous coverage. Interactive Supervised Classification was also employed to expedite the maximum likelihood classification process [77].

In this study, the measurement of leachate dispersion, crucial for assessing the functionality of a final disposal site (FDS), particularly when operating as open dump or controlled site lacking proper infrastructure, is emphasized. The research utilizes the ArcGIS 10.X Groundwater Toolset to track groundwater flow direction and movement, along with spatial and temporal variations in the plume of conservative contaminant (Ion Chloride Cl-). This tool provides a simplified approach to determine leachate dispersion, aiding in the evaluation of FDS performance. Although this work is only able to highlight the anthropogenic actions that occurred in the study area which were empirically modeled, the results provide useful information for the process of decision-making. The analysis of distances and restrictions also showed that the final disposal site (FDS) partially followed Mexican regulations. The GIT tools and techniques used in this paper allowed for visualizing the movement and spatial scope of the most important subproducts generated by MSW, such as landfill gas and leachate. Changes that occurred in the regional context, which are difficult to identify with the naked eye, such as LULC or LST, were also detected.

Additionally, the use of these tools entailed advantages such as simple procedures, use of free and easily accessible data, and the possibility of replicating in other places [76].

5.8.4 Urban Landfill Assosa Town, Ethiopia

The study [78] utilizes GIS and Multi-Criteria Evaluation (MCE) to identify potential landfill sites for Assosa town, located in Benishangul-Gumuz, Ethiopia, as it faces the challenge of waste management due to rapid urbanization and inadequate infrastructure, necessitating the selection of suitable landfill sites for environmental protection and public health. In this study, eight criteria, including settlement areas, drainage, groundwater wells, roads, airport, socially valuable areas, land use/cover, and slope, were assessed using the Analytic Hierarchy Process (AHP). The utilized eight datasets were sourced from both primary (collected *via* field survey with GPS tools) and secondary (websites, literature, government institutions) sources. Groundwater well data was from Assosa city water authority, while satellite imagery and elevation data were acquired from USGS Earth Explorer. The criteria layers were standardized and overlaid in GIS to determine suitability. This study offers preliminary insights for further detailed analysis, aiding authorities in enhancing solid waste management in the region, emphasizing the importance of integrating GIS and MCE in urban planning and environmental management. It employs GIS spatial analysis techniques like scanning, georeferencing, digitizing, buffering, and overlay analysis, integrating them with Multi-Criteria Evaluation (MCE) using Analytic Hierarchy Process (AHP). A range of input datasets are processed to derive weighted criteria, producing a thematic map of potential landfill sites. The suitability analysis showed that 31.95% of this site is unsuitable while 36.54% is highly suitable for the purpose of landfill siting. Pairwise comparison using AHP indicates consistent weightage, with the most important criteria being distance to settlement area (26.9%), followed by socially valuable area (21.3%), distance to road network (12.7%), distance to drainage (9.9%), land use/land cover (16.3%) and distance to airport site (6.8%). The study aims to propose landfill sites while adhering to environmental standards and minimizing costs. Results show 10.15% unsuitable, 5.79% highly suitable, and 59.53% less suitable areas for landfill siting. GIS-based analysis aids decision-makers in waste management and public health protection. Further hydrogeological investigations are recommended for final site selection.

A similar study for Dejen town in Ethiopia was also conducted. For the rapidly growing Dejen, proper waste management is crucial. Using ArcGIS

10.5 and AHP techniques, a suitability index map for waste disposal was created, considering ten factors. The study found 7.9% of the study area is highly suitable, 46.4% being suitable, 22.2% less suitable, 6.5% moderately suitable, and 17% unsuitable for waste disposal. This map assists local administrators in selecting appropriate sites and emphasizes the importance of public awareness, reuse, recycling, and using waste for energy to maintain environmental health [79].

5.8.5 Landfill in Vietnam

The study [81] aimed to identify optimal landfill sites in Thach That district, Hanoi, using GIS and MCDA. Nine criteria were evaluated, including proximity to water, historical sites, rural areas, the waste collection points, land use and slopes. Using a Boolean logic model and AHP, it was found that 56.81% (10,641.40 hectares) of area is unsuitable, 42.01% (7,867.60 hectares) is less suitable, and 1.18% (220.89 hectares) is suitable. Despite some suitable sites being far from communes, the study provides a scientific basis for landfill site selection in Vietnam [80].

Thach That district, northwest of Hanoi, spans 20°56′45′ to 21°06′10′ N latitude and 105°23′30′ to 105°38′22′ E longitude. The district has three terrain types: low mountainous (29%), semi-mountainous (46%), and plain (25%). Elevation ranges from 5m to 15m. It features a tropical monsoon climate with mean annual temperature of 23.4°C, 1,628mm of precipitation, and 1,680 hours of sunshine annually. The population is 253,786 with an economic growth rate of 13.6%. The district includes 22 communes, 1 town, and over 50 craft villages. Data was processed using QGIS3, MicroStation V8i, and the EPSG:5897 coordinate system

MCDA assigns weights to criteria based on their significance. This study used Boolean Logic, Analytic AHP, and Weighted Linear Combination (WLC) methods for identifying optimal landfill sites [81]. Boolean Logic converts information from input maps into binary format (0 for false, 1 for true) [82], indicating suitability for a landfill. It uses Equation 1 to categorize sites based on criteria, resulting in a binary map with suitable and unsuitable areas. This method is useful for initial screening and straightforward when deterministic rules apply, producing a map that clearly differentiates between satisfactory and unsatisfactory locations [83, 84].

$$\mathbf{C = \{1 \text{ if class } A > \text{ or } < X\} \text{ and } C = \text{class } A > \text{ or } < X}$$

In locating landfills using Boolean logic, buffer zones of 3,000m around urban areas, 100m around main roads, and 1,000m around industrial zones were established.

A suitability map for selecting the landfill sites in Thach That district of Hanoi by using GIS and MCDA, identified 10,641.40 hectares (56.81%) as being unsuitable, 7,867.60 hectares (42.01%) as less suitable, and only a small portion of 220.89 hectares (1.18%) as suitable. Suitable locations are in Canh Nau, Huu Bang, Di Nau, Dai Dong, Phung Xa, Cam Yen, and Tien Xuan communes, with the largest areas available in Canh Nau and Di Nau. Two areas are too small for landfills but three are suitable for small-sized landfills and four for medium-sized ones. Despite some sites being 20 km from certain communes, the distances are comparable to the current landfill. This entire research highlights the scientific and social importance of proper landfill site selection in developing countries and provides valuable insights for Hanoi's waste management planning [80].

5.8.6 Landfill in Saudi Arabia

This study [85] utilized MCDA to assess and to recommend the suitable landfill locations in Najran, KSA. Through use of remote sensing data and ArcGIS software, nine thematic layers were prepared accordingly, encompassing aspects like drainage density, groundwater depth, and proximity to residential and protected areas. Weightings for these layers were determined *via* expert surveys and AHP. The AHP and FAHP methods were used in calculating final weightings for each criterion, ensuring objectivity and transparency in decision-making. Each layer was subdivided into five subclasses with local weightings ranging from one to five. Pixels were assigned integer values from one to five based on suitability, with five being most suitable and one being less suitable. The overlaying of thematic layers and summing weighted scores was later used to generate the final suitability map. Five main components in landfill location determination are outlined in subsequent sections, covering criteria definition, constraint classification, weighting, thematic layer preparation, and decision-making processes. Sensitivity testing was conducted to validate the weightings, leading to the creation of a landfill suitability index (LSI) map. The map categorized areas into five suitability classes, with 18.0% deemed most suitable and 18.2% suitable for landfill. The study's findings offer valuable guidance for Najran's municipality for regional planners and decision-makers in future for landfill site selection endeavors, accompanied by recommendations to refine the suitability map further.

5.8.7 Solid Waste Case Study of Ghana

To address the challenges of Municipal Solid Waste (MSW) disposal, particularly in Bolgatanga Municipality, Ghana, this study utilized Multi-Criteria Evaluation (MCE) and GIS analysis to assess landfill site suitability based on various criteria. Proximity to water bodies, major roads, and distance from towns were key considerations. Areas within 164m of water bodies were unsuitable, while those 328m away were most suitable, while those farther away were considered more suitable. Similarly, proximity to major roads within 332m while 665m away was most suitable. It influenced suitability, with closer distances being less favorable. Slope gradient and soil type were also assessed, with specific ranges deemed suitable or unsuitable. A model builder process in ArcGIS was utilized to integrate these criteria, producing a suitability index map. From this, potential sites were identified and refined, selecting a site with optimal characteristics, away from water bodies and major roads, and suitable in terms of slope and soil type. This selected site offers logistical advantages and lower management costs compared to the current site, highlighting the importance of careful landfill site selection for environmental and economic considerations [86].

5.8.8 Site Selection for Solid Waste, Ecuador

Azuay Province in Ecuador comprises 15 cantons, including Cuenca, the most populous. Cantonal governments are responsible for Integrated Solid Waste Management (ISWM). Cuenca has the highest urban per capita, while Sevilla de Oro leads in this aspect. Two Final Disposal Sites (FDSs) serve multiple cantons, while others rely on open dumps or landfills. Some cantons face long distances to reach FDSs. Azuay lacks Transfer Stations (TSs) in its ISWM system. The study aimed to identify the most suitable locations for setting up Transfer Stations (TSs) in Azuay Province in Ecuador, using three-level hierarchy structure. The authors utilized the AHP technique with GIS-MCDA methodology, since they are popular for site location issues due to cost-effectiveness and easy scalability. The approach converted complex decision-making into simpler hierarchal systems, integrating qualitative and quantitative attributes into a geographical database. Using multi-criteria decision analysis, it also assesses technical, environmental, economic, and social factors. The research identifies potential TS locations in Paute, Sigsig, and Nabón, serving nine out of fifteen cantons. Overall, the results showed that 79.12% of study area is deemed restrictive for TS placement, highlighting the need for sustainable and integrated solutions in MSW management. While economic factors often dominate, this research highlighted environmental, social, and technical aspects in terrain suitability assessment. Future research could explore fuzzy sets theory for uncertainty management and probabilistic sensitivity testing for decision-making validation [87].

5.8.9 Geospatial Waste Collection, Pakistan

Approximately 50-70% of total waste management cost is being spent on collection, making route optimization vital for cost savings. This study demonstrates the effectiveness of GIS-based route optimization in the twin cities, showing improvements in route lengths and travel times. The findings suggest extending this approach to other Pakistani cities for better waste collection optimization. This research served as a major decision support tool for waste management, aiding lower fuel consumption, workload distribution and scheduling, and benefiting municipal administrations in efficient solid waste management operations. Additionally, future enhancements could include incorporating additional spatial data for better modeling.

The existing solid waste collection routes in Islamabad span 356 km in 7.5 hours, while in Rawalpindi, they cover 393 km in 8.3 hours annually. After optimization, proposed routes reduce travel distances to 294 km in 5.8 hr for Islamabad and 358 km in 7.2 hr for Rawalpindi annually. This results in potential savings of 10,642 km and 345 hr per year for Rawalpindi and 19,406 km and 531 hours per year for Islamabad. These savings are achieved through decreased mileage and overall travel time, leading to reduced fuel and maintenance costs and lower emissions.

A comprehensive database spanning 2005 to 2012 was collated in collaboration with the municipality, encompassing both static and dynamic waste management data. Static datasets covered population density, waste bin positions and types, road networks, proper service zones and vehicle capacities. These dynamic datasets included the current waste generation rates, vehicle routing systems, and also traffic data. To optimize collection processes, various data were generated, including study area boundaries, satellite imagery, urban plans, population density distributions, land use data, and road network information. A spatial geodatabase was constructed to integrate and analyze this data, ensuring compatibility with existing municipal data and facilitating the generation of optimal collection routes. Non-spatial data such as travel times, vehicle speeds, capacity of bins, and collection schedules were updated through the use of field surveys for accuracy. The proposed study focuses on optimizing transportation management within the Vehicle Routing Problem (VRP) framework to enhance public service operations and management efficiency. GIS technology is employed to handle spatial data, facilitating importation, storage, management, analysis, and visualization of relevant data [88].

5.8.9.1 GIS and MCD in Pakistan

The study conducted in Peshawar district, Pakistan, emphasizes the importance of appropriate landfill site selection (LSS) for an effective municipal solid waste (MSW) management. Utilizing a sophisticated MCDA approach within a geographical information system (GIS), the study incorporates various economic, social and environmental criteria. Pairwise comparisons and the Analytical Hierarchy Process (AHP) techniques were employed to rank the criteria, followed by Weighted Linear Combination (WLC) which was used to identify the most suitable landfill site. Results indicate that around 11.4% to 16.6% of the study area is highly suitable for landfill construction, while approximately 20.8% is unsuitable. The study recommended that the aim of landfill sites should be to minimize public health risks and environmental degradation, offering a scientific basis for MSW disposal in urbanized areas.

The study utilized eleven datasets from various departments to assess suitable landfill sites which cover environmental, social, and economic factors. Data sources included groundwater data from the Public Health Engineering Department Peshawar, soil texture data from IGIS NUST/ Soil Survey of Pakistan, surface water, railway lines, and airport points from topographic sheets of the Survey of Pakistan, aspect and slope from ALOS DEM, and LULC settlements data from Landsat 8. Road lines were digitized from OpenStreetMap, and land value data were obtained from District Administration Office Peshawar. The MCDA with the AHP technique in ArcGIS 10.5 was employed to generate thematic maps for each criterion, followed by a weighted linear combination (WLC) to identify final suitable landfill sites [88].

5.8.9.2 Plastic Waste Disposal, Delhi

Plastic waste, a major component of urban MSW, requires sustainable management, currently lacking in India's policy framework. This study also used the Multi-Criteria Decision Analysis (MCDA) to evaluate the waste disposal options for plastic waste in Delhi. A panel consisting of nine members from IIT Delhi assessed seven disposal methods (Landfill, Recycling, Incineration, Pyrolysis, and combinations of these) against environmental, health, financial, and legislative criteria. The panel later concluded that a combination of recycling and incineration were the best options, balancing reduced air pollution and economic benefits from recycling with decreased land and water pollution from incineration. Recycling alone was preferred over incineration alone due to its broader benefits. Landfilling was the least preferred due to its contribution to land and water pollution, despite its low cost and ease of implementation. Pyrolysis was also not favored due to perceived hazards and impracticality in India, despite the existence of clean technologies like Plasma Pyrolysis Technology (PPT). The study suggests MCDA is effective for robust decision-making in waste management but notes the need for future sensitivity analysis in order to test the robustness of the criteria weightings [89].

5.8.9.3 *Solid Waste Collection Class 2 Indian City Using GIS*

Illegal dumping is a significant problem in urban solid waste management and poses a large number of challenges for municipal authorities. The study [90] examines Rampurhat Municipality's waste management inefficiencies, revealing a 44.60% labor shortfall and a lack of required vehicles for secondary waste transportation. The resulting illegal dumping shows a cluster pattern. The mean distance of 78.88 meters exists between sites. Analysing the factors helped identify two key factors influencing dump site frequency, significantly affecting the northern area. According to the 2020-2021 MSW report by the Central Pollution Control Board (CPCB) [91], India generates more then 160,038.9 tons per day (TPD) of solid waste. Of this, about 152,749.5 TPD is collected, achieving a 95.4% collection efficiency. However, only 50% (79,956.3 TPD) of this waste is processed, while 18.4% (29,427.2 TPD) is landfilled, leaving 31.7% (50,655.4 TPD) unaccounted for. The problem of state-to-state disparities in solid waste management are significant, with processing rates ranging from 0% to 85%. Madhya Pradesh leads in environmental solid waste management with a score of 76.75, followed by Goa, Chandigarh, and others. States like Arunachal Pradesh, Assam, Bihar, and West Bengal perform poorly. All the geospatial analysis was carried out in ArcGIS 10.5 software, whereas IBM SPSS Statistics v25 has been considered for the statistical analysis section.

Effective waste management is hindered by a 44.60% shortage in required human resources and a 14% vehicle deficiency for door-to-door collection, alongside a 95% and 72% vehicle shortage for secondary transportation. This leads to 150 illegal dump sites. The study used correlation analysis and PCA, identifying seven variables influencing illegal dumping, with road density, population density, and pond density being significant factors. The composite index model allowed to validate by regression analysis with an

R^2 of 0.874, highlighted specific wards with high illegal dumping incidents [90].

5.9 Role of GIS for Planning of Waste Disposal

5.9.1 Sustainable Agriculture Waste Management

Agriculture significantly contributes to major global anthropogenic emissions, including waste and greenhouse gases. To mitigate all these impacts, the role of adopting a circular economy for agricultural waste management is essential. MCDA methods offer a comprehensive approach for evaluating the treatments and for better valorization options by supporting long-term and economically viable as well as ecologically sound and socially acceptable policies. This study [92] reviews the MCDA application in agricultural waste management. It significantly highlights its growing use in decision-making and stakeholder involvement. The research is part of a broader Next-Generation-EU-integrated program by the National Research Centre for Agricultural Technologies.

The article critically reviews studies using Multi-Criteria Decision Analysis (MCDA) in agricultural waste management. The review also highlights:

1. MCDA Methods: There were 15 different MCDA methods identified in 58 papers. The Analytic Hierarchy Process (AHP) was most common (29 applications), followed by ELECTRE, TOPSIS, and the weighted sum method (with three applications each). MCDA was often part of a multi-methodological framework (31 papers).
2. Technologies: Commonly applied technologies include anaerobic digestion, composting, and thermal treatment, primarily producing biofertilizers, biogas, energy, and fertilizers.
3. Stakeholder Involvement: Most studies involved stakeholders through interviews or surveys to identify and weigh criteria. AHP, sometimes combined with Life Cycle Assessment (LCA), was frequently used alongside participatory methods.

The study notes that future research will test AHP with LCA to assess sustainability of innovative technologies in AGRITECH project, aiming for a multidimensional, repeatable, and expandable evaluation of agricultural waste management technologies.

5.9.2 Healthcare Waste Disposal

Rapid urbanization in South Africa has led to increased exposure to waste sites, with households' median distance decreasing substantially from 2008 to 2015. Proximity to nearest waste sites within 5 km was also associated with higher risks of heath conditions like asthma, diabetes, tuberculosis and depression, even after adjusting for the socioeconomic factors. This highlights the urgent need to mitigate health risks by reducing waste site numbers and sizes. The study from the South African National Income Dynamics Study (SA-NIDS) between 2008 and 2015 involving 32,255 participants helped to find significant health risks associated with living near waste sites. The prevalence of households within 5 km of the waste sites increased from 2.8% in 2008 to 35.5% in 2015, with more than half (53.4%) within 10 km by 2015. Adjusted relative risks for living within 5 km of the waste site were significantly higher for tuberculosis (1.18), asthma (1.41), diabetes (1.25), and depression (1.08). This study also highlights the urgent need for sustainable waste management policies to mitigate adverse health outcomes and to uphold human rights, aligning with the Sustainable Development Goals (SDGs). The findings underscore the importance of balancing economic development with environmental protection to ensure the well-being of the communities [93].

5.9.3 Waste-to-Energy Technology Review

Enhanced Waste Management (EWM) and Enhanced Landfill Mining (ELFM) aim to sustainably manage landfill waste by converting it into materials (Waste-to-Product, WtP) or energy (Waste-to-Energy, WtE). This paper gives insights into thermochemical technologies like incineration, gasification, pyrolysis, and plasma technologies for the processing of municipal solid waste (MSW) and refuse-derived fuel (RDF). It highlights plasma gasification or vitrification as a promising method for recovering both materials and energy, noting its proven feasibility and expected improvements in process control and the efficiency to enhance commercial viability. Enhanced Landfill Mining (ELFM) aims to valorize landfill waste as materials (WtP) and energy (WtE) while reducing CO_2 and pollutant emissions. Plasma gasification/vitrification achieves significant volume reduction which is over 5:1 for ash and more than 50:1 for solid waste. This process is socially accepted as a recovery method rather than disposal.

Factors affecting landfill mining potential include waste characteristics, landfill age, and regional location. Belgium's tests show feasibility in converting excavated MSW into RDF similar to fresh MSW. Advanced thermochemical technologies, sensitive to feedstock variations, benefit from homogeneous waste inputs for stable operations and high-quality outputs. Plasma technologies are emerging as promising options for combined energy and material recovery, although more comparative data with conventional methods like incineration are needed. Future commercial viability depends on continued support from regulatory, economic, and socio-political factors [94].

5.9.4 Municipal Waste to Energy, Bangladesh

Bangladesh, facing increased urban energy demands and waste generation, can benefit from incineration of municipal solid waste in order to produce electricity. A proposed steam power plant is suggested using municipal solid waste in Chattogram and Dhaka shows potential capacities of 83 MW and 169 MW, respectively. Energy and exergy efficiencies are predicted at 32% and 27%, with potential CO_2 emission reductions of at least 1 MMT in city of Dhaka and about 0.5 MMT in Chattogram. Ecological efficiency exceeds 87%, with the cost of electricity remaining lower than that of the Narayanganj plant [95].

5.9.5 Waste-to-Energy Plant, India

Waste-to-energy (WtE) technologies offer an excellent solution by converting MSW into a form of renewable energy sources like electricity and heat in an environmentally friendly manner. This review clearly outlines the challenges of MSW management; it also highlights the major health implications, discusses WtE opportunities and technologies, and evaluates the current status of WtE initiatives in India. Additionally, it offers recommendations for enhancing solid waste management practices in an Indian context, providing valuable insights for scholars, authorities, researchers, and stakeholders that are involved in MSW management [96].

5.9.6 Waste-to-Energy Plant, China

The paper presents a unique methodology by combining the HFL-BWM-ANP and double parameters TOPSIS, for selecting the optimal sites for the waste-to-energy (WTE) incineration plants. It constructs a comprehensive evaluation index system with five dimensions and thirteen sub-criteria, blending quantitative calculations with qualitative expressions using HFLTS. The proposed methodology accurately determines criteria weights, considering hesitation and randomness in comparisons, and leverages double parameters in TOPSIS; it is to rank alternative sites, factoring in decision-makers' risk appetites and preferences. A numerical case which demonstrates feasibility, sensitivity analysis showcases the robustness, and comparative analysis confirms effectiveness. This approach effectively addresses WTE incineration plant site selection challenges, enhancing overall decision consistency and computational efficiency. Its applicability extends to other decision problems under uncertainty, like WTE technology selection and green supply chain management. Future research should be focused on heterogeneous evaluation index systems, also on incomplete information, and conflict analysis in group decision-making. Overall, the proposed methodology presented offers a promising decision-making framework in addressing complex site selection problems in waste management and related fields [97].

5.10 Conclusions

The role of AI in environmental remediation is significant and promising. AI can help to improve the efficiency and effectiveness of environmental remediation processes by providing accurate and real-time monitoring of environmental pollutants, predicting potential risks and identifying the most appropriate remediation strategies. One of the key advantages of using AI in environmental remediation is its ability to process large amounts of data from different sources and provide insights that would be difficult to obtain through traditional methods. AI can analyze data from satellite imagery, sensors, and other sources to identify areas that are most contaminated and prioritize remediation efforts. AI can also help to optimize remediation strategies by simulating different scenarios and predicting the outcomes of different approaches. This can help to reduce costs, minimize environmental impact, and ensure that the remediation process is as effective as possible. AI has the potential to revolutionize the field of environmental remediation by providing new tools and approaches to address environmental challenges. In spite of having so many advantages, there is often a lack of quality data on the extent and nature of environmental pollution, which can limit the effectiveness of AI models and algorithms. In addition, the use of AI in environmental remediation raises ethical concerns around issues such as data privacy, algorithmic bias, and the displacement of human workers.

References

1. Basso, F., Bove, E., Dumontet, S., Ferrara, A., Pisante, M., Quaranta, G., Taberner, M., Evaluating environmental sensitivity at the basin scale through the use of geographic information systems and remotely sensed data: an example covering the Agri basin (Southern Italy). *Catena*, *40*, 1, 19–35, 2000.
2. Castka, P., Searcy, C., Fischer, S., Technology-enhanced Auditing in Voluntary Sustainability Standards: The Impact of COVID-19. *Sustainability*, 2020, 12, 1–24, 2020, doi: 10.3390/su12114740.
3. Castoldi, N., Stein, A., Bechini, L., Agri-environmental assessment of extractable soil phosphorus at the regional scale. *NJAS Wageningen J. Life Sci.*, *56*, 4, 325–343, 2009.
4. Chowdhary, K.R., *Fundamentals of Artificial Intelligence*, pp. 603–649, Springer India, New Delhi, 2020.
5. Delavar, M.R., Gholami, A., Shiran, G.R., Rashidi, Y., Nakhaeizadeh, G.R., Fedra, K., Hatefi Afshar, S., A novel method for improving air pollution prediction based on machine learning approaches: a case study applied to the capital city of Tehran. *ISPRS Int. J. Geo-Inf.*, *8*, 2, 99, 2019.
6. Deng, K., Fang, T., Feng, H., Peng, H., Löwenstein, L., Hameyer, K., Hierarchical eco-driving and energy management control for hydrogen powered hybrid trains. *Energy Convers. Manage.*, *264*, 115735, 2022.
7. Engel-Cox, J.A., Hoff, R.M., Haymet, A.D.J., Recommendations on the use of satellite remote-sensing data for urban air quality. *J. Air Waste Manage. Assoc.*, *54*, 11, 1360–1371, 2004.
8. Everitt, J.H., Yang, C., Escobar, D.E., Webster, C.F., Lonard, R.I., Davis, M.R., Using remote sensing and spatial information technologies to detect and map two aquatic macrophytes. *J. Aquat. Plant Manage.*, *37*, 2, 71–80, 1999.
9. Feizizadeh, B. and Blaschke, T., Examining urban heat island relations to land use and air pollution: Multiple endmember spectral mixture analysis for thermal remote sensing. *IEEE J. Sel. Top. Appl. Earth Obs. Remote Sens.*, 6, 3, 1749–1756, 2013.
10. Fortuna, M.E., Simion, I.M., Gavrilescu, M., Sustainability in environmental remediation. *Environ. Eng. Manage. J. (EEMJ)*, *10*, 12, 2011.
11. Geissdoerfer, M., Savaget, P., Bocken, N.M., Hultink, E.J., The Circular Economy–A new sustainability paradigm? *J. Clean. Prod.*, *143*, 757–768, 2017.
12. Gholizadeh, A., Saberioon, M., Ben-Dor, E., Borůvka, L., Monitoring of selected soil contaminants using proximal and remote sensing techniques: Background, state-of-the-art and future perspectives. *Crit. Rev. Environ. Sci. Technol.*, *48*, 3, 243–278, 2018.
13. Gómez-Sal, A., Belmontes, J.A., Nicolau, J.M., Assessing landscape values: a proposal for a multidimensional conceptual model. *Ecol. Modell.*, *168*, 3, 319–341, 2003.

14. Haupt, S.E., Pasini, A., Marzban, C. (Eds.), *Artificial Intelligence Methods in the Environmental Sciences*, Springer Science & Business Media, 2008.
15. Hunsaker, C.T. and Carpenter, D.E. (Eds.), *Environmental Monitoring and Assessment Program: Ecological Indicators*, Atmospheric Research and Exposure Assessment Laboratory, Office of Research and Development, US Environmental Protection Agency, 1990.
16. Jensen, J.R., Ramsey III, E.W., Holmes, J.M., Michel, J.E., Savitsky, B., Davis, B.A., Environmental sensitivity index (ESI) mapping for oil spills using remote sensing and geographic information system technology. *Int. J. Geogr. Inf. Syst.*, *4*, 2, 181–201, 1990.
17. Jerlov, N.G., *Marine Optics*, Elsevier, 1976.
18. Kaplan, G., Aydinli, H.O., Pietrelli, A., Mieyeville, F., Ferrara, V., Oil-Contaminated Soil Modeling and Remediation Monitoring in Arid Areas Using Remote Sensing. *Remote Sens.*, *14*, 10, 2500, 2022.
19. Kirk, J.T., Volume scattering function, average cosines, and the underwater light field. *Limnol. Oceanogr.*, *36*, 3, 455–467, 1991.
20. Liu, W., Zhao, J., Du, L., Padwal, H.H., Vadivel, T., Intelligent comprehensive evaluation system using artificial intelligence for environmental evaluation. *Environ. Impact Assess. Rev.*, *86*, 106495, 2021.
21. MacMillan, R., Automated knowledge-based classification of landforms, soils and ecological spatial entities, 2004, DOI:10.13140/RG.2.1.1122.9048. Retrieved April, 24, 2017.
22. Magazzino, C., Mele, M., Sarkodie, S.A., The nexus between COVID-19 deaths, air pollution and economic growth in New York state: Evidence from Deep Machine Learning. *J. Environ. Manage.*, *286*, 112241, 2021.
23. Marwa, M., Salhi, B., Jarboui, A., Environmental audit and environmental disclosure quality. *Sci. Ann. Econ. Bus.*, *67*, 1, 93–115, 2020.
24. Nazarova, K., Hotsuliak, V., Miniailo, V., Nezhyva, M., Mysiuk, V., Accounting, analysis and environmental audit as an imperative of the development of green economy in the state's economic security system. *E3S Web Conf.*, 166, 13003, 2020, EDP Sciences.
25. Oikonomidis, D., Dimogianni, S., Kazakis, N., Voudouris, K., A GIS/remote sensing-based methodology for groundwater potentiality assessment in Tirnavos area, Greece. *J. Hydrol.*, *525*, 197–208, 2015.
26. Rahman, M.R., Shi, Z.H., Chongfa, C., Soil erosion hazard evaluation—an integrated use of remote sensing, GIS and statistical approaches with biophysical parameters towards management strategies. *Ecol. Modell.*, *220*, 13-14, 1724–1734, 2009.
27. Rahman, M.R., Shi, Z.H., Chongfa, C., Assessing regional environmental quality by integrated use of remote sensing, GIS, and spatial multi-criteria evaluation for prioritization of environmental restoration. *Environ. Monit. Assess.*, *186*, 6993–7009, 2014.

28. Ramadas, M. and Samantaray, A.K., Applications of remote sensing and GIS in water quality monitoring and remediation: a state-of-the-art review, in: *Water Remediation*, pp. 225–246, 2018.
29. Rummell, A.J., Leon, J.X., Borland, H.P., Elliott, B.B., Gilby, B.L., Henderson, C.J., Olds, A.D., Watching the Saltmarsh Grow: A High-Resolution Remote Sensing Approach to Quantify the Effects of Wetland Restoration. *Remote Sens.*, *14*, 18, 4559, 2022.
30. Sagan, V., Peterson, K.T., Maimaitijiang, M., Sidike, P., Sloan, J., Greeling, B.A., Adams, C., Monitoring inland water quality using remote sensing: Potential and limitations of spectral indices, bio-optical simulations, machine learning, and cloud computing. *Earth Sci. Rev.*, *205*, 103187, 2020.
31. Shi, T., Guo, L., Chen, Y., Wang, W., Shi, Z., Li, Q., Wu, G., Proximal and remote sensing techniques for mapping of soil contamination with heavy metals. *Appl. Spectrosc. Rev.*, *53*, 10, 783–805, 2018.
32. Shi, T., Yang, S., Zhang, W., Zhou, Q., Coupling coordination degree measurement and spatiotemporal heterogeneity between economic development and ecological environment—Empirical evidence from tropical and subtropical regions of China. *J. Clean. Prod.*, *244*, 118739, 2020.
33. Singh, A., Satapathy, S.C., Roy, A., Gutub, A., AI-based mobile edge computing for IoT: Applications, challenges, and future scope. *Arabian J. Sci. Eng.*, 1–31, 2022.
34. Smith, C.J. and Wong, A.T., Advancements in artificial intelligence-based decision support systems for improving construction project sustainability: a systematic literature review. *Informatics*, 9, 2, 43, 2022, May, MDPI.
35. Somvanshi, S.S., Vashisht, A., Chandra, U., Kaushik, G., Delhi air pollution modeling using remote sensing technique, in: *Handbook of Environmental Materials Management*, pp. 1–27, 2019.
36. Srinivasa Rao, Y. and Jugran, D.K., Delineation of groundwater potential zones and zones of groundwater quality suitable for domestic purposes using remote sensing and GIS. *Hydrol. Sci. J.*, *48*, 5, 821–833, 2003.
37. Styles, D., Schoenberger, H., Galvez-Martos, J.L., Environmental improvement of product supply chains: Proposed best practice techniques, quantitative indicators and benchmarks of excellence for retailers. *J. Environ. Manage.*, *110*, 135–150, 2012.
38. Takahashi, T., Matsushita, K., Nishimura, T., Community actions against anticommons of forests in contemporary Japan: Case studies of former common forests. *J. For. Res.*, *26*, 1, 68–74, 2021.
39. Duncan, B.N., Lamsal, L.N., Thompson, A.M., Yoshida, Y., Lu, Z., Streets, D.G., ... & Pickering, K.E., A space-based, high-resolution view of notable changes in urban NOx pollution around the world (2005–2014). *J. Geophys. Res. Atmos.*, 121, 2, 976–996, 2016.
40. Tian, C., Peng, J., Zhang, W., Zhang, S., Wang, J., Tourism environmental impact assessment based on improved AHP and picture fuzzy PROMETHEE II methods. *Technol. Econ. Dev. Econ.*, *26*, 2, 355–378, 2020.

41. Vimal, S., Robinson, Y.H., Kadry, S., Long, H.V., Nam, Y., IoT based smart health monitoring with CNN using edge computing. *J. Internet Technol.*, *22*, 1, 173–185, 2021.
42. Yang, J., Zhang, J.C., Zhuang, J.Y., Multi-resources data-based environment assessment of the Grand Canal in Beijing with CEAI model. *Procedia Environ. Sci.*, *13*, 660–669, 2012.
43. Yaseen, Z.M., An insight into machine learning models era in simulating soil, water bodies and adsorption heavy metals: Review, challenges and solutions. *Chemosphere*, *277*, 130126, 2021.
44. Yuen, K.F., Wang, X., Ma, F., Li, K.X., The psychological causes of panic buying following a health crisis. *Int. J. Environ. Res. Public Health*, *17*, 10, 3513, 2020.
45. Zhang, Z., Lu, W.X., Zhao, Y., Song, W.B., Development tendency analysis and evaluation of the water ecological carrying capacity in the Siping area of Jilin Province in China based on system dynamics and analytic hierarchy process. *Ecol. Modell.*, *275*, 9–21, 2014.
46. Zhao, L., Dai, T., Qiao, Z., Sun, P., Hao, J., Yang, Y., Application of artificial intelligence to wastewater treatment: A bibliometric analysis and systematic review of technology, economy, management, and wastewater reuse. *Process Saf. Environ. Prot.*, *133*, 169–182, 2020.
47. Zamani Joharestani, M., Cao, C., Ni, X., Bashir, B., Talebiesfandarani, S., PM2. 5 prediction based on random forest, XGBoost, and deep learning using multisource remote sensing data. *Atmosphere*, 10, 7, 373, 2019.
48. Zhu, X., Wan, Z., Tsang, D.C., He, M., Hou, D., Su, Z., Shang, J., Machine learning for the selection of carbon-based materials for tetracycline and sulfamethoxazole adsorption. *Chem. Eng. J.*, *406*, 126782, 2021.
49. Câmara, G., Monteiro, A.M., Fucks, S.D., Carvalho, M.S., Spatial analysis and GIS: a primer. Image Processing Division, National Institute for Space Research (INPE), Brasil, 2004.
50. Maguire, D.J., An overview and definition of GIS, in: *Geographical Information Systems: Principles and Applications*, vol. 1, no. 1, pp. 9–20, 1991.
51. Eastman, J.R., Multi-criteria evaluation and GIS, in: *Geographical information Systems*, vol. 1, no. 1, pp. 493–502, 1999.
52. Zlaugotne, B., Zihare, L., Balode, L., Kalnbalkite, A., Khabdullin, A., Blumberga, D., Multi-criteria decision analysis methods comparison. *Rigas Tehniskas Universitates Zinatniskie Raksti*, 24, 1, 454–471, 2020.
53. Breunig, M., Bradley, P.E., Jahn, M., Kuper, P., Mazroob, N., Rösch, N., Jadidi, M., Geospatial data management research: Progress and future directions. *ISPRS Int. J. Geo-Inf.*, 9, 2, 95, 2020.
54. Wu, C., Zhu, Q., Zhang, Y., Du, Z., Ye, X., Qin, H., Zhou, Y., A NOSQL–SQL hybrid organization and management approach for real-time geospatial data: A case study of public security video surveillance. *ISPRS Int. J. Geo-Inf.*, 6, 1, 21, 2017.

55. Hu, F., Xu, M., Yang, J., Liang, Y., Cui, K., Little, M.M., Yang, C., Evaluating the open source data containers for handling big geospatial raster data. *ISPRS Int. J. Geo-Inf.*, 7, 4, 144, 2018.
56. Reshetyuk, Y., *Terrestrial laser scanning: Error sources, self-calibration and direct georeferencing*, VDM Verlag Dr. Muller, 2009.
57. Stefanakis, E., Web mercator and raster tile maps: two cornerstones of online map service providers. *Geomatica*, 71, 2, 100–109, 2017.
58. Kang, T.W. and Hong, C.H., A study on software architecture for effective BIM/GIS-based facility management data integration. *Autom. Constr.*, 54, 25–38, 2015.
59. Hor, A.H., Sohn, G., Claudio, P., Jadidi, M., Afnan, A., A semantic graph database for BIM-GIS integrated information model for an intelligent urban mobility web application. *ISPRS Ann. Photogramm. Remote Sens. Spatial Inf. Sci.*, 4, 89–96, 2018.
60. Egenhofer, M.J., Reasoning about binary topological relations, in: *Symposium on Spatial Databases*, pp. 141–160, Springer Berlin Heidelberg, Berlin, Heidelberg, 1991, August.
61. Kovacs-Györi, A., Ristea, A., Havas, C., Mehaffy, M., Hochmair, H.H., Resch, B., Blaschke, T., Opportunities and challenges of geospatial analysis for promoting urban livability in the era of big data and machine learning. *ISPRS Int. J. Geo-Inf.*, 9, 12, 752, 2020.
62. Gorelick, N., Hancher, M., Dixon, M., Ilyushchenko, S., Thau, D., Moore, R., Google Earth Engine: Planetary-scale geospatial analysis for everyone. *Remote Sens. Environ.*, 202, 18–27, 2017.
63. Kamilaris, A. and Ostermann, F.O., Geospatial analysis and the internet of things. *ISPRS Int. J. Geo-Inf.*, 7, 7, 269, 2018.
64. Jing, C., Wang, S., Wang, M., Du, M., Zhou, L., Sun, T., Wang, J., A low-cost collaborative location scheme with GNSS and RFID for the Internet of Things. *ISPRS Int. J. Geo-Inf.*, 7, 5, 180, 2018.
65. Kamilaris, A. and Pitsillides, A., The impact of remote sensing on the everyday lives of mobile users in urban areas, in: *2014 Seventh International Conference on Mobile Computing and Ubiquitous Networking (ICMU)*, 2014, January, IEEE, pp. 153–158.
66. Puiu, D., Barnaghi, P., Tönjes, R., Kümper, D., Ali, M.I., Mileo, A., Fernandes, J., Citypulse: Large scale data analytics framework for smart cities. *IEEE Access*, 4, 1086–1108, 2016.
67. Choukolaei, H.A., Ghasemi, P., Goodarzian, F., Evaluating the efficiency of relief centers in disaster and epidemic conditions using multi-criteria decision-making methods and GIS: A case study. *Int. J. Disaster Risk Reduct.*, 85, 103512, 2023.
68. Aksha, S.K., Resler, L.M., Juran, L., Carstensen Jr., L.W., A geospatial analysis of multi-hazard risk in Dharan, Nepal. *Geomatics Nat. Hazards Risk*, 11, 1, 88–111, 2020.

69. Jena, R., Pradhan, B., Naik, S.P., Alamri, A.M., Earthquake risk assessment in NE India using deep learning and geospatial analysis. *Geosci. Front.*, 2021, May 1, https://doi.org/10.1016/j.gsf.2020.11.007.
70. Sohail, S.S., Javed, Z., Nadeem, M., Anwer, F., Farhat, F., Hussain, A., Madsen, D.O., Multi-criteria decision making-based waste management: A bibliometric analysis. *Heliyon*, 2023.
71. Coelho, L.M.G., Lange, L.C., Coelho, H.M.G., Multi-criteria decision making to support waste management: a critical review of current practices and methods. *Waste Manage. Res.*, 35, 1, 3–28, 2017.
72. Torkayesh, A.E., Rajaeifar, M.A., Rostom, M., Malmir, B., Yazdani, M., Suh, S., Heidrich, O., Integrating life cycle assessment and multi criteria decision making for sustainable waste management: key issues and recommendations for future studies. *Renew. Sustain. Energy Rev.*, 168, 112819, 2022.
73. Duan, Z., Scheutz, C., Kjeldsen, P., Trace gas emissions from municipal solid waste landfills: A review. *Waste Manage. (Oxford)*, 119, 39–62, 2021.
74. Sliusar, N., Filkin, T., Huber-Humer, M., Ritzkowski, M., Drone technology in municipal solid waste management and landfilling: A comprehensive review. *Waste Manage. (Oxford)*, 139, 1–16, 2022.
75. Shen, S., Chen, Y., Zhan, L., Xie, H., Bouazza, A., He, F., Zuo, X., Methane hotspot localization and visualization at a large-scale Xi'an landfill in China: effective tool for landfill gas management. *J. Environ. Manage.*, 225, 232–241, 2018.
76. Araiza-Aguilar, J.A., Rojas-Valencia, M.N., Nájera-Aguilar, H.A., Gutiérrez-Hernández, R.F., García-Lara, C.M., Analysis of a Municipal Solid Waste Disposal Site: Use of Geographic Information Technology Tools for Decision Making. *ISPRS Int. J. Geo-Inf.*, 12, 7, 280, 2023.
77. Hu, Q., Wu, W., Xia, T., Yu, Q., Yang, P., Li, Z., Song, Q., Exploring the use of Google Earth imagery and object-based methods in land use/cover mapping. *Remote Sens.*, 5, 11, 6026–6042, 2013.
78. Kebede, Y.S., Alene, M.M., Endalemaw, N.T., Urban landfill investigation for managing the negative impact of solid waste on environment using geospatial technique. A case study of Assosa town, Ethiopia. *Environ. Challenges*, 2021, August 1, https://doi.org/10.1016/j.envc.2021.100103.
79. Zewdie, M.M. and Yeshanew, S.M., GIS based MCDM for waste disposal site selection in Dejen town, Ethiopia. *Environ. Sustain. Indic.*, 18, 100228, 2023, ISSN 2665-9727, https://doi.org/10.1016/j.indic.2023.100228.
80. Optimal Solid Waste Landfill Site Identification Employing GIS-Based Multi-Criteria Decision Analysis Within the Thach That District, Hanoi, Vietnam. *Int. J. Geoinformatics*, 2024, https://doi.org/10.52939/ijg.v20i1.3021.
81. Mvula, R.L.S., Mundike, J., Nguvulu, A., Spatial suitability analysis for site selection of municipal solid waste landfill using hybrid GIS and MCDA approach: The case of Kitwe, Zambia. *Sci. Afr.*, 21, e01885, 2023, ISSN 2468-2276.

82. Shahabi, H., Keihanfard, S., Ahmad, B.B., Amiri, M.J.T., Evaluating Boolean, AHP and WLC methods for the selection of waste landfill sites using GIS and satellite images. *Environ. Earth Sci.*, 71, 9, 4221–4233, 2013, https://doi.org/10.1007/s12665-013-2816-y.
83. Ghose, M., Dikshit, A., Sharma, S., A GIS based transportation model for solid waste disposal – A case study on Asansol municipality. *Waste Manage. (Oxford)*, 26, 11, 1287–1293, 2006, https://doi.org/10.1016/j.wasman.2005.09.022.
84. Shafei, S., Vahdati, H., Sedghi, T., Charmin, A., CBMIR System Based on Matrix Weighting Framework and Linear Transformation with KNN. *Electrica*, 22, 2, 258–266, 2022, https://doi.org/10.54614/electrica.2022.210130.
85. Elkhrachy, I., Alhamami, A., Alyami, S.H., Landfill Site Selection Using Multi-Criteria Decision Analysis, Remote Sensing Data, and Geographic Information System Tools in Najran City, Saudi Arabia. *Remote Sens.*, 15, 15, 3754, 2023, https://doi.org/10.3390/rs15153754.
86. Ampofo, S., Issifu, J.S., Kusibu, M.M., Mohammed, A.S., Adiali, F., Selection of the final solid waste disposal site in the Bolgatanga municipality of Ghana using analytical hierarchy process (AHP) and multi-criteria evaluation (MCE). *Heliyon*, 9, 8, e18558, 2023, https://doi.org/10.1016/j.heliyon.2023.e18558.
87. Cobos-Mora, S.L., Guamán-Aucapiña, J., Zúñiga-Ruiz, J., Suitable site selection for transfer stations in a solid waste management system using analytical hierarchy process as a multi-criteria decision analysis: a case study in Azuay-Ecuador. *Environ. Dev. Sustain.*, 25, 1944–1977, 2023, https://doi.org/10.1007/s10668-022-02134-8.
88. Ali, I., Islam, A., Ali, S.M., Adnan, S., Identification and selection of suitable landfill sites using GIS-based multi-criteria decision analysis in the Peshawar District, Pakistan. *Waste Manage. Res.: J. Sustain. Circ. Econ.*, 41, 3, 608–619, 2022, https://doi.org/10.1177/0734242x221124069.
89. Bhagat, S., Bhardawaj, A., Mittal, P., Chandak, P., Akhtar, M., Sharma, P., Evaluating Plastic Waste Disposal Options in Delhi Using Multi Criteria Decision Analysis. Department of Civil Engineering, Indian Institute of Technology Delhi, Hauz Khas, New Delhi, India.
90. Mondal, S. and Mandal, B., Assessment of urban solid waste management in a Class II Indian city using geospatial and statistical approaches: A case study of Rampurhat municipality. *Waste Manage. Bull.*, 1, 4, 74–92, 2024, https://doi.org/10.1016/j.wmb.2023.09.003.
91. Dixit, A., Singh, D., Shukla, S.K., Assessment of human health risk due to leachate contaminated soil at solid waste dumpsite, Kanpur (India). *Int. J. Environ. Sci. Technol.*, 21, 909–924, 2024, https://doi.org/10.1007/s13762-023-04868-y.
92. Lombardi, P. and Todella, E., Multi-Criteria Decision Analysis to Evaluate Sustainability and Circularity in Agricultural Waste Management. *Sustainability*, 15, 20, 14878, 2023, https://doi.org/10.3390/su152014878.

93. Tomita, A., Cuadros, D.F., Burns, J.K., Tanser, F., Slotow, R., Exposure to waste sites and their impact on health: a panel and geospatial analysis of nationally representative data from South Africa, 2008–2015. *Lancet Planet. Health*, 4, 6, e223–e234, 2020, https://doi.org/10.1016/s2542-5196(20)30101-7.
94. Ferdoush, M.R., Aziz, R.A., Karmaker, C.L., Debnath, B., Limon, M.H., Bari, A.M., Unraveling the challenges of waste-to-energy transition in emerging economies: Implications for sustainability. *Innov. Green Dev.*, 3, 2, 100121, 2024, https://doi.org/10.1016/j.igd.2023.100121.
95. Adnan, A., Mahmud, S., Uddin, M.R., Modi, A., Ehsan, M.M., Salehin, S., Energy, Exergy, Exergoeconomic, and environmental (4E) analyses of thermal power plants for municipal solid waste to energy application in Bangladesh. *Waste Manage. (Oxford)*, 134, 136–148, 2021, https://doi.org/10.1016/j.wasman.2021.08.006.
96. Chand Malav, L., Yadav, K.K., Gupta, N., Kumar, S., Sharma, G.K., Krishnan, S., Rezania, S., Kamyab, H., Pham, Q.B., Yadav, S., Bhattacharyya, S., Yadav, V.K., Bach, Q.V., A review on municipal solid waste as a renewable source for waste-to-energy project in India: Current practices, challenges, and future opportunities. *J. Clean. Prod.*, 277, 123227, 2020, https://doi.org/10.1016/j.jclepro.2020.123227.
97. Luo, C., Ju, Y., Santibanez Gonzalez, E.D., Dong, P., Wang, A., The waste-to-energy incineration plant site selection based on hesitant fuzzy linguistic Best-Worst method ANP and double parameters TOPSIS approach: A case study in China. *Energy*, 211, 118564, 2020, https://doi.org/10.1016/j.energy.2020.118564.
98. Aslam, S., Rehman, R., Asad, M., Linking environmental management practices to environmental performance: the interactive role of environmental audit. *J. Environ. Manag.*, 14, 99–119, 2020.
99. Dhedheya, I., Dube, S. S., Sanders, I., A framework for the adoption of Geographic Information Systems (GIS) in service delivery: A case of Harare city council, Zimbabwe. *The Wisconsin Journal of Arts and Sciences*, 43, 2023.
100. Zhou, M., Zuo, H., Xiao, P., Zhang, X., RETRACTED ARTICLE: Mountain air pollution evaluation and tourism brand building based on remote sensing image processing. *Arab. J. Geosci.*, 14, 18, 1851, 2021.

6

Process Engineering and Economical Aspects

Priyanka Singh[1]*, Meena Choudhary[1] and Ram Bhajan Sahu[2]

[1]NIMS Institute of Allied Medical Science and Technology, NIMS University Rajasthan, Jaipur, Rajasthan, India
[2]Department of Paramedical, Apex School of Paramedical Science, Apex University, Jaipur, Rajasthan, India

Abstract
The informal sector generally includes small-scale manufacturing products and low provision of services. Their entrepreneurs have no registered license and do not pay taxes according to government policy. They mainly involved the waste recycling activities for screening of recyclable and reusable materials from disposed solid waste items. These activities depicted the informal sector as labour-intensive, low earning, unrecorded and unregulated work. They are either operated as micro enterprises (up to 10 group members) and small enterprises (20-30 group) with low capital investment. These entrepreneurs applied their innovative skill to improve market values of collected solid waste items. The informal sectors for solid waste management systems generally operate at a low standard due to lack of funds, and availability of low-skilled labour force. The major challenges in informal sectors are inappropriate solid waste collection points and improper disposal in open dumps. These improper disposal activities are considered as backward, unhygienic, and highly unregulated informal systems. On the other hand, modern waste management follows the R3 principle (reducing the disposal of items, reuse of wasted items and recycling process) and acts as an efficient recycling system for these products to move up the waste hierarchy.

This chapter will provide information for an effective sustainable role of informal waste recycling and modern waste management in developing countries. The local economy, policies and their characteristics for concerns of informal solid waste management has been also discussed. Enzymatic, fermentative approach and valorisation processes for bioconversion of biodegradable waste product into

**Corresponding author*: priyay20@gmail.com

Priyanka Singh, Pooja Agarwal and V. Vivekanand (eds.) Solid Waste Management: Challenges, Sustainability and Advancements, (183–212) © 2026 Scrivener Publishing LLC

value added industrial products through modern waste management process have been simultaneously discussed.

Keywords: Recycle, solid waste managment, sutainable approach, valorization, bioconversion, modern waste managmen, value added product, informal sector

6.1 Informal Sector Recycling in Solid Waste Management

Waste picking is the major section of informal waste recycling system in a developing country. Generally poor, illiterate and unemployed social groups are being involved for waste picking for small income generation on a daily basis [37]. Informal recyclers generally includes minority-based social groups all over the world like Egypt (Zabbaleen), Mexico (Pepenadores, Catroneros and Buscabotes), Colombia (Cartoneros, Traperos, Chatarreros), Ecuador (Chamberos), Costa Rica (Buzos) and Argentina (Cirujas) [13, 38]. Itinerant waste pickers have generally sorted dry recyclable items from different types of solid waste collected from householders and finally transported to recycling shop (Table 6.1).

Table 6.1 Different target sites for generation of types of solid waste material.

Sources	Waste generation site	Types of waste
Domestics	Individual houses, apartments, villas	Food scraps, plastics, cloths, glass, metals, ashes
Commercial sites	Shopping mall, market, hotels, restaurants, offices	Paper, wood, plastics, food scraps, glass, metals, papers, corrugated boxes
Institutional centres	Colleges, schools, universities, government offices, medical care centres	Paper, cloths, wood, plastics, food scraps, glass, metals, papers, corrugated boxes
Public Facilities	Street cleaning, park, beaches, landscaping, recreation areas	Plant trimmed parts, plastics, bottles, general wastes from recreation areas

They are also investing capital to arrange vehicles apart from their labour to carry these waste materials. This mode of informal recycling has been popularized in most developing countries like China, Bangkok, and India [30]. Some street waste pickers from economically developing countries are collecting the waste from open dump areas nearby construction sites [12]. They also collect secondary raw materials from mixed waste thrown on the streets or from communal bins before collection. There are some authorized persons from municipal waste collection to collect secondary raw materials by vehicles and transport them to disposal sites. Informal recycling activities have important consequences for income generation for financially weak persons. The secondary materials collected through informal recycling are generally traded to local industries, including craftsmen and artisans. A chain of intermediate dealers often exists between the waste pickers and end-users. These dealers may include intermediate processors, brokers and wholesalers and are mainly involved in both formal and informal sector activities [18, 28, 40, 56]. In a hierarchy recycling system, the higher secondary raw material will be traded for conversion into value-added items as shown in Figure 6.1.

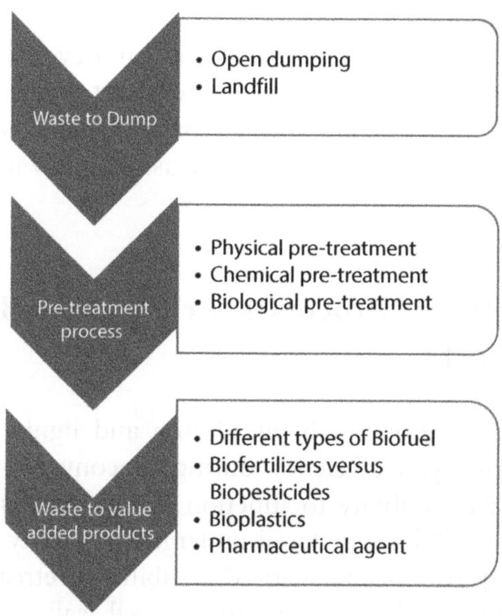

Figure 6.1 Different steps for vaporization process for bioconversion of biological solid wastes into value-added products.

Informal recycler activities are restricted to the base of the secondary materials trade hierarchy which significantly reduces their potential income. Individual waste pickers are the most vulnerable as they do not have an organised supportive network. Family-organised activities including women, children and elders has commonly been conducted for dump scavenging and their family members have been exposed to increased health risks. Family-organised activities provide a level of social and economic support for these families but their children are affected by not getting a proper education. The training program for informal recyclers into municipal sectors has effectively upgraded their ability to use collected waste materials (like plastics, paper, cardboard, aluminium, steel, other metals, glass and textiles) as value-added items [28, 37, 55]. Organic wastes enriched with nutritive and bioenergetic components are commonly used as livestock fodder, soil improvers and fuel [21]. The recycling activity depends on income levels, existence of local and national markets, and the level of financial and regulatory governmental intervention. These value-added products have low income due to their low position in the trade hierarchy for recycled materials [45].

Informal recycling systems can give significant economic benefits for developing countries. They provide a steady, reliable supply of secondary raw materials for the local manufacturing industry to manufacture low-cost, affordable products [28, 55].

The informal waste recycling systems exist in developing countries to reduce the cost of formal waste management systems as they reduce the quantity of waste for collection. These economic benefits are achieved at no direct cost to the taxpayers and are also associated with social benefits [37].

6.2 Pretreatment Processes for Partially Hydrolysing Solid Waste

The higher concentrations of hemicellulose and lignin in biomass are reducing the hydrolysis efficiency during bioconversion because they obstruct the enzyme's ability to function *in situ* process. To overcome this problem, the cellulosic material undergoes a variety of pretreatment techniques to improve its enzymatic digestibility. Pretreatment has been commonly used to modify the structure of cell wall to improve hydrolytic enzyme access to sugar polymers. The majority of hemicellulose has been typically hydrolysed using water and acidic pretreatment techniques, whereas the amount of monomeric sugar has been enhanced by lowering pH pre-treatment. Alkaline pretreatment has been used for partial denaturation of lignin and carbohydrates ester during the extraction process. These deacetylated hemicellulose are being effectively hydrolysed with cellulase, pectinase hydrolase enzyme [15].

The rigidity of cell walls might be due to the presence of a higher percentage of cellulose in the cell wall structure. Hemicellulose as a plant's structural and reserve material has also contributed to rigidity of cell wall structure. The pretreatment process generally includes physical, biological, and chemical approaches. The following specifications for the pretreatment approach should be taken into account:

1) The ability of bioconversion of cellulose-based raw materials into specific sugars upon hydrolysis.
2) The process should be standardised to avoid deterioration or depletion of carbohydrates.
3) The toxic by-products should not be produced during the bioconversion process.
4) The process should be cost-effective and highly efficient.

The pretreatment process has been commercially applied for production of many value-added products in several researches [32].

6.2.1 Physiochemical Pretreatment Process

Physical pretreatment process results in improvement in surface area, particle sizes, crystallinity properties, and biomass polymerization. Lignocellulosic materials could be reduced with reduction of cellulose crystallinity by incorporation of chipping, milling, and grinding processes. After grinding or milling, the materials were reduced up to 10 to 30 mm sizes with a range of 0.2 to 2 mm. Vibratory ball has reduced the cellulose structure more effectively than using a regular ball, and it has also improved the digestibility of cellulose [27]. These physical pretreatments have the major limitation of their higher rate of energy consumption, which makes lignocellulosic biomass processing more expensive. A chemical pretreatment process is generally applied for conversion of lignocellulosic structure into different generation of biofuel [60]. The specific sorting unit is used for digestion of waste products collected from the textile industry [26]. This sorting system has been used firstly to press out fasteners, buttons, and other non-textile product parts from textile trash by following the processes of disinfection and dusting. The remaining textile waste products has been further treated within a conveyor and passed through a spectrum lamp by sending delta signal to the robot. In this process, two parallel-arranged disc knives are permanently mounted above the table and a conveyor belt is fixed for the supply of textile products underneath the knives. It has also a specific middle guide chute for fixing the fastener-zipper, side chutes for the function of disc knives, a front

Table 6.2 Merits and demerits of different pretreatment methodologies.

Pretreatment methodologies	Merits	Demerits
Mechanical process	The crystalline structure of cellulose has been disintegrated.	It consumes high power and energy
Pyrolysis	It produced maximum percentage of gaseous products and liquids.	• Requires large temperature levels to run, produces ash as a by-product
Acid Hydrolysis	Hemicellulose content has been hydrolysed to xylose and other sugars. It changed the structural form of lignin.	• Very expensive • Secretion of harmful by-products • Chances to corrode instruments by exposure to acid
Alkaline hydrolysis	It degraded efficiently hemicellulose and lignin, and enables access to a larger surface area.	• Require of maximum residence time • Imbedding of irretrievable salts
Organic solvents	Hydrolyzation of hemicellulose content and lignin.	• Not cost-effective • Require additional processes like evaporation and condensation
Biological process	Disintegrates lignin and hemicellulose content. It requires small amount of energy.	• Very low degree of hydrolysis process under standard condition

pressing roller for the supply of the textile product, and a rear roller for its removal characterise. The merit and demerit of the chemical treatment process is described in Table 6.2.

They generally include the following chemical agents for pretreatment process.

6.2.2 Acid Based Pre-Treatment Process

Acidic pretreatment has been applied as the most effective technique to solubilize hemicellulose, or increase availability of cellulose for hydrolysis. The key steps include the precipitation of lignin, the hydrolysis of hemicellulose, and condensation process. Pretreatment process with strong acids results in the release of glucose molecules from hydrolysed cellulosic materials in higher yield at low temperature [7]. Sulphuric acid has affected cellulosic structure to a greater extent and enables the bioconversion of these partially hydrolysed molecules into glucose derivatives in the presence of macerozymes [24, 57]. The quantity of cellulose and lignin in partially treated lignocellulosic materials has been increased by sulphuric acid. It effectively induced random dissolution of glycosidic linkages and effective removal of hemicellulose.

Diluted acids have been demonstrated to be a very helpful solution to the problem of excessive expenditures without sacrificing the effectiveness of the process. Prior to the materials being hydrolysed, diluted-acid hydrolysis has also been utilised to effectively treat them. It has been demonstrated that the use of dilute sulphuric acid (H_2SO_4) for pretreatment can lead to remarkable reaction rates for hydrolysis of cellulose. This pretreatment process with use of dilute acid has reduced the production of inhibitors as by-product [4]. Sulphuric (H_2SO_4) and phosphoric (H_3PO_4) acids are generally preferred due to their mild character and high productivity rate for biofuel by hydrolysing lignocellulose. Hydrochloric acid is also used in some cases to recover different types of other value-added products. The cellulosic hydrolysis with nitric acid (HNO_3) also help in the production of higher concentration of glucose. These two acids (phosphoric acid and nitric acids) are more expensive than sulphuric acid. Acid media generally hydrolysed hemicellulose structure of lignocellulosic materials into specific sugar [54]. This partial hydrolysis of cellulosic structure results in the disruption of xylosidic bonds and the breaking down of acetyl ester bunches. The disintegration of lignin has increased the level of cellulose contents and therefore increased the crystallinity properties of pretreated biomaterials. The condition of pretreatment has been mainly affected by size of the molecules, reaction duration, acidity fixation, fluid-to-strong ratio, and temperature. Temperatures (120°C to 210°C) and weights enable response times to be completed instantly or in a very short amount of time under mild acids [59]. The pretreatment process is generally carried out in two stages: (i) the majority of hemicelluloses are solubilized in the first stage using a weaker acidic substrate; (ii) in the second stage, residual hemicellulose has been treated with a stronger acid agent for complete hydrolysis of cellulose. These pretreatment strategies enable the release of a larger number of free glucose molecules. Protein stacking and acidic fixation has mainly influenced the glucose output more at optimum reaction temperature.

Concentrated sulphuric (65–86% w/v), hydrochloric (41%), or phosphoric (85% w/w) acids are used in this pretreatment approach against lignocellulosic biomass (with 5–10% moisture) at optimum temperature and pressures. This loosens the biomass materials with deionized water and finally results in the separation of free cellulosic hydrolysates at moderate temperatures (70–121°C). Acetic acid is further added to the pretreated material and the blend is fermented to stop the reaction and separate the solids from the lignin-containing supernatant. They are further washed prior to enzymatic hydrolysis [7].

The pretreatment efficiency is mainly affected by duration, temperature, and ration of acidic/biomass. Depending on the H_2SO_4 concentration, two licenced cycle arrangements based on the Arkenol cycle and Biosulfurol cycle are also engaged. The Arkenol cycle involves pretreatment (decrystallization of the biomass) at temperatures below 50°C with concentration of acid (70–77%)/ biomass (10%) with production rate 1.25 times. The biomass is treated with 70% acidic medium in the Biosulfurol cycle after addition of dry CO_2 in the fermenter. This is followed by the pretreated slurry being diluted with water. Prior to the ageing process, the acidic medium has been separated from the biomass slurry using layers, and after ageing, it has been partially separated in an anaerobic digester. The Biosulfurol cycle has major advantage of capacity to fractionate various biomasses with high ethanol yields, low temperature treatment, low formation of some products, and not-requirement of specific catalysts. The recovery of sugar and polyester from textile wastes has been carried out after treatment with enzymatic hydrolysis and acid hydrolysis (H_3PO_4). The highest recovery yield (79%) was reported after carrying pretreatment process for 7 hours with 85% phosphoric acid at 50°C.

6.2.3 Alkaline Pretreatment

These alkaline pretreatment methodologies mainly hydrolyse the intermolecular ester bonds of Xylan and some other components [10]. Ammonia (NH_3) and other bases including sodium hydroxide, potassium hydroxide, and calcium hydroxide are frequently utilised in the alkaline pretreatment of lignocellulosic materials. Sodium hydrates are responsible for partial hydrolysis of textile fibres by digesting cellulose fibres of textile. NaOH (8–10%) solubilised cellulose with low to moderate degree of polymerization will directly hydrolyse functional groups of cellulose rather than rupturing the inter- or intramolecular bonds within molecules. The supplementation of urea and NaOH has efficiently hydrolyzed cellulosic fibres with a higher rate in comparison to NaOH treatment. The urea hydrates act as hydrogen donors/receptor between solvent molecule by hydrolysing intra- and interhydrogen bonds within cellulosic polymers in textile fibres. This will prevent reassembling of cellulose molecules and improve the dissolution of cellulose at molecular level. The higher rate of hydrolysis of cellulose has been observed after enzymatic pretreatment followed by treatment with a NaOH/urea solution. Enzymatic pretreatment has improved the solubility of cellulose in cold NaOH/urea solution (30% to 65%) [63].

This alkaline biomass pre-treatment reaction results in an increase in surface area, a decrease in cellulose crystallinity, and a reduction in the degree of polymerization. This methodology has effectively hydrolysed biomaterials with low lignin concentration. This process is generally preferred as a low-cost process due to use of inexpensive chemicals, non-hard reaction conditions, and effective lignin, Xylan, and biomass fractionation probability elimination. The main drawbacks are lengthy cycle time and difficulty of neutralising the posttreatment combination.

6.2.4 Use of Oxidizing Agents

An oxidising agent for pretreatment process for lignocellulosic materials generally includes ozone, hydrogen peroxide. This process primarily entails the extraction of lignin from the biomass structure and improves cellulose accessibility [49]. Cane bagasse endured this pretreatment procedure using hydrogen peroxide, which significantly boosted its susceptibility to enzymatic hydrolysis. Additionally, moist oxidation and alkaline hydrolysis were carried out on wheat straw (20 g straw/l, 170°C, 5–10 min), yielding an 85% glucose production rate. Fibre recovery is a potential method of recycling fibre, where fabrics are being hydrolysed into unwrapped strands that may be twisted back into yarn. Fibre reclamation process has been carried out to extract fibres from textile waste and reuse them into another specific biomaterial. Recycling of post-consumer trash is more challenging than recycling pre-consumer waste due to the presence of specific contaminants and differences in wear and tear conditions of these cloths. Cellulose resists undergoing molecular change in standard solvent due to exhibiting strong intermolecular hydrogen bonds within cellulose molecules [44]. These types of cotton fabrics were dissolved in solvent of ionic liquid 1-butyl-3-methylimidazolium acetate and dimethyl sulfoxide (DMSO). The artificial cellulose fibres are further regenerated using wet spinning method [33]. The post-consumer cotton textile wastes have been completely dissolved in the ionic liquid 1,5-diazabicyclo [4.3.0] non-5-enium acetate ([DBNH][OAc]) to form continuous filaments. Pretreatment with acid-enzyme hydrolysis followed by blending the waste cotton with birch pre-hydrolysed kraft pulp has improved the degree of polymerization in cellulosic fibre. The heterogeneous raw materials showed variation in molar mass distribution and degree of polymerization. These techniques improved tensile strength of cellulosic fibre up to 870 MPa across fibres, which is higher than the tenacity of man-made cellulosic fibres and native cotton [6].

6.3 Biological Approach for Pretreatment Process

The application of various rot fungus is a safe and eco-friendly process known as biological therapy. This method of lignin extraction from lignocellulosic biomass is becoming more and more popular since, despite the significant lignin degradation, it does not need a lot of energy. Microorganisms including brown, white, and soft-rot fungi are utilised in biological pretreatment techniques to break down lignin and hemicellulose in waste materials. While brown rots typically concentrate on cellulose alone, white and soft rots primarily focus on both lignin and cellulose. Many lignin-degrading enzymes, including lactase and peroxidase, are responsible for the lignin breakdown induced by white-rot fungus; although the hydrolysis rate is relatively tiny in their pretreatment methods, biological pretreatment offers the advantage of low energy consumption with few limits [46].

6.3.1 Fermentation Strategies

Various industrial enzymes have been produced by solid state fermentation processes. This type of fermentation promotes the growth of conidiophores in particular as well as the growth of other fungal species on solid substrates, including ascomycetes, basidiomycetes, and deuteromycetes [11]. Fungal origins account for the majority of industrial enzymes. In submerged fermentation as well as solid state fermentation, fungal organisms express themselves differently at their genetic level. Solid state fermentation has not been used for large-scale manufacturing due to its weak repeatability of findings and difficulties to standardise methods. The fermentative approaches have been severely affected by high temperature and longer fermentation time. Aeration and evaporation with water loss may resolve these issues. The development of techniques to make solid state fermentation a manageable process has been somewhat successful because of advancements in fermentation technology as shown in Figure 6.2. A non-air flow box (NAB) has been designed using moisture-permeable flour-permeable membrane.

Enzyme synthesis occurs quickly and with great repeatability when water vapour exits through the membrane, keeping the substrate dry. This promotes uniform culture development. Reduced effluent output and little downstream processing are two benefits of solid-state fermentation. It also offers great volumetric productivity and high product concentration. However, volatile organic compounds (VOC) including CH_4, N_2O, and NH_3 were found to be released in research using organic waste for the solid-state fermentation regime in the manufacture of enzymes [36].

The most popular technique for producing enzymes on a big scale is submerged fermentation. A stirred tank reactor, one kind of submerged fermentation utilised to produce enzymes using lignocellulose as a substrate, is shown in Figure 6.3. It uses a water-based medium in the reactor, which allows for agitation and aeration inside the vessel while also assisting in regulating the pH and temperature. Within the reactor, a sterile environment may be maintained with a lower risk of contamination. A homogeneous environment is maintained during submerged fermentation, and improved control over process parameters is made possible by the reactor.

The inclusion of detrimental elements such as butylated hydroxytoluene, hydrogen peroxide, and metal ions, which cause oxidative stress, might compromise the entire process even if it aids optimisation studies and the uniform delivery of nutrients and oxygen to the developing microbe [51].

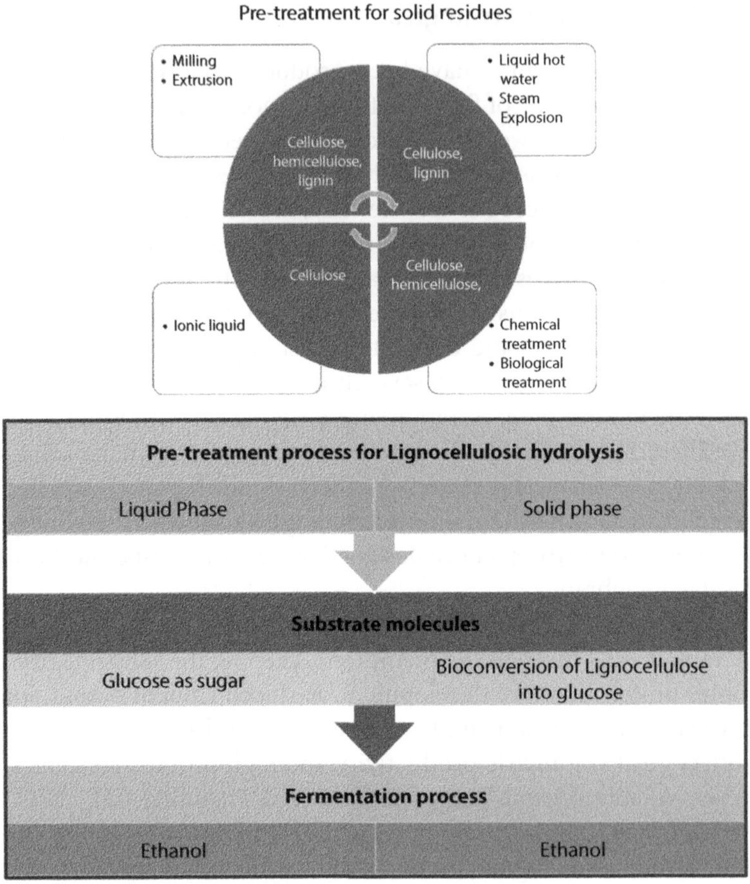

Figure 6.2 Role of pretreatment and fermentation approaches for bioconversion of lignocellulosic biomass to ethanol.

Numerous filamentous fungal species are utilised in the synthesis of diverse enzymes and bioactive substances. While submerged fermentation is utilised in the industry to cultivate these microorganisms, several studies demonstrate that this approach is not optimal. Due to the need for big, complex equipment, submerged fermentation is not as cost-effective as solid-state fermentation. In research, comparing the manufacture of biodiesel, it was found that submerged fermentation needed 78% more capital than the solid-state method.

Biomass pretreatment and the use of microorganisms appropriate to the required enzyme are the two main steps in the manufacture of enzymes from lignocellulosic substrates. In certain instances, a crucial phase in this

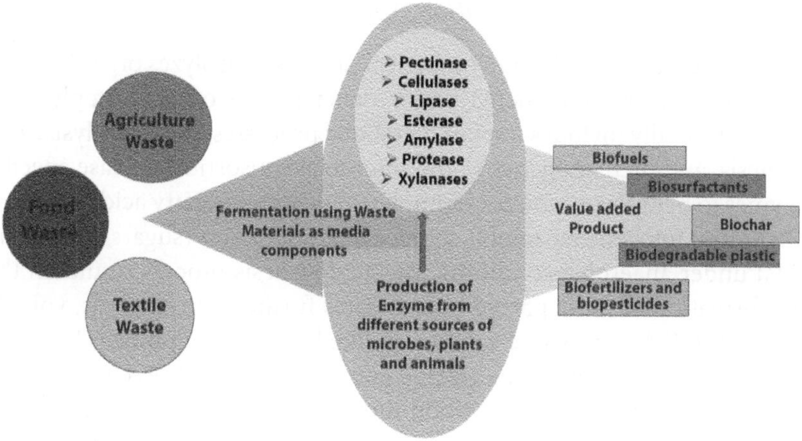

Figure 6.3 Enzyme mediated biotransformation of solid waste materials into value-added products.

process that yields a hydrolysed containing fermentable sugars is the enzymatic hydrolysis of prepared biomass. The release of two primary reducing sugars, xylose and glucose, is caused by the action of enzymes (cellulose and hemicellulose). Mannose, galactose, and arabinose are among the other sugars released. This method of producing α-amylase was used by Rajagopalan and Krishnan [48] with hydrolysed sugar cane bagasse. Nevertheless, the kind of microbial digestion of the biomass determines whether or not the enzymatic hydrolysis step is included.

In order to decompose the polymeric structure of lignocellulose, certain bacteria create enzymes. Under such circumstances, the economic viability of the production process is compromised by the needless step of employing enzymes to predigest lignocellulose. In recent times, researchers have been more interested in sequential solid state and submerged fermentation processes. With this method, lignocellulosic biomass is utilised as the substrate for a solid-state pre-culture fermentation.

To get the moisture level down to 70%, nutritional broth is added to the biomass. Then, according to the solid biomass concentration, spores are introduced in particular ratios. After the fermentation medium is supplied with bigger volumes of nutritional medium containing increased quantities of glucose, the solids state fermentation is carried out for 24 hours, with the remainder of the process taking place under submerged circumstances. This method was described by Cunha *et al.* [19] when they used sugarcane bagasse as a substrate and *Aspergillums niger* to produce cellulose.

6.3.1.1 *Anaerobic Fermentation Approaches*

Anaerobic digestion (AD) as a biological process hydrolyzes organic materials into smaller components under deficiency of oxygen supply. This process normally includes a set of metabolic processes like hydrolysis, acetogenesis, and methanogenesis. The microbial consortium release specific enzymes to hydrolyze complex biomolecules (proteins, fatty acids, and carbohydrates) into smaller soluble molecular components (sugars and amino acids) under anaerobic condition. The acidogenesis process results in the transformation of these partial hydrolysates further into alcohols, volatile fatty acids (VFA), inorganic substances like hydrogen sulphide (H_2S), carbon dioxide (CO_2), ammonia (NH_3), and hydrogen (H_2). Acetogenesis as the final stage has produced carbon dioxide, acetate, and hydrogen from these partially hydrolysed products [52]. Biogas as a primary by-product of the anaerobic digestion (AD) process is mainly composed of carbon dioxide and methane (50–75%). These by-products are preferred as natural fertilizers due to the presence of nutritive components like potassium, nitrogen, and phosphorus. The quality of these digested products could be varied according to the type of waste product [52]. The dark fermentation process results in the secretion of hydrogen as a by-product of anaerobic digestion. The production of hydrogen has been mainly affected by fermentation conditions like temperature, configuration of reactor, pH, and fermentation period. Short chain aliphatic carboxylic acids (lactic acid, valeric, propionic, caproic, acetic acids) has been also secreted during acidogenesis and acetogenesis processes. Succinic acid has been also secreted by this fermentation process which are in great demand in industrial activities like the production of pharmaceuticals, ink, and polymers [22].

Solid waste products are pretreated prior to carry anaerobic fermentation process for optimum size reduction, sorting, and sterilisation purposes. This pretreatment process has increased the production of biogas yield at lower operating expenses [23]. The pretreatment process mainly includes physical, chemical, and biological treatment methodologies to improve the hydrolytic stage of bioconversion. The mechanical pretreatment process has enhanced the crystalline characteristics of lignocellulosic material. This pretreatment process has major challenges of reduced biogas output and VFA accumulation due to over-reduction of particle size. The presence of large number of lignocellulosic components could inhibit the bioconversion of solid waste products into value-added products. A thermal process has also been employed for breakage of lignocellulosic materials into cellulosic fibres. High temperature hydrolysed the lignin component by allowing assimilation of cellulosic fibres [52]. The chemical process includes ozone, acids and alkalis for hydrolysis of lignocellulosic components in municipal solid waste products.

These anaerobic digestion processes have specific disadvantages like low product quality, inefficient processes, and interference from intermediates. Researchers have mainly conducted studies to examine the potential for efficient electron transfer process to enhance yield of methane in bacteria [42]. The production of methane by the syntrophic metabolism of the fermentative intermediates has been aided by direct transfer of electron between bacteria and methanogenic archaea. In this case, biotic components have contributed the electron transport pathway based on cytochromes. These direct electron transfers could be improved by the introduction of abiotic elements such as conductive carbon compounds. Magnetite, graphite powder has improved the methanation rate in anaerobic digesters [42]. Current research has also shown that addition of biochar can boost the direct electron transfer process, support growth of biofilms and increase electrical conductivity for the methane generation process. These bichars can speed up the lag period in the methanogenesis stage and accelerate substrate degradation rates [2].

6.3.1.2 Aerobic Digestion Process

Organic carbon has been hydrolysed and stabilised during the composting process to improve its productivity as fertiliser. Thermophilic, mesophilic and early activation are important stages of the composting process [22]. Mesophilic microbes first hydrolysed the waste products at higher composting pile's temperature. The mesophilic bacteria become less common for environmental temperature greater than 40°C. Thermophilic microorganisms are more likely to grow at this higher temperature. During the thermophilic phase of organic waste decomposition, bacteria use oxygen and secrete ammonia and carbon dioxide as by-products. The contaminated organisms and viability of the seeds present in the organic solid waste can be destroyed up to temperature 55–70°C [22, 34]. Aeration and uniform mixing should be required throughout the process to increase the rate of decomposition process in solid waste products. The maturation phase has started after the depletion of decaying organic portion in solid waste products and decreasing the temperature in the composting pile. The organic matter has been stabilised and humidified during the maturation phase to improve humic qualities in compost [22].

The pyrolysis process results in the conversion of solid waste products into three major products: biochar (carbon enriched), bio-oil (liquid obtained after condensation of volatile organic matter, and non-condensable gases (methane, hydrogen, carbon monoxide. Biochar has significant application in environmental sectors like wastewater treatment, soil remediation, carbon dioxide capture, and energy storage worldwide. The pyrolysis process generally involves catalytic process and heating techniques after adding feedstocks like sludge, coal, or polymers [25]. The pyrolysis process has been modified using catalytic hydrodeoxygenation under hydrogen-rich atmosphere for production of aromatic hydrocarbons. This study has involved use of several catalyst like metal oxides, zeolites, and inorganic salts to increase the yield of bio-oil at a large scale [25]. Solar- and microwave-assisted pyrolysis have been used as alternative methods against electrically assisted heating process [25].

6.3.2 Enzymatic Approaches

Enzymes as catalyst accelerate the bioconversion process and transformation of organic matter in solid waste products and therefore played a vital role for solid waste management, as shown in Figure 6.3.

Enzyme-mediated waste treatment results in the transformation of waste materials into value-added products like bioethanol, butanol, and biogas [41]. The mixture of enzymes of cellulases, cellobio-hydrolase, endoglucanase, β-glucosidase has been commonly used for hydrolysis of cellulose of newspaper-based solid waste into easily digestible glucose monomers [35, 39]. These enzymes hydrolyzed the cellulose-based waste product into ethanol through yeast cell. In this process, the pretreatment process was carried out using laccase enzyme prior to hydrolysis in the presence of the above mixture of enzymes for bioethanol production.

Textile waste can be produced at all stages of the manufacture process like spinning, dyeing, knitting, weaving, and finishing [14]. These wastes can be categorised as cellulose, protein, and synthetic fibers on the basis of their composition. The enzymatic hydrolysis of pretreated textile waste has recovered glucose and polyester [31] after being treated with cellulase and β-glucosidase under specific optimum condition.

Plastics can be efficiently degraded using specific plastic degrading enzymes under standard condition [65]. Some Esterase enzymes isolated from marine organism had the ability to hydrolyze polylactic acid (PLA), polyhydroxybutyrate (PHB), polycaprolactone (PCL), and poly (butylene succinate-co-butylene adipate) (PBSA) in plastic substrates. Cutinase isolated from leaf-branch compost can degrade PCL; and polyethylene terephthalate (PET). PHB depolymerase can efficiently degrade PHB plastics.

Organic food waste (FW) with specific polysaccharide like cellulose, starch can be converted into bioethanol through enzymatic hydrolysis by cellulase and amylase [47, 53]. Therefore, we should consider cost feasibility for use of enzymatic approach for management of solid waste product into value-added products. The bioconversion process mainly includes the conversion of cellulose to glucose, glucose to alcohol, starch to invert sugar, and dilute waste stream to methane by anaerobic digestion [17].

Lactose as a major food waste can be processed to glucose and galactose through enzymatic approach by treating with β-galactosidase [1]. These lactose molecules could be modified as sweeteners after the fermentation process through immobilized β-galactosidase from Aspergillus oryzae [3]. β-Galactosidase isolated from *Kluyveromyces lactis* has been extensively used in industries for degradation of lactose [1]. Genetically engineered β-galactosidase has been extensively used for lactose hydrolysis [61]. These polysaccharide waste products has been transformed into value-added products such as sweeteners, bioplastics, biofuels, and prebiotics after treating with amylases, cellulases, and xylanases [1, 5].

These enzymatic approaches have major limitations in the industrial sector due to involvement of high-cost decomposition techniques for feedstock and expensive purification strategies. Bioethanol production has been improved by degradation of solid waste using genetically improved strains of *Saccharomyces cerevisiae*. Bioethanol, produced by the fermentative enzymatic approaches, has been preferably used as a fuel source in automobiles and clean-burning fuel [20]. It is commonly used as different beverages in the brewing industry and as a solvent in different industrial sectors. The degradation of textile wastes by cellulase- and β-galactosidase has provided the recovery of glucose and polyesters as value-added products. Glucose is currently used as sweetener, thickener, and moisturizing agent in the food and beverage industry. The polyester in the textile industry has been degraded by a mixture of enzymes of esterase (extracted from *Pseudozyma antarctica*), cutinase (isolated from Pichia pastoris), and PHB depolymerase [29].

Amylase and cellulase are commercially used for enzymatic saccharification of food waste to produce bioethanol at a higher rate. Microbial consortia are more preferred than pure culture with efficiency of secretion of enzymes for degrading food waste into value-added products. The degradation of food waste is generally carried out using consortium of cellulolytic bacteria like *Bacillus thuringiensis* and *B. licheniformis* and amylolytic bacteria (*Brevibacillus borstelensis* and *B. licheniformis*) [62].

6.4 Role of Enzyme Immobilization and Genetic Engineering in Solid Waste Management

Enzyme immobilization involves the attachment or confinement of enzymes to a solid support or matrix. This method has increased reuse and stability of these immobilized enzymes, and therefore offered several advantages for solid waste management. Immobilized enzymes onto solid supports provide protection against harsh environmental conditions such as temperature and pH fluctuations, thereby increasing their stability and prolonging their activity. This technique has improved the overall cost-effectiveness, and reduced generation of waste from enzyme disposal [64]. Cellulase immobilized in alginate-chitosan beads exhibited 83% activity and increased the production of erythritol from crop wastes. This improvement rate of production confirmed that immobilized enzymes are eco-friendly and economically feasible method. The immobilization process has been accelerated in the presence of cost-effective carriers like eggshells, coffee grounds, and brown onion skins, [16]. These carrier matrices have improved the enzyme activity, specificity, and degradation of solid wastes. The immobilized enzymes can be designed to target specific waste components to facilitate their breakdown and conversion into simpler product molecules. Immobilized enzymes could be efficiently separated from treated waste by filtration, centrifugation, or their affinity for certain materials. Enzyme immobilization has been carried out by different methodologies like adsorption, entrapment, covalent binding, encapsulation, and cross-linking. Each method has its advantages and limitations, and the choice of immobilization technique depends on the specific application and properties of the enzymes and waste materials involved.

Solid wastes from the agriculture and food sector are good sources of polysaccharides, and the bioconversion of these types of waste can be improved by the immobilization of enzymes [16]. Tang *et al.* [58] have studied the immobilization of enzyme/microorganism complexes for the degradation of microplastics. The authors affirmed that metal nanoparticle-enzyme complexes enhance microplastic biodegradation by producing photogenerated radicals that promote polymer oxidation and encourage the development of bacterial consortia for biological degradation, thereby securing the enzymes, enhancing their endurance, and absorbing water for hydrolysis. Carbon particle-enzyme aggregates allow enzyme immobilization on a carbon-based substrate or matrix by covalent bonding, adsorption, entrapment, encapsulation, or a combination of the above methods, which is enhanced by the development of cross-links among the enzymes. These complexes have been found to boost the effectiveness of microplastics degradation and facilitate the recycling of enzymes. Enzymes fused with polymer-binding modules, hydrophobins, new nanoparticles, and peptides are further forthcoming nanoparticles and/or enzymatic technologies for solid waste management [58]. Halloysite nanotubes grafted with 3-aminopropyltriethoxysilane have been employed in the development of nano-biocatalysts using cellulase and acetyl cholinesterase [43]. The immobilization enabled the enzyme to tolerate pH and temperature efficiently and also enhanced the storage time. They claimed that the biocatalyst could be reused for five cycles. Cel @ f-Hal has been utilized to degrade solid wastes such as bamboo leaf, orange peel, bagasse, blotting paper, and wheat bran, whereas AChE @ f-Hal has been used as a bio-electrochemical sensor for malathion [43]. Again, Fe_3O_4 @ Fe (0) (obtained by NaBH4 reduction of Fe_3O_4 nanoparticles) immobilized laccase was used to augment the degradation of the insecticide methoxychlor [8, 9] immobilized esterase from Bacillus subtilis on halloysite nanotubes and studied its dibutyl phthalate degradation efficiency. Moreover, the potential for DBP degradation by esterase and lipase enzymes derived from diverse microorganisms, as well as the immobilization of the best performing enzymes, was explored in this study. Immobilization of cutinase in Mg/Al-stratified double hydroxide structures provides thermal insulation. The effectiveness of cutinase immobilization was determined to be efficient, as was its ability to be released in an appropriate medium. Cutinase's thermal stability was greatly increased following immobilization, as evidenced by a 6-fold enhanced half-life at 90°C relative to free enzyme and an elevated functional retention (>60 %) following short exposure to temperatures as high as 200°C. Furthermore, a poly (butylene succinate-co-adipate) film manufactured with 5% immobilized cutinase decomposed entirely within 24 h [50]. Hence, in the context of solid waste management, enzyme immobilization can be an effective tool for enhancing the degradation or transformation of both inorganic and organic solid waste materials.

6.5 Economical Aspects of Solid Waste Management

The informal waste and recycling sector in cities generally comprises two distinct subdomains:

(i) Informal service sector: it consists of micro-enterprise informal service providers (ISPs) which enable individuals to earn money for removal of waste, excreta, litter, and dirt from households and street.

(ii) Informal valorisation sector: It consists of individuals, co-operatives, and micro-enterprises from the extractive resource industry. Its main activity is to identify and separate valuable materials from the waste stream.

Valorisation of waste products in the informal private recycling sector are generally carried out by two processes:

(a) Personal or commercial reuse: The collected waste materials are generally used for transformation into beautiful designs for decoration of the household and also used as agricultural or business inputs. These types of value-added products are generally sold in secondhand shops and flea markets.

(b) Reuse with repair: In this case, waste collected is repaired and sold in the market. A recycling process is first conducted by collecting separately and/or identifying, sorting, processing, storing and trading materials into the global industrial value chain. Organics valorisation is further conducted as a second step by collecting separately or sorting and processing kitchen, garden, commercial, agricultural and animal wastes and paper, and marketing it as animal feed or compost.

Both these domains of informal sectors are significant aspects in the private sector. They consist of individuals, small or medium-sized enterprises (MSMEs), and families working as enterprises. Private sectors are currently active in participation in the formal solid waste sector and face challenges of varying institutional and financial arrangements as shown in Figure 6.4.

The finance of public-private partnerships has been utilized for transport, handling, recovery and disposal of materials. In many cities, a contractual or franchise agreement has been created between city authorities and formal solid waste collection. Transport companies have been established for collection of street waste and perform other services for households and businesses. These informal sector workers generally provide some portion of required services, supplement formal contracts, and fill small niches.

Waste materials recycled by informal valorisation sector are sold directly into the market through the industrial value chain. Organic waste products have been supplied for animal feeding operations and composting processes after a proper formal recovery process. Recycling of materials recovered by informal sector pickers has been supplied extensively to the formal sector factories or exporters. Municipal employees are also involved in separation of recyclable materials from waste products and sell them unofficially to informal sector dealers. Cooperatives constituted by informal sector workers may undertake some (formal) work in association with municipal authority for informal recycling of these organic waste products.

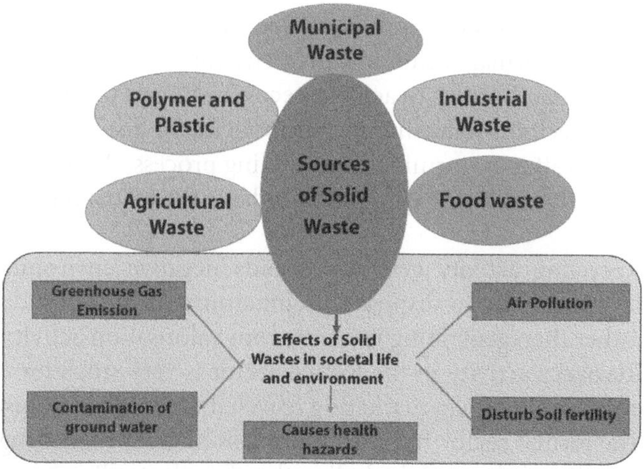

Figure 6.4 Different sources of solid waste products and their impact on social life and environment.

6.6 Cost and Effectiveness of Informal SWM Activities

The formal sector basically works in service provision for waste collection and disposal and has not achieved high recycling rates. The informal sector primarily works both for recovery and recycling of valuable materials. The operational costs are often higher for informal sectors than the formal sector; this results in high revenues from materials at lower cost with maximum net benefit.

Informal valorisation businesses have been mainly involved in the extraction, processing and selling of waste materials with high intrinsic value. These informal valorisation activities along the entire value chain have a high degree of profit. Service revenues are different from net cost calculation due to involvement of recovery cost covered by fees, taxes or government subsidies. Therefore, their actual cost recovery depends on political or socioeconomic considerations, and fees may be recovered from various parties. The informal sector saves the formal authorities a great deal of money in many cities. In the informal sector, materials are directly collected door-to door and therefore reduce expenses for collection, transport and disposal. Transport issues could be resolved by directly recycling the materials collected from the waste stream. The recovery of useful items at a disposal site has directly reduced the disposal cost. Recyclers should utilize the waste items at the earliest possible stage in the industrial sector to get minimum exposure from hazardous waste products. Valorisation activity of the informal sector will give environmental benefits and will allow municipal authorities to achieve recovery targets to a greater extent.

The formal sector is following specific policies for disposal of hazardous waste materials after performing the recycling process. Recycling activities of formal selective collection programmes have the following limitations:

- Recycling activity generally avoids negative environmental effects due to disposal of contaminated waste products rather than generating revenue from valorisation activity.
- Recovery activity in the formal sector is very small for efficient logistics than marketing recycled materials in industry.
- The formal sector has significant less idea for marketing of recycling industry, which directly affect the trading activity.

These studies will cause global environmental impacts for both formal and informal waste sectors. Carbon impacts determine the economic value of emissions, which will be reduced by the recycling and valorisation activity in the formal and informal sectors. These recovery activities will reduce the decomposition and methane formation in landfills, and therefore improve the carbon footprint of waste management. These carbon benefits from recovery activities will occur due to avoiding proper disposal, reducing extraction of raw materials and recovery of secondary raw materials for the production cycle. This activity will use less energy for the recycling process than production processes with primary raw materials. These benefits depend on the equipment used for recovery and type of recovered material with different potentials for reducing greenhouse gas (GHG) emissions. The rate of recovery is better in the informal sector than formal recovery, because of less use of the transport system for waste collection and use of reduced fossil energy. There are significant differences between the informal service sector and the valorisation occupations. The informal sector relies on service occupations with waged or performance-based labour, whereas valorisation occupations are associated with other resource extraction industries.

The earning scale for individuals through full-time informal valorisation is generally higher than minimum wage. The role of informal recovery activity always varies for income and maintenance of family. Women mainly earn more livelihoods through recovery activity of informal waste and valorisation sectors in most Indian cities. Women are mainly involved in informal valorisation process rather than valorisation process in the formal sector. In the valorisation sector, women mainly worked as pickers, sorters and traders of materials, at the level of junk shops or as initiators of community-based enterprises.

Workers in recycling activity have suffered from health issues due to exposure to contaminated waste materials. Proper management is always required to improve health and safety conditions, but it may affect the profitability of the enterprises. Informal enterprises are commonly unregistered, competitive and cause exploitation. Therefore, they would prefer to be recognised as formal businesses. Many policies show a stronger integration of the informal sector for solid waste system in cities. These policies have potential to increase the informal sector revenues, reduce the cost of the formal sector and total solid waste system in a city. The integration of the informal sector has contributed to the reduction of greenhouse gas (GHG) emissions and increased the quantity of recycled materials. The informal sector recovery activities using human or animal traction represent less energy consumption, and therefore cause lower carbon impacts in comparison to those of the formal sector. The integration of the informal sector in solid waste management activities causes higher recovery of materials at lower environmental cost.

6.7 Recommendations for Specific Policy and Practices

Policies are required to improve performance and equitability of waste management practices with potential benefits for the informal sector entrepreneurs. General information for policies mainly include the understanding of the informal sector and developing a feasible integration scenario. The economic and technical performance of these sectors should work together with their socioeconomic profile. Understanding and measuring of the informal valorisation activities provide a baseline for recycling activities to meet national or international recovery goals.

The workers in informal sectors should have awareness of industrial supply chain policy to hand over recovered biomaterials. The recovery goals for these resources are set by government and just need to be documented. Municipal (or NGO) activities are also conducted to set up recycling processes for poor persons. Careful investigation shows a lower degree of informal valorisation activity, a weak supply chain, and no valorisation for geographic or demographic areas. In this situation, consultation with informal entrepreneurs could result in legitimising of existing efforts and supporting informal sectors to diversify their business models.

Workers in these activities are facing serious issues of health and safety conditions; and negative environmental effects from recovery processes. They are also facing transaction costs in terms of harassment, bribery, criminalisation; and monopoly structures in the industrial value chain at the informal sector. These informal entrepreneurs always need support for organization of market relationships for longer-distance logistics. Informal workers require facilitation, advocacy, leadership, inter-agency co-operation and an interdisciplinary team to find solutions. The municipality has started schemes with NGOs to organise sourcing, horizontal linkages, and sociopolitical protection measures for management of workers in informal sectors. They create a portfolio of low-threshold formalisation measures for regulating facilitation of improvements. Light regulation can provide key data of workers in informal sectors, like street address, identity card or bank account.

Local authorities develop structural relationships between the solid waste system and the valorisation sector at a formal and informal level.

They require the appointment of representatives by waste pickers to speak on their behalf or the organisation of waste pickers in associations or co-operatives. Reverse institutional and systemic linkages have driven out the informal sector in order to pay for themselves.

6.8 Conclusions

This study has critically analysed the parameters for informal and formal sectors in terms of recycling and valorisation aspects. The informal sector showed more effectivity in valorisation than the formal sector. They are facing major challenges like low-energy-fuel collection and treatment technologies creating significant carbon benefits, and non-motorised approaches. Interventions and research have supported the design and procurement of low-energy and small-scale collection, transfer, processing and disposal systems. The incorporation of existing informal recycling systems into the operations of formal MSWM will bring significant benefits. The planning strategies of MSWM should be documented with effective informal collection and recycling systems.

References

1. Andler, S.M. and Goddard, J.M., Transforming food waste: how immobilized enzymes can valorize waste streams into revenue streams. *npj Sci. Food*, 2, 19, 2018, https://doi. org/10.1038/s41538-018-0028-2.
2. Arenas, C.B., Meredith, W., Snape, C.E., Gomez, X., Gonzalez, J.F., Martinez, E.J., Effect of char addition on anaerobic digestion of animal by-products: evaluating biogas production and process performance. *Environ. Sci. Pollut. Res.*, 27, 24387–24399, 2020, https://doi:10.1007/s11356-020-08828-8.
3. Arndt, E.A. and Wehling, R.L., Development of hydrolyzed and hydrolyzed-lsomerized syrups from cheese whey ultrafiltration permeate and their utilization in ice cream. *J. Food Sci.*, 54, 880–884, 1989, https://doi.org/10.1111/j.1365-2621.1989.tb07904.x.
4. Arora, A., Nandal, P., Singh, J., Verma, M.L., Nanobiotechnological Advancements in Lignocellulosic Biomass Pretreatment. *Mater. Sci. Energy Technol.*, 3, 308–318, 2020, doi: 10.1016/j.mset.2019.12.003.
5. Arya, P.S., Yagnik, S.M., Rajput, K.N., Panchal, R.R., Raval, V.H., Valorization of agro-food wastes: Ease of concomitant-enzymes production with application in food and biofuel industries. *Bioresour. Technol.*, 361, 127738, 2022, https://doi.org/10.1016/j. biortech.2022.127738.

6. Asaadi, S., Hummel, M., Hellsten, S., Härkäsalmi, T., Ma, Y., Michud, A., Sixta, H., Renewable high-performance fibers from the chemical recycling of cotton waste utilizing an ionic liquid. *ChemSusChem*, 9, 3250–3258, 2016.
7. Asgher, M., Bashir, F., Iqbal, H.M.N., A Comprehensive Ligninolytic Pretreatment Approach from Lignocellulose Green Biotechnology to Produce Bio-Ethanol. *Chem. Eng. Res. Des.*, 92, 8, 1571–1578, 2014, doi: 10.1016/j.cherd.2013.09.
8. Bai, H., Yang, Y., Yuan, H., Liu, X., Ni, C., Preparation of Fe3O4@ Fe (0) immobilized enzyme to enhance the efficient degradation of methoxychlor. *Environ. Sci. Pollut. Res.*, 30, 1, 917–929, 2023, https://doi.org/10.1007/s11356-022-22265-9.
9. Balci, E., Rosales, E., Pazos, M., Sofuoglu, A., Sanromán, M.A., Immobilization of esterase from Bacillus subtilis on Halloysite nanotubes and applications on dibutyl phthalate degradation. *Environ. Technol. Innov.*, 30, 103113, 2023, https://doi.org/ 10.1016/j.eti.2023.103113.
10. Ballesteros, L.F., Michelin, M., Vicente, A.A., Teixeira, J.A., Cerqueira, M.A., *Lignocellulosic Materials and Their Use in Bio-Based Packaging*, 2018.
11. Batche, H., Valenzuela-Solano, C., Hernandez-Martinez, R., Production of Conidiospores in Solid-State Fermentation of Trichoderma Harzianum and T. asperellum Isolated from Grapevine in Baja California; Phytopathology: St. Paul, MN, USA, p. 179, 2014.
12. Bernache, G., The environmental impact of solid waste management: The case of Gaudalajara metro area. *Resour. Conserv. Recycl.*, 39, 3, 223–237, 2003.
13. Berthier, H.C., Garbage, work and society. *Resour. Conserv. Recycl.*, 39, 3, 193–210, 2003.
14. Bidu, J.M., Njau, K.N., Rwiza, M., Van der Bruggen, B., Textile wastewater treatment in anaerobic reactor: Influence of domestic wastewater as co-substrate in color and COD removal. *S. Afr. J. Chem. Eng.*, 43, 112–121, 2023, https://doi.org/ 10.1016/j.sajce.2022.10.007.
15. Brethauer, S. and Studer, M.H., Biochemical Conversion Processes of Lignocellulosic Biomass to Fuels and Chemicals - A Review. *Chimia*, 69, 572–582, 2015, doi: 10.2533/chimia.2015.572.
16. Budžaki, S., Velić, N., Ostojčić, M., Stjepanović, M., Rajs, B.B., Šereš, Z., Strelec, I., Waste management in the agri-food industry: The conversion of eggshells, spent coffee grounds, and brown onion skins into carriers for Lipase immobilization. *Foods*, 11, 409, 2022, https://doi.org/10.1016/j.indcrop.2021.114351.
17. Carroad, P.A. and Wilke, C.R., Enzymes and microorganisms in food industry waste processing and conversion to useful products: A review of the literature. *Resour. Recovery Conserv.*, 3, 165–178, 1978, https://doi.org/10.1016/0304-3967 (78)90003-3.
18. Coad, A., Solid waste collection that benefits the urban poor—suggested guidelines for municipal authorities under pressure. Prepared for the Collab-

orative Working Group on solid waste management in low- and middle-income countries by the SKAT Foundation, St Gallen, Switzerland, 2003, http://www.skat-foundation.org/.
19. Cunha, F.M., Esperanca, M.N., Zangirolami, T.C., Badino, A.C., Farinas, C.S., Sequential solid-state and submerged cultivation of Aspergillus niger on sugarcane bagasse for the production of cellulase. *Bioresour. Technol.*, 112, 270–274, 2012.
20. Demirbaş, A., Ethanol from cellulosic biomass resources. *Int. J. Green Energy, 1*, 1, 79–87, 2004.
21. Dulac, N., The organic waste flow in integrated sustainable waste management. Tools for decision-makers. Experiences from the Urban Waste Expertise Programme, WASTE, The Netherlands, 2001, http://www.waste.nl/page/526.
22. Richard, E., Hilonga, A., Machunda, R., Njau, K., A review on strategies to optimize metabolic stages of anaerobic digestion of municipal solid wastes towards enhanced resources recovery. *Sustain. Environ. Res.*, 1, 1–13, 2019, https://doi.org/10.1186/ s42834-019-0037-0.
23. Fan, Y.V., Klemes, J.J., Lee, C.T., Perry, S., Anaerobic digestion of municipal solid waste: energy and carbon emission footprint. *J. Environ. Manage.*, 223, 888–897, 2018, https://doi.org/10.1016/j.jenvman.2018.07.005.
24. Ferreira da Silva, I., Reis Fontinelle Souto, L., Collins, S.R.A., Elliston, A., de Queiroz, J.H., Waldron, K.W., Impact of Hot Water and Alkaline Pretreatments in Cellulosic Ethanol Production from Banana Pseudostem. *Bioenerg. Res.*, 13, 4, 1159–1170, 2020, Doi: 10.1007/s12155-020-10123-w.
25. Wang, G., Dai, Y., Yang, H., Xiong, Q., Wang, K., Zhou, J., Li, Y., Wang, S., A Review of Recent Advances in Biomass Pyrolysis. *Energy Fuels*, 2020, https://doi.org/10.1021/acs.energyfuels.0c03107.
26. Girfanova, L.R. and Abdyrasulova, R.R., Improvement of textile waste sorting processes. *IOP Conf. Ser. Earth Environ. Sci.*, 666, 22–27, 2021.
27. Guerrero, L.A., Maas, G., Hogland, W., Factors Influencing the Performance of Waste Management Systems, 2012.
28. Haan, H.C., Coad, A., Lardinois, I., Municipal waste management: Involving micro-and-small enterprises. Guidelines for municipal managers, International Training Centre of the ILO, SKAT, WASTE, Turin, Italy, 1998, http://www.skat-foundation.org/ publications/waste.htm.
29. Kaushal, J., Khatri, M., Arya, S.K., Recent insight into enzymatic degradation of plastics prevalent in the environment: A mini-review. *Clean. Eng. Technol.*, 2, 100083, 2021.
30. Li, S., Junk-buyers as linkage between waste sources and redemption depots in urban China: The case of Wuhan. *Resour. Conserv. Recycl.*, 36, 4, 319–335, 2002.
31. Li, X., Hu, Y., Du, C., Lin, C.S.K., Recovery of glucose and polyester from textile waste by enzymatic hydrolysis. *Waste Biomass Valorization*, 10, 3763–3772, 2019, https://doi. org/10.1007/s12649-018-0483-7.

32. Łukajtis, R., Rybarczyk, P., Kucharska, K., Konopacka-Łyskawa, D., Słupek, E., Wychodnik, K., *et al.*, Optimization of Saccharification Conditions of Lignocellulosic Biomass under Alkaline Pre-treatment and Enzymatic Hydrolysis. *Energies*, 11, 4, 886, 2018, Doi: 10.3390/en11040886.
33. Ma, Y., Rosson, L., Wang, X., Byrne, N., Upcycling of waste textiles into regenerated cellulose fibres: Impact of pretreatments. *J. Text. Inst.*, 111, 630–638, 2020.
34. Vaverkova, M.D., Adamcová, D., Winkler, J., Koda, E., Petrželová, L., Maxianova, A., Alternative method of composting on a reclaimed municipal waste landfill in accordance with the circular economy: benefits and risks. *Sci. Total Environ.*, 723, 137971, 2020, https://doi.org/10.1016/j.scitotenv.2020.137971.
35. Mattedi, A., Sabbi, E., Farda, B., Djebaili, R., Mitra, D., Ercole, C., Cacchio, P., Del Gallo, M., Pellegrini, M., Solid-State Fermentation: Applications and Future Perspectives for Biostimulant and Biopesticides Production. *Microorganisms*, 11, 6, 1408, 2023.
36. Maulini-Duran, C., Abraham, J., Rodríguez-Pérez, S., Cerda, A., Jiménez-Peñalver, P., Gea, T., Gaseous emissions during the solid state fermentation of different wastes for enzyme production at pilot scale. *Bioresour. Technol.*, 179, 211–218, 2015.
37. Medina, M., Scavenger cooperatives in Asia and Latin America. *Resour. Conserv. Recycl.*, 31, 1, 51–69, 2000.
38. Medina, D., A short history of scavenging. *Comp. Civilizations Rev.*, 42, 2000.
39. Nakanishi, A., Kuroda, K., Ueda, M., Direct fermentation of newspaper after lactase-treatment using yeast codisplaying endoglucanase, cellobiohydrolase, and β-glucosidase. *Renew. Energy*, 44, 199–205, 2012, Https: //doi.org/10.1016/j. renene.2012.01.078.
40. Neamatalla, M., Zabbaleen Environment & Development Program, Cairo, Egypt. The Mega-Cities Project, Publication MCP-018D, 1998, http://www.megacitiesproject.org/publications/pdf/mcp018d.pdf.
41. Nicell, J.A., Environmental applications of enzymes. *Interdiscip. Environ. Rev.*, 3, 1, 14–41, 2001.
42. Namal, O.O., Investigation of the effects of different conductive materials on the anaerobic digestion. *Int. J. Environ. Sci. Technol.*, 17, 473–482, 2020, https://doi.org/10.1007/s13762-019-02498-x.
43. Pandey, G., Tharmavaram, M., Khatri, N., Rawtani, D., Halloysite nanotubes as nano-support matrix to tailor cellulase and acetylcholinesterase-based "nanobiocatalysts" for waste degradation and electrochemical sensing. *Appl. Clay Sci.*, 234, 106852, 2023, https://doi.org/10.1016/j.clay.2023.106852.
44. Pinkert, A., Marsh, K.N., Pang, S., Staiger, M.P., Ionic liquids and their interaction with cellulose. *Chem. Rev.*, 109, 6712–6728, 2009.
45. Porter, C.R., *The Economics of Waste*, Resources for the Future (RFF) Press, Washington, DC, 2002.

46. Pramanik, S.K., Suja, F.B., Zain, S.M., Pramanik, B.K., The Anaerobic Digestion Process of Biogas Production from Food Waste: Prospects and Constraints. *Bioresour. Technol. Rep.*, 8, 100310, 2019, doi: 10.1016/j.biteb.2019.100310.
47. Puthur, E.J., Singh, G., Jos, T., Microbial mediated waste management and bioenergy production, in: *Utilization of Waste Biomass in Energy, Environment and Catalysis*, CRC Press, 2022.
48. Rajagopalan, G. and Krishnan, C., α-Amylase production from catabolite derepressed Bacillus subtilis KCC103 utilizing sugarcane bagasse hydrolysate. *Bioresour. Technol.*, 99, 3044–3050, 2008.
49. Ravindran, R. and Jaiswal, A.K., A Comprehensive Review on Pretreatment Strategy for Lignocellulosic Food Industry Waste: Challenges and Opportunities. *Bioresour. Technol.*, 199, 92–102, 2016, doi: 10.1016/j.biortech.2015.07.106.
50. Romano, A., Rosato, A., Totaro, G., Zanaroli, G., Celli, A., Sisti, L., Recycling by-design of plastic through formulation with thermally protected enzymes in layered double hydroxide structures. *J. Clean. Prod.*, 384, 135517, 2023, Doi: 10.106/j.jclepro.2022.135517.
51. Roukas, T., The role of oxidative stress on carotene production by Blakeslea trispora in submerged fermentation. *Crit. Rev. Biotechnol.*, 36, 424–433, 2016.
52. Wainaina, S., Awasthi, M.K., Sarsaiya, S., Chen, H., Singh, E., Kumar, A., Ravindran, B., Awasthi, S.K., Liu, T., Duan, Y., Kumar, S., Zhang, Z., Taherzadeh, M.J., Resource recovery and circular economy from organic solid waste using aerobic and anaerobic digestion technologies. *Bioresour. Technol.*, 301, 122778, 2020, https:// doi.org/10.1016/j.biortech.2020.122778.
53. Salimi, E., Saragas, K., Taheri, M.E., *et al.*, The Role of Enzyme Loading on Starch and Cellulose Hydrolysis of Food Waste. *Waste Biomass Valorization*, 10, 3753–3762, 2019, https://doi.org/10.1007/s12649-019-00826-3.
54. Samuel, R., Pu, Y., Foston, M., Ragauskas, A.J., Solid-state NMR Characterization of Switchgrass Cellulose after Dilute Acid Pretreatment. *Biofuels*, 1, 85–90, 2010, doi: 10.4155/BFS.09.1710.4155/bfs.09.17.
55. Scheinberg, A., Financial and economic issues in integrated sustainable waste management. Tools for decision-makers. Experiences from the Urban Waste Expertise Programme, WASTE, The Netherlands, 2001a, http://www.waste.nl/page/525.
56. Scheinberg, A., Micro- and small enterprises in integrated sustainable waste management, WASTE, The Netherlands, http://www.waste.nl/page/524, 2001b.
57. Shimizu, F.L., Monteiro, P.Q., Ghiraldi, P.H.C., Melati, R.B., Pagnocca, F.C., Souza, W.D., *et al.*, Acid, Alkali and Peroxide Pretreatments Increase the Cellulose Accessibility and Glucose Yield of Banana Pseudostem. *Ind. Crops Prod.*, 115, 62–68, 2018, doi: 10.1016/j.indcrop.2018.02.024.

58. Tang, K.H.D., Lock, S.S.M., Yap, P.-S., Cheah, K.W., Chan, Y.H., Yiin, C.L., Ku, A.Z.E., Loy, A.C.M., Chin, B.L.F., Chai, Y.H., Immobilized enzyme/ microorganism complexes for degradation of microplastics: A review of recent advances, feasibility and future prospects. *Sci. Total Environ.*, 832, 154868, 2022.
59. Tao, L., Aden, A., Elander, R.T., Pallapolu, V.R., Lee, Y.Y., Garlock, R.J., *et al.*, Process and Technoeconomic Analysis of Leading Pretreatment Technologies for Lignocellulosic Ethanol Production Using Switchgrass. *Bioresour. Technol.*, 102, 24, 11105–11114, 2011, doi: 10.1016/j.biortech.2011.07.051.
60. Taylor, M., Alabdrabalameer, H., Skoulou, V., Choosing Physical, Physicochemical and Chemical Methods of Pre-treating Lignocellulosic Wastes to Repurpose into Solid Fuels. *Sustainability*, 11, 13, 3604, 2019, Doi: 10.3390/ su11133604.
61. Velikodvorskaya, G.A., Tikhonova, T.V., Gurvits, I.D., Karyagina, A.S., Lavrova, N.V., Sergienko, O.V., Tashlitskii, V.N., Lunina, N.A., Lunin, V.G., Chimeric lactase capable of spontaneous and strong immobilization on cellulose and development of a continuous-flow system for lactose hydrolysis at high temperatures. *Appl. Environ. Microbiol.*, 76, 24, 8071–8075, 2010.
62. Wang, Q., Awasthi, M.K., Ren, X., Zhao, J., Recent advances in composting of organic and hazardous waste: a road map to safer environment, in: *Biosynthetic Technology and Environmental Challenges, Energy, Environment, and Sustainability*, S.J. Varjani, *et al.* (Eds.), pp. 307–329, Springer, Singapore, 2018, https://doi.org/10.1007/ 978-981-10-7434-9.
63. Wang, Y., Zhao, Y., Deng, Y., Effect of enzymatic treatment on cotton fiber dissolution in NaOH/urea solution at cold temperature. *Carbohydr. Polym.*, 72, 178–184, 2008.
64. Yaashikaa, P.R., Devi, M.K., Kumar, P.S., Advances in the application of immobilized enzyme for the remediation of hazardous pollutant: A review. *Chemosphere*, 299, 134390, 2022, https://doi.org/10.1016/j.chemosphere.2022.134390.
65. Zhu, B., Wang, D., Wei, N., Enzyme discovery and engineering for sustainable plastic recycling. *Trends Biotechnol.*, 40, 1, 22–37, 2022, https://doi.org/10.1016/j. tibtech.2021.02.008.

7

Valorization of Biomass Wastes for Energy Production

Ankit Tripathi[1], Anurag[1], Aarti[1], Subhalaxmi Pradhan[1]* and A. K. Jain[2]

[1]Department of Chemistry, School of Basic Sciences, Galgotias University, Greater Noida, UP, India

[2]Department of Chemistry, School of Sciences, ILM University, Greater Noida, Uttar Pradesh, India

Abstract

The energy landscape is changing globally to include more renewable and sustainable sources. The need for alternative energy sources has attracted more attention due to climate change, depletion of fossil resources, and environmental issues. A potential solution to these problems that also increases energy security and rural development is the production of biofuels from biomass wastes. Fossil fuel consumption has a major impact on air pollution and greenhouse gas emissions, which exacerbates global climate calamities. Concerns regarding long-term energy sustainability are also raised by the finite supply of fossil fuels. In this case, producing bioenergy from biomass wastes is a creative solution with substantial positive effects on the environment and the economy. Biomass waste is a renewable resource that may be grown and exploited sustainably. It comprises organic waste, forestry by-products, and agricultural residues. A greener and more sustainable way to utilize biomass waste's energy potential is to convert it into biodiesel, bio alcohols, biogas, syngas, and biohydrogen. These biofuels can reduce greenhouse gas emissions and reliance on fossil fuels by either completely replacing or blending with conventional fuels in transportation, industry, and energy production. In this chapter, we will discuss how to combine cutting-edge technology, interdisciplinary collaborations, and sustainable practices to address the technological, economic, and environmental concerns associated with producing biofuel from biowastes.

***Keywords*:** Biomass, sustainable energy, biodiesel, syngas, biohydrogen, biorefinery approach

**Corresponding author*: subhalaxmi.pradhan@galgotiasuniversity.edu.in

Priyanka Singh, Pooja Agarwal and V. Vivekanand (eds.) Solid Waste Management: Challenges, Sustainability and Advancements, (213–250) © 2026 Scrivener Publishing LLC

Abbreviations

MSW	Municipal Solid Waste
RT	Residence time
DME	Dimethyl Ether
HTC	Hydrothermal Carbonization
HTL	Hydrothermal Liquefaction
HTG	Hydrothermal Gasification
SCWG	Supercritical Water Gasification
CHP	Combined Heat and Power
GHG	Greenhouse Gas
LCA	Life Cycle Assessment

7.1 Introduction

Biomass wastes encompass a wide range of organic materials, including agricultural residues, forestry residues, food waste, and municipal solid waste. These materials, often considered as by-products or waste streams, can be converted into valuable energy through various technological processes. The valorization of biomass involves harnessing the energy content within these organic materials to generate electricity, heat, and biofuels. This process not only reduces the reliance on non-renewable energy sources but moreover mitigates the effects on the environment associated with the disposal of biomass wastes [1].

Several technologies play a crucial role in the valorization of biomass for energy production. Biomass can be combusted directly to produce heat or converted into biofuels such as biogas, bioethanol, and biodiesel through thermochemical and biochemical processes [2]. Anaerobic digestion, pyrolysis, gasification, and fermentation are among the key techniques employed in biomass valorization, each offering unique advantages based on the characteristics of the feedstock and desired energy outputs [2].

The valorization of biomass waste not only provides a sustainable and green energy source but also presents economic opportunities for communities and industries. By converting biomass wastes into energy, regions can reduce their dependence on finite fossil fuels, create local jobs, and establish a circular economy by closing the loop on organic waste streams [3]. Additionally, the utilization of biomass for energy aligns with the global transition towards a low-carbon and resilient energy infrastructure.

This review explores the various aspects of biomass waste valorization for energy production, including the available technologies, environmental benefits, economic considerations, and potential challenges. By understanding and promoting the valorization of biomass, we can contribute to a more sustainable and diversified energy portfolio, fostering a greener future for upcoming generations [3]. Biofuels offer an environmentally acceptable energy source that can be utilized as a potential substitute in the transportation sectors as they may be generated anywhere in the world where biomass is present in sufficient amount. This helps to ensure the green sustainability and growth of forest-agriculture and associated industries with it [3].

Biomass is a viable resource for producing the energy needed for the sustainable growth of human society because of its availability, low cost, and composition. Production of renewable energy has drawn more attention and will be crucial to our efforts to become carbon neutral in the future. In this regard, generating energy from renewable sources is crucial to reducing reliance on fossil fuels and, thus, mitigating the risks associated with climate change at the moment [1]. In that regard, biomass valorization presents countless opportunities for the production of renewable energy to fulfil the demands of the modern industry, the standards of a post-petroleum society that is carbon neutral, and the sustainability of the energy and environmental systems [1].

In an agro-industry producing animal feed, the experiments compare the performance of mixing wheat straw and maize stalk with forestry wood to that of using 100% forestry woody pellets as the reference fuel. Different pellets, or mixtures of woody and herbaceous biomass, were made and tested in a fixed-bed reactor to examine their burning behaviour. Moreover, agro-industries' capacity to produce blended pellets that meet ISO 17225-6 requirements for non-woody pellets for the energy [3].

7.2 Classification and Potential of Biomass Waste

Biomass waste refers to organic materials that are generated from living organisms or their by-products and can be used as a source of energy or other valuable products. The classification and potential of biomass waste are crucial aspects in understanding and harnessing the diverse range of organic materials available for sustainable resource utilization. Each category offers unique characteristics and challenges, demanding tailored approaches for efficient valorization [4]. Biomass waste can be broadly categorized into different types on the basis of their availability in nature as follows in Figure 7.1.

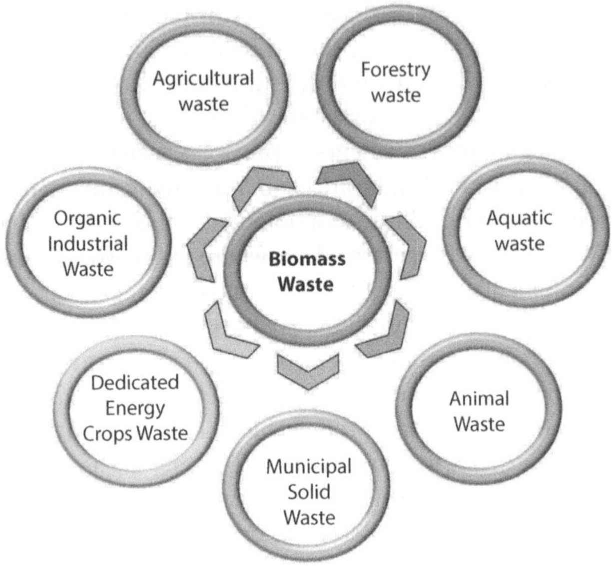

Figure 7.1 Classification of biomass wastes.

7.2.1 Agricultural Waste

Agricultural wastes include crop residues (stalks, husks, cobs and leaves), various types of straw (wheat straw, oat straw, barley straw, and rice straw, etc.) etc. There are numerous circumstances to use rural (agricultural) resources on available lands without hindering the generation of nourishment (food), fibres, or timberland items. Agrarian crop buildups, which incorporate the leaves and stalks, are inexhaustible, assorted, and broadly disseminated over the joined together states. Cases incorporate corn stover (stalks, clears out, cobs and husks), oat straw, wheat straw, grain straw, rice straw, and sorghum stubble. The deal of these buildups to a nearby biorefinery moreover speaks to a beneficial circumstance for agriculturists to produce extra wage [5]. Utilising biomass from agricultural waste is expected to be a significant alternative energy source that significantly reduces greenhouse gas emissions. The wastes that are left over from each stage of agricultural production, known as agricultural residues, are utilised as sustainable biomass resources to produce bioenergy [6].

7.2.2 Dedicated Energy Plants/Crops Waste

Dedicated energy plants/crops are the crops which are not used in feeding (i.e., non-food plants and trees) and can be developed on minimal land (the land which is not appropriate for conventional plants/crops like soybeans and corn) particularly to supply biomass. Depending on the lifetime of the crops, these can be divided into two common classes as herbaceous and woody [7].

- **Herbaceous vitality plants/crops** – These are lasting grasses (the plant/crops that have a lifetime for more than 2 years, i.e., a long time, almost 10-20 years) that are collected yearly after planting 2-3 years to gain proper efficiency. They incorporate bamboo, miscanthus, switchgrass, tall fescue, sweet sorghum, wheatgrass, kochia and others [8].
- **Short-rotation woody plants/crops** – These are the fast-growing timbers (hardwood trees) which are collected inside 5-8 years of seeding. They incorporate cross-breed poplar, eastern cottonwood, half-breed willow, black walnut, green fiery remains, silver maple, sycamore, and sweetgum. Most of these species can offer assistance progress water and soil quality, enhance natural life environment corresponding to yearly crops, differentiate modes of pay as well as progress in general cultivate efficiency [9].

7.2.3 Forestry Waste

Forest biomass feedstocks/build-ups drop into two classes: timberland residues/buildups left after accumulating timber/hardwood trees (including tops, appendages and separated trees/plants and their components that would be something else unmerchantable) or entire-tree biomass collected expressly for biomass/buildups [10]. Dead, infected, ineffectively shaped, and other unmerchantable plants/trees are frequently cleared out within the woods taking after timber collect. These woody wastes are collected to utilize in producing bioenergy, leaving sufficient behind, however, to supply territory as well as keep up hydrologic highlights and legitimate nutrient. There are, moreover, openings to create utilization of abundant biomass that are grown up on millions of acres of woodlands. Collecting over the top woody biomass can decrease the chance of bugs and fire as well as help in timberland reclamation, efficiency, essentialness, and flexibility. On an average, only almost 50% of the harvested wood is used for value-added purposes and the remains are unused that goes to waste [11]. This biomass may be collected for bioenergy without adversely affecting the well-being and soundness of timberland environmental structure and work.

7.2.4 Aquatic Biomass Waste

Aquatic biomass wastes (aquatic plants and algae, etc.) as feedstocks/build-ups for bioenergy and biorefinery refers to a different bunch of exceedingly beneficial living beings that incorporate macroalgae (seaweed), microalgae and cyanobacteria (known as "blue-green algae"). Many utilize daylight and supplements to make biomass, which consists of some key components—including proteins, lipids and carbohydrates—which can be changed over and updated to a variety of biogases, biofuels and some other items. Based on the strain, algae can develop by utilizing new, brackish or saline water from groundwater, surface water sources or seawater. Moreover, they can develop in water from the second-use sources, such as treated mechanical wastewater; civil, aquaculture or rural wastewater; or delivered water created from gas and oil penetrating operations [5].

7.2.5 Animal Waste

Animal wastes transcendently incorporate fertilizers from dairy animals, pigs, and chickens. Animal wastes are toxins of expanding concern both to the public and to administrative bodies since they have the potential to contaminate both earth and groundwater. Animal squanders can be utilized as a root of biomass-dependent transformation operations, especially in bio-fertilizer and bioenergy generation. The foremost general sources consolidate dairy shed emerging (consisting of pee, wash water, dunk, waste nourishments and remaining deplete), dairy manures, poultry litter (a blend of manures, spilled feed, water, bedding texture and quills), renderings as well as other wastes from creatures wrapping up procedures. Poultry litter and dairy manure are among the first generally utilized animal squanders in biomass-based generation chains all through the world [7].

7.2.6 Organic Industrial Waste

Organic industrial wastes incorporate a broad range of natural materials acquired from commercial and mechanical operation. These organic squanders, either solid or liquid, are an extraordinary substrate for biogas production by making utilization of anaerobic assimilation. Anaerobic absorption of organic mechanical squander is quickly picking up notoriety around the world. Various industries create liquid and solid squanders that are appropriate for anaerobic absorption, such as nourishment preparing, pharmaceuticals, organic chemicals, paper fabricating and tannery industries. A few of the squanders can be troublesome to handle as a sole substrate, but those may be biochemically degraded in combination with sewage sludge or fertilizer. The combined absorption of particular squanders is called co-digestion [12].

In later decades, the fast-growing nourishment and refreshment preparing industry has surprisingly extended significantly all over the world. Solid squanders incorporate scraps and peelings from vegetables and fruit, nourishment that does not meet quality control guidelines, fiber and pulp from starch and sugar extraction, coffee grounds and channel sludges. Liquid squanders are created by washing fruit, meat and vegetables, brightening vegetables and fruit, poultry and fish, pre-cooking meats, cleaning and handling with operations [12].

The paper and pulp industry is viewed as one of the profoundly contaminating industries and expends a huge quantity of vitality and water in different unit processes. It produces a huge amount of more polluted coloured wastewater. The wastewater released by these industries is profoundly a mixture and heterogeneous blend because it consists of compounds from wood or other crude materials, prepared compound as well as chemicals shaped amid handlings [13].

7.2.7 Municipal Solid Waste (MSW)

Municipal solid waste resources incorporate mixed commercial and private trash, such as yard trimmings, paper and paperboard, plastics, rubber, leather, materials, item packaging, grass clippings, furniture, clothing, bottles, nourishment scraps, daily papers, appliances, paint, and batteries. This comes from our homes, schools, hospitals, and businesses. MSW, by producing bioenergy through biorefineries creates an opportunity to diminish commercial and residential waste by redirecting significant volumes from landfills to the biorefineries [5].

MSW is ordinarily burned especially at waste-to-energy industrial units that utilize the heat from that burning residue to create steam for heating or creating power buildings. In 2022, 63 U.S. power plants created almost 12.8 billion kilowatt-hours of electricity and power from burning almost 26.6 million tons of combustible municipal solid wastes (MSW) for power generation [14]. Biomass materials accounted for around 61% of the weight of the combustible MSW and for almost 45% of the power created. The leftover portion of the combustible MSW was non biomass fabric, basically plastics. Many huge landfills also produce power by utilizing the methane gas that is produced from breaking down biomass in landfills.

Producing biomass energy as electricity is the only reason to burn municipal solid waste. Burning squander too diminishes what would likely be buried in landfills. Waste-to-energy industrial units/plants decrease 2,000 pounds of rubbish to fiery remains that weigh almost 300 - 600 pounds, and they diminish the amount of squander by almost 87% [14].

7.2.8 Potential of Biomass Waste

The potential of biomass waste lies in its abundance and renewable nature. Agricultural residues, for instance, are generated annually during crop harvesting, providing a readily available and often underutilized resource [2]. Forestry residues, comprising branches and wood scraps, are by-products of timber harvesting and processing. Municipal solid waste, rich in organic content, presents an opportunity to divert substantial volumes from landfills and convert them into valuable energy or bio-based products. Furthermore, organic industrial waste streams, such as those from the food industry, can be harnessed for energy production and resource recovery.

A huge potential exists for these industrial wastes to be anaerobically processed to create biogas, or matured to create ethanol, and a few commercial illustrations of waste-to-energy conversion as of now exist in totally different parts of the world. The utilization of biogas innovation is expanding quickly for mechanical wastewater treatment. Over 30 types of industry have been identified as having wastewaters amenable to anaerobic absorption treatment, counting processors of refreshments, chemicals, nourishment, meat, drain, pulp and paper, and pharmaceutics.

The versatility of biomass waste extends beyond its use for traditional combustion. Advanced conversion technologies such as thermochemical (i.e., pyrolysis, gasification, liquefaction, etc.) and biochemical (alcoholic fermentation, anaerobic digestion and photobiological hydrogen production, etc.) offers opportunities to transform biomass waste into biogas, biofuels and bio-based chemicals [15]. These processes allow for a more efficient and sustainable use of biomass wastes, contributing to the development of circular economies.

In summary, the classification of biomass waste into distinct categories highlights the diverse sources of organic materials available for utilization. The potential of biomass waste lies not only in its abundance but also in the variety of environmentally friendly and economically viable applications. As technology advances, there is a growing and demanding recognition of the importance of harnessing biomass waste for energy production and the development of bio-based products, contributing significantly to the paradigm shift towards a more sustainable and circular approach to resource management [1]. Sometimes, sustainability is categorized in these three types: weak sustainability (economic value-centred view), strong sustainability (nature-centred view) and absurdly strong sustainability, which assumes that the exploitation and utilisation of ecosystems should be terminated [16].

7.3 Thermochemical Conversion Processes

Fast industrialization as well as the enhancement of the living residencies measures of society depends intensely on the consistent reservoir of vitality, with the utilization of fossil powers (coal, petroleum, common gas) responsible for 80% of the vitality need around the world [17]. A critical sum of this vital energy is utilized to create electrical power that the developed world anticipates being accessible 24/7, while the intemperate utilization of fossil fuels causes resource consumption and generates air contamination and worldwide warming (global warming), making irreversible damage to the environment [18].

Hence, to ease worldwide natural issues and fulfil the energy necessities of current modern society, a range of renewable and green vitality assets have come frontward as maintainable alternatives to conventional fossil powers [19]. Right now, the renewable and green energy operations of commercial importance are sun, wind and hydroelectric based [20].

Both wind energy and sun-based energy depend on the climate. Hydroelectric energy comes with its associating vitality capacity, a storage of water at a high height (elevation) and the related potential vitality is the extreme vitality source from which the generation of hydroelectricity is operated. The above techniques present various issues [20]. In solution of these problems, thermochemical conversion processes are evolved essentially. Thermochemical conversion processes complement existing vitality assets by convincing an economical and flexible approach to energy generation, waste treatment plants, carbon neutrality and climate mitigation, and economic advancement. They play a crucial part in addressing energy and natural challenges and cultivating development and versatility within the energy sector [20].

Thermochemical conversion processes include a set of chemical reactions that change biomass, waste materials, or fossil fuels into vitality, fuels, chemicals, or other profitable items through the application of heat. These processes typically occur in the absence of oxygen (anaerobic) or with limited oxygen supply. Thermochemical conversion can be broadly categorized into various techniques, each with its own set of advantages and challenges. Some common thermochemical conversion processes are described below.

7.3.1 Pyrolysis

- Pyrolysis is a process in which thermal decay takes place in absence of O_2 or in the presence of a much smaller amount of O_2 to change biomass into fluid (bio-oil), gases and solid charcoal at elevated temperatures [21].
- Depending on TGA (thermal gravity analysis) analysis of biomass, there are three phases for an ordinary pyrolysis process that happens. The primary phase (pre-pyrolysis) happens between 120-200 °C with a minute noticed loss in weight, when a few internal modifications, such as the appearance of free radicals, bond breakage and the arrangement of carbonyl groups (-CO, -CHO) is observed with a corresponding discharge of lesser quantity of carbon monoxide (CO), carbon dioxide ($CO_{2)}$ and water (H_2O). The second phase is the principal pyrolysis procedure in which solid deterioration happens, associated with a significant loss in weight from the elementary fed biomass. Due to the further cleavage of C–O and C–H bonds, the final phase is the persistent char devolatilization [21].

Based on the residence time and temperature in reaction, pyrolysis can be sub-divided as fast pyrolysis, flash pyrolysis, and slow/conventional moderate pyrolysis.

7.3.1.1 Fast Pyrolysis

Fast pyrolysis is a process in pyrolysis having elevated heating rate (at temp. 400–550 °C) and brief residence time (~ 0.5–10 s). It especially supports the arrangement of fluid items, but hinders the generation of solid chars. The fluid items (bio-oils) are made up of an aqueous/liquid phase which consists of a few light organo-oxygen compounds having low molecular weight, and a non-aqueous phase (tar) which incorporates a number of aromatic organic compounds (which are insoluble) of high molecular weight. Bio-oil, the major item produced from fast pyrolysis, may be a potential fluid fuel that can be effortlessly put away and transported [21].

7.3.1.2 Flash Pyrolysis

Flash pyrolysis is a variety of fast pyrolysis that works at even higher temperatures (ordinarily over 700 °C; 550–750 °C) and amazingly short residence time (less than a moment; ~ 0.1–0.5s). The fast heating and nearly immediate vaporization of the biomass waste create a high production of bio-oil and gases, making flash pyrolysis appropriate for biofuel generation and chemical synthesis [22].

7.3.1.3 Slow/Conventional Moderate Pyrolysis

A slow or conventional moderate pyrolysis process, with a generally long vapour RT (residence time, RT ~ Several hours to days) and low heating rate (HR ~ 0.1–2 s at temp. T ~<500 °C), has been utilized to synthesize charcoal for many years. The product, charcoal, may be utilized in broad areas, from household cooking and heating to chemical or metallurgical utilization as the crude material to produce chemicals, activated carbon, fireworks, absorbents, soil conditioners and pharmaceuticals [21].

Each sort of pyrolysis process has its advantages and restrictions based on factors such as feedstock characteristics, desired item yields, energy necessities, and process adaptability. The choice of the suitable pyrolysis method depends on the particular application and goals, whether it is biochar generation, bio-oil extraction, or biofuel synthesis.

7.3.2 Gasification

Gasification is a process in which thermal conversion of biomass waste takes place and carbonaceous material is changed over a synthetic gas or gaseous product under a gas stream containing managed oxygen rate at elevated temperatures (almost 550–1000 °C). A syngas or product gas is composed of carbon monoxide (CO) and hydrogen (H_2) with some amount of carbon dioxide (CO_2), nitrogen (N_2), water (H_2O), methane (CH_4) and higher hydrocarbons. The presence of a gasifying reagent (air or oxygen, etc.) is necessary for conducting the biomass gasification. It produces a final product gas which can be transformed into different types of chemicals through Fisher Tropic ways and different sorts of fuels used in transport and vehicles such as DME, methanol and so on. Biomass gasification is the foremost proficient technique of converting biomass wastes into energy and heat by means of turbines, engines and boilers. Naqvi *et al.* (2023) suggested that power and heat may be produced effectively during gasification by utilizing fuels having low energy density, for example, lignocellulosic biomass etc. [22].

Some typical reactions which are processed in gasification are as follows:

$$C_nH_mO_p \rightarrow CO_2\uparrow + H_2O + CH_4\uparrow + CO\uparrow + H_2\uparrow + (C_2 - C_5) \quad (7.1)$$

$$n\,CO + n\,H_2O \leftrightarrow n\,H_2 + n\,CO_2 \quad (7.2)$$

$$C + O_2 \rightarrow CO_2 + \text{Heat} \quad (7.3)$$

$$C + \tfrac{1}{2}\,O_2 \rightarrow CO + \text{Heat} \quad (7.4)$$

$$H_2 + \tfrac{1}{2}\,O_2 \rightarrow H_2O \quad (7.5)$$

$$C + H_2O \rightarrow CO + H_2 \quad (7.6)$$

$$C + 2\,H_2O \rightarrow CO_2 + 2\,H_2 \quad (7.7)$$

$$C + CO_2 \rightarrow 2\,CO \quad (7.8)$$

$$C + 2\,H_2 \rightarrow CH_4\uparrow \quad (7.9)$$

$$CO + 3\,H_2 \leftrightarrow CH_4\uparrow + H_2O \quad (7.10)$$

$$C + H_2O \rightarrow \tfrac{1}{2}\,CH_4\uparrow + \tfrac{1}{2}\,CO_2\uparrow \quad (7.11)$$

Pyrolysis process (reaction 7.1) is the first step in biomass gasification. In this process, tars as well as various volatile hydrocarbon gases having some amount of hydrogen (H_2) from biomass stocks released during the gasification processes. The water–gas shift reaction (reaction 7.2) has also a high importance due to its critical role in generating hydrogen. After that, a number of reactions occur in a gasifier as oxidation reactions (in presence of oxygen) (reactions 7.3–7.5), principal gasification reactions (water–gas reactions; reactions 7.6 & 7.7), methanation reactions (reactions 7.9–7.11), etc. Oxidation reactions 7.3 & 7.4 are highly exothermic which generates sufficient heat to dry feedstocks, to break bonds in chemical compounds in pyrolysis process of biomass and to retain elevated temperatures in gasifiers for gasification reactions. Reaction 10 (methanation) proceeds slowly in the absence of any catalysts at low temperatures. Both reactions 7.2 & 7.10 occur in either direction based on the specific pressure, temperature as well as the reactant concentrations in the process. Through these reactions, it can be noticed that the syngas is a mixture of gases which is mainly made up of H_2, CH_4, CO_2, CO and H_2O vapour [21].

7.3.3 Hydrothermal Treatments

Using high-temperature, high-pressure water or steam, biomass materials are transformed into valuable products through the hydrothermal treatment of biomass wastes. The environmental advantages of this process as well as its potential to yield beneficial products like biofuels (biocrude oils), charcoal (hydro char), syngas and biochemicals, are making it more and more popular. Depending on variety of biomass being utilized and the desired result, different pressure and temperature settings may be used. Hydrothermal treatment of biomass wastes can be categorized as follows [22].

Hydrothermal Carbonization (HTC) or Wet Carbonization
Hydrothermal Carbonization (HTC) changes biomass into a carbon-rich material called hydro char or biochar having properties comparable to low–rank coal. The process involves heating biomass in the presence of water at sufficient temperatures (typically between 180 - 250°C) and pressure (2 – 10 MPa) [22].

Hydrothermal Liquefaction (HTL): Hydrothermal Liquefaction (HTL) involves converting biomass into liquid items such as bio-oil or biocrude oil through a combination of high temperature (typically 250-400 °C) and pressure (5 – 35 MPa) in the presence of water or a solvent.

Hydrothermal Gasification (HTG) or Supercritical Water Gasification (SCWG): HTG converts biomass into syngas (a mixture of hydrogen-H_2, methane – CH_4, carbon dioxide – CO_2 and carbon monoxide - CO) by subjecting it to high temperatures (375 - 750°C) and pressure (> 22 MPa) in a water-rich environment. Compared to conventional gasification procedures carried out at 800 °C, undesirable compounds like tar can be eliminated due to the high solubility of SCW (supercritical water) in the HTG process [22].

Hydrothermal treatments have many advantages like resource utilization, energy recovery, waste reduction and biochar production, etc. Hydrothermal treatment of biomass wastes is a useful method for handling waste management issues and producing useful and value-added products and renewable energy (bioenergy production). This treatment is also a promising technique to achieve the safe treatment and utilization of food waste [23].

7.3.4 Torrefaction

Torrefaction, also known as mild pyrolysis (which carries out at relatively low temperatures of 200–300°C and residence time; RT ~ 10 – 120 min, in a low oxygen region or inert atmosphere), is a valuable preliminary treatment to remove the low-weight organic volatile components and the moisture content in biomass, which produces a biochar or hydrophobic solid fuel having high calorific value, flowability as well as enhanced grindability [24]. As in pyrolysis, torrefaction can generate bio-oil, biogas and biochar. Among these, biochar is the principal product produced during torrefactions [25]. The bio-oil includes water, phenols, acids, alcohols, ketones, aldehydes, esters and some other contents, where H_2O is the most abundant product. A few portions of biogas including CH_4, H_2, CO, CO_2 and C_2H_4 are produced amid torrefaction, which is mainly dominated by carbon dioxide (CO_2) [26]. In summary, torrefaction of biomass waste gives a better route for the conversion of biomass waste into a high-quality solid fuel with improved potentials for energy production, waste utilization, and environmental benefits.

Key steps involved in torrefaction:

- **Drying stage:** Initially, the biomass is heated to eliminate moisture, which lowers the amount of energy needed for further stages [27].
- **Devolatilization stage:** Volatile substances, including hemicellulose and some lignin constituents, are liberated from the biomass, aiding in the remaining solid's mass loss and raising its energy density [27].
- **Carbonization stage:** A partial carbonization of the residual solid biomass produces a substance with a higher mass percentage of carbon and energy [27].

Torrefaction of biomass wastes produces torrefied biomass which has properties like higher energy density, improved storage and handling, enhanced combustion characteristics, improved fuel characteristics such as lower ignition temperature and higher heating value, palletisation potential, etc.

7.3.5 Combustion

- During the combustion process, biomass is burned through a sequence of chemical processes in excess air and at high temperature nearly above 900°C. Carbon present in biomass is converted into CO_2 (carbon dioxide) during combustion, and H_2 is turned into H_2O (water) through oxidation. Due to the high reactivity of the fuel and char as well as the high fuel volatility, biomass is regarded as an essential combustion source. Combustion converts chemical energy accumulated in the biomass waste to heat, mechanical power or electricity, etc. [22].
- Any sort of biomass can be burned but combustion is only possible for biomass that has above 50% moisture content. The combustion process has some drawbacks. Biomass hardly occurs naturally in a burning that is suitable. Most of the time, some pretreatment is necessary, such as drying, chopping, grinding, etc., which has an impact on energy and cost [28].
- Combustion has some advantages such as renewable energy sources, reduces carbon emissions, helps in waste management and supports energy production sector, etc. In contrast to fossil fuels, which release sequestered carbon, the combustion of biomass releases carbon dioxide (CO_2), but over time, this emission is reabsorbed by new biomass growth, making biomass combustion carbon neutral [22].

Some applications of this process are as follows:

- **Heat Generation:** In boilers used for hot water supply, process heating, and space heating in homes, businesses, and industries, biomass combustion is a popular method of producing heat.
- **Electricity Generation:** By burning biomass, biomass power plants use gas or steam turbines to produce electricity, which

helps to meet renewable energy targets and maintain system stability.

- **Combined Heat and Power Unit (CHP):** CHP systems combine the burning of biomass with the production of electricity, maximizing energy efficiency by putting the electricity and heat outputs to use in different ways.

7.4 Bioethanol from Lignocellulose

Plant biomass that is high in cellulose, hemicellulose, and lignin is known as cellulosic biomass. Lignocellulosic biomass makes up a sizable amount of the biomass that is available on earth among all other types of biomass [29]. Alvira *et al.* stated that the most researched feedstock for the synthesis of biofuel was lignocellulosic biomass [29]. However, a number of conditions, including excessive humidity and contaminants, make direct conversion processes—like burning biomass—not economically or environmentally viable. To convert biomass feedstock into value-added products, raw biomass is mostly converted through physical [30], biological [31], chemical [32], biochemical [33] or thermochemical pathways [34]. Typically, lignocellulosic biomass is composed of 10–50% hemicellulose, 9–80% cellulose, and 5–35% lignin. Small amounts of extractives and minerals make up the remaining portion. Bioethanol production from lignocellulosic materials can be classified into four groups based on resource

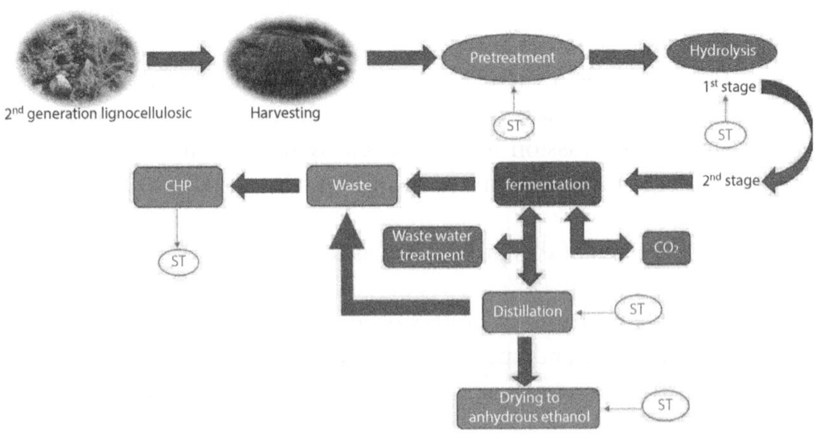

Figure 7.2 Second-generation biomass-to-ethanol production (ST: steam addition), combined heat and power unit (CHP).

type: (1) forestry residues, (2) municipal solid waste, (3) wastepaper, and (4) crop residue resources. The literature reports several articles on the use of various lignocellulosic wastes such as rice straw [35], corn stalks [36], switchgrass [37], palm residue [38], etc. The process of producing bioethanol from lignocellulose, which is represented in Figure 7.2, typically involves several steps.

7.4.1 Pretreatment

Pretreatment is considered as one of the most expensive processing steps in converting biomass into fermentable sugars [39]. Lignocellulosic biomass is treated to break down the complex structure and make cellulose more accessible. Common pretreatment methods include steam explosion, acid hydrolysis, alkaline treatment, and enzymatic pretreatment.

7.4.1.1 Hydrolysis

Hydrolysis of lignocellulosic biomass is done in two stages. In the first stage, enzymes or acids are used to break down cellulose and hemicellulose into simple sugars (such as glucose, xylose, arabinose, mannose and galactose). Enzymatic hydrolysis is often preferred because it is more specific and efficient. The biomass's freed cellulose is transformed to glucose in the second hydrolysis stage. This process is once more aided by concentrated or diluted acid, or ideally by cellulase enzymes. Alternatively, the glucose can be produced independently in a reactor or acquired externally from businesses. The metabolism of cellulose and hemicellulose can be demonstrated by the reaction of glucan (for hexose), which is given in Equation 7.12 and xylan (for pentose) with water as given in Equation 7.13.

$$(C_6H_{10}O_5)_n + nH_2O \rightarrow nC_6H_{12}O_6 \tag{7.12}$$

$$(C_5H_8O_4)_n + nH_2O \rightarrow nC_5H_{10}O_5 \tag{7.13}$$

7.4.1.2 Fermentation

Current methods for converting lignocellulose to ethanol are still in the experimental and demonstration stages, in contrast to the established technologies used to convert disaccharides and starch to ethanol. ETEK (Sweden) [40], Iogen Corporation (Canada) [41] and NREL (USA) [42] have built pilot plants capable of producing several hundred thousand liters of ethanol per year. The biological process of fermentation occurs when different microorganisms, such as bacteria, yeast, or fungus, transform hexoses and pentoses into ethanol. The metabolic reaction for hexose (C6) and pentose (C5) is as follows

$$C_6H_{12}O_6 \rightarrow 2C_2H_5OH + 2CO_2 \tag{7.14}$$

$$3C_6H_{10}O_5 \rightarrow 5C_2H_5OH + 5CO_2 \tag{7.15}$$

The obtained sugars are fermented by microorganisms (usually yeast, such as Saccharomyces cerevisiae) to produce ethanol. Some microorganisms can ferment both glucose and xylose, while others may require additional engineering for better xylose utilization. Ethanol is separated from the fermentation broth through distillation or other separation techniques. Additional steps, such as dehydration, may be employed to increase the ethanol concentration.

7.4.2 Waste Treatment

Residual materials, such as lignin and unfermented sugars, are often treated or utilized for other purposes. Lignin can be used for heat and power generation, as a feedstock for chemicals, or for soil improvement. Challenges in lignocellulosic bioethanol production include the recalcitrance of lignocellulose, the cost of enzymes for hydrolysis, and the need for robust microorganisms capable of efficient fermentation. Research is ongoing to optimize each step of the process and make lignocellulosic bioethanol more economically viable and sustainable. Additionally, advancements in biotechnology, metabolic engineering, and synthetic biology aim to improve the efficiency of microorganisms in converting lignocellulosic sugars to bioethanol and broaden the range of feedstocks that can be used.

7.5 Biodiesel from Biomass

Renewable fuels such as biodiesel can be made from a variety of biomass sources. The primary feedstocks for biodiesel production include vegetable oils, animal fats, and used cooking oils. Transesterification is the procedure that turns these feedstocks into biodiesel. Because vegetable oils produce less pollution and are renewable, they are a more appealing option for producing biodiesel than traditional petroleum diesel fuel [43]. Below is a basic overview of how biodiesel can be produced from biomass.

7.5.1 Feedstock Selection

Over the course of several generations, several raw material kinds have been employed. Figure 7.3 presents the various raw materials utilized in the production of first-, second-, third-, and fourth-generation biodiesel [44]. It can be financially advantageous to use wastewater as a source of nutrients, but there are a number of things to take into account, including public opinion, legal restrictions, and technological viability [45]. There are two types of vegetable oils: edible and non-edible. Because edible oils eventually compete with food items, using them has been a major source of stress. Additionally, some non-edible oils are available, offering advantages including abundance, fluid mobility, excellent biocompatibility, reduced sulfur and odor content, and no discernible detrimental effects on crops intended for human consumption. Larger-scale biodiesel production can be achieved by using the producing feedstocks, such as animal fats, fish oil, fat oil, microalgae, and so on. The fourth generation of biodiesel is produced using photobiosolar fuel and electric fuel. Solar energy is converted into biodiesel utilizing raw ingredients to create solar biofuel [46].

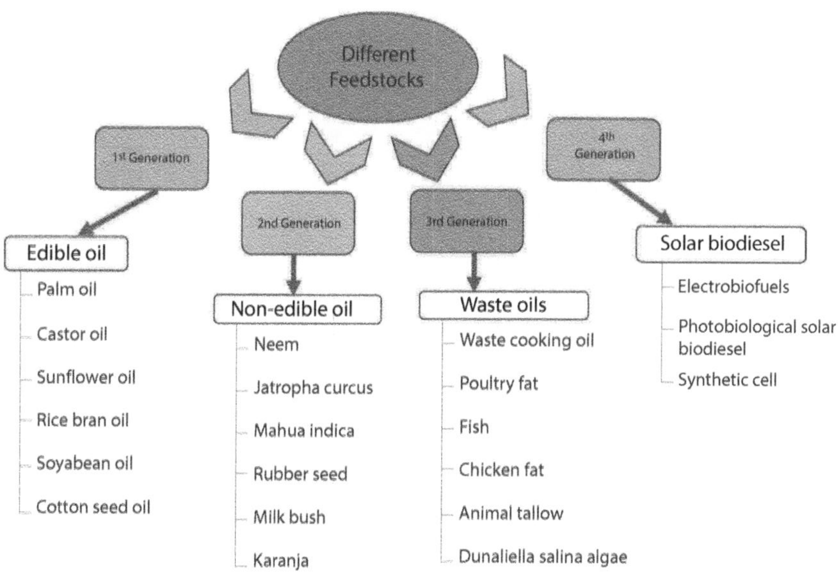

Figure 7.3 Different generation feedstocks for biodiesel.

Vegetable oils are frequently utilized as feedstocks for the manufacturing of biodiesel, including canola, palm, and soybean oil and used cooking oils and animal fats can be used for biodiesel production.

7.5.2 Preprocessing

If the feedstock is a triglyceride (as is the case with most vegetable oils and animal fats), it needs to undergo preprocessing. This typically involves removing impurities, moisture, and free fatty acids from the feedstock [43].

7.5.3 Transesterification

Three catalysts are used in the transesterification process to produce biodiesel: bases, acids, and enzymes. Base-catalyzed transesterification has become a standard commercial method because, in contrast to acid-catalyzed transesterification, it can produce considerable yields of fatty acid methyl ester quickly and under extremely moderate reaction conditions [47]. Enzyme catalysts need to be refined before being used in a commercial context, despite the fact that they are very effective and environmentally friendly [47]. The two-step esterification and transesterification conversion process is commonly used to make biodiesel. The key

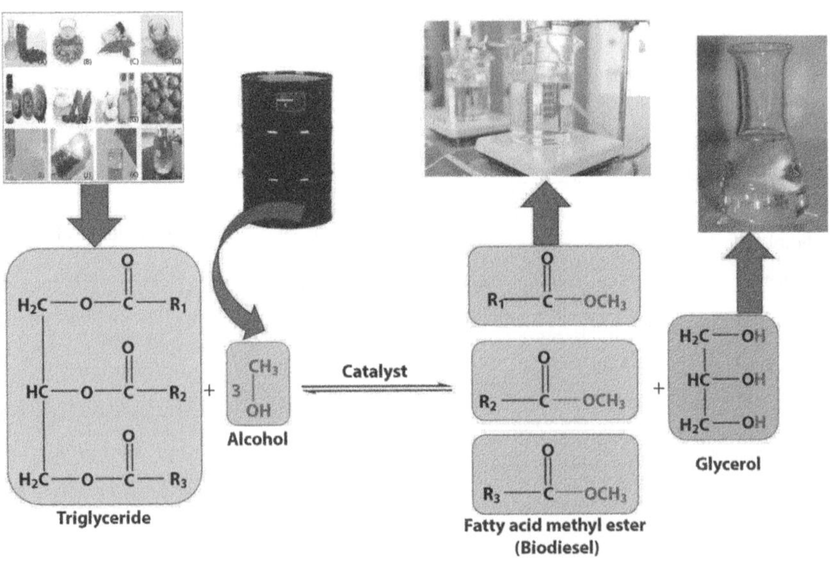

Figure 7.4 Transesterification reaction for biodiesel production.

step in biodiesel production is transesterification, where the triglycerides in the feedstock react with an alcohol (usually methanol or ethanol) in the presence of a catalyst (such as sodium hydroxide or potassium hydroxide). This reaction produces biodiesel (methyl or ethyl esters) and glycerol as a by-product [48]. The chemical reaction can be represented in Figure 7.4 as follows:

7.5.4 Separation and Purification

After transesterification, the biodiesel and glycerol phases need to be separated. Typically, this is done through settling, and then further purification processes are employed to remove any remaining impurities. A lot of washing of biodiesel is carried out with water to get rid of extra catalyst and other contaminants. The biodiesel is dried to remove any leftover water content after washing.

7.5.5 Quality Control

The procedures of quality control are carried out to guarantee that the biodiesel satisfies the standard parameters necessary for fuel. Testing for characteristics like viscosity, acidity, flash point, cloud point, pour point, cetane number and oxidation stability may be necessary for this. It's important to note that the sustainability of biodiesel production depends on the type of feedstock used. While biodiesel from waste oils and fats is generally considered more environmentally friendly, the use of certain feedstocks like palm oil has raised concerns about deforestation and habitat destruction [49]. Additionally, ongoing research is exploring advanced techniques and alternative feedstocks, such as algae and certain types of bacteria, to increase the production of biodiesel's effectiveness and environmental impact.

7.6 Syngas from Biomass

Syngas, or synthesis gas, can be produced from biomass through a process called biomass gasification. Thermochemical conversion of biomass into a mixture of gases, mostly carbon monoxide (CO), methane (CH_4), and hydrogen (H_2), as well as other trace gases, is known as biomass gasification. Heat production, power generation and the synthesis of chemicals and fuels are just a few uses for this syngas [50]. Below is a general overview of the biomass gasification process.

7.6.1 Feedstock Preparation

Biomass feedstocks such as wood, agricultural residues, or energy crops are prepared and dried to reduce moisture content.

7.6.2 Gasification

The prepared biomass is subjected to high temperatures in a gasifier, usually in the presence of regulated air or oxygen levels, occasionally steam, or a mix of these. The lack of complete combustion distinguishes gasification from traditional combustion processes [51].

7.6.3 Syngas Formation

The complex organic components in biomass are broken down into simpler gases by the high temperature, mainly carbon monoxide, hydrogen, and methane. The variety of biomass, the gasification technique and the gasifier design are some of the variables that affect the composition of syngas.

7.6.4 Gas Cleanup

The raw syngas may contain impurities such as tars, particulates, and trace contaminants. Gas cleanup processes are employed to remove these impurities and ensure the syngas meets the specifications required for the intended application.

7.6.5 Syngas Utilization

The cleaned syngas can be utilized for different purposes, including power generation, heat production and chemical synthesis. Syngas can be used in internal combustion engines or gas turbines to generate electricity. Syngas can be burned for heat production in industrial processes or for district heating. The syngas components (CO and H_2) can be used as building blocks for the synthesis of chemicals and fuels. Biomass gasification has both environmental and energy-related benefits. It allows for the utilization of green and renewable biomass resources to produce a versatile energy carrier (syngas) that can replace or supplement traditional or conventional fossil fuels. It's important to note that various technologies and gasification methods exist, each with its own advantages and challenges. The choice of technology depends on factors such as the type of biomass feedstock, desired syngas composition, and the specific application requirements [52].

7.7 Biohydrogen from Biomass

Biohydrogen production from biomass involves the conversion of organic materials, such as harvesting/agricultural residues, organic waste or energy plants/crops, into hydrogen gas through biological processes. There are several pathways for biohydrogen production, and among them, dark fermentation and photo fermentation are two commonly studied methods.

The fundamental chemical process that produces hydrogen is described as follows [53].

$$2H+ + 2e- \leftrightarrow H_2 \tag{7.16}$$

7.7.1 Dark Fermentation

Dark fermentation is a biological process that occurs in the absence of light. It involves the anaerobic (absence of oxygen) breakdown of organic compounds by certain bacteria to produce hydrogen gas [54]. This approach is unusual in that it doesn't require light for fermentation, has rapid cell development, and can be economically implemented because garbage may be utilized as a substrate [55]. Various types of bacteria, such as Clostridium and Enterobacter species, are capable of dark fermentation. These bacteria typically ferment sugars and other organic compounds to produce hydrogen, organic acids, and carbon dioxide. The pH range, temperature, type and composition of the substrate, organic load, and hydraulic retention duration all affect this fermentation process [56, 57]. The growth of microorganisms and modifications in metabolism are caused by pH, which is a crucial component in fermentation. This shift has been noted by numerous scientists to occur when the pH falls to 4.5 or below. The reaction's production of volatile fatty acids is also to blame for the pH shift. The microbial population and the rate of H_2 (Hydrogen) production rely on the pH range being maintained [58]. Depending on the type of microbe, different temperature ranges have varying significance. These temperature ranges are further separated into categories based on mesophilic, ambient, thermophilic, hyperthermophilic, and severe thermophilic values, which are 15–30°C, 30-49°C, 50–64°C, and > 80°C, respectively [59]. The organic materials used as feedstock for dark fermentation can include sugars, starches, and complex organic substances found in biomass.

7.7.2 Photofermentation

Photofermentation is another biological process, but it requires the presence of light. Certain photosynthetic bacteria, such as purple non-sulfur bacteria (e.g., *Rhodobacter* spp.), can utilize organic compounds to produce hydrogen in the presence of light [54]. In photofermentation, organic substrates like carbohydrates are oxidized to produce hydrogen, carbon dioxide, and other byproducts. Light energy is captured by pigments in the bacteria, and this energy is used to drive the hydrogen production process. The reactions of hydrogen production can be given as follows [53].

$$C_6H_{12}O_6 + 12H_2O + \text{light energy} \rightarrow 12H_2 + 2CO_2 \tag{7.17}$$

$$CH_3COOH + 2H_2O + \text{light energy} \rightarrow 4H_2 + 2CO_2 \tag{7.18}$$

$$COOHCH_2CHOHCOOH + 3H_2O + \text{light energy} \rightarrow 6H_2 + 4CO_2 \tag{7.19}$$

$$CH_3CH_2COOH + 4H_2O + \text{light energy} \rightarrow 7H_2 + 3CO_2 \tag{7.20}$$

$$CH_3CH_2CH_2COOH + 6H_2O + \text{light energy} \rightarrow 10H_2 + 4CO_2 \tag{7.21}$$

Dark fermentation is simpler and less expensive than photofermentation since photon energy processing devices are not required [60]. Both dark fermentation and photofermentation have their advantages and challenges. Dark fermentation is often faster and can use a wider range of feedstocks, but it is less efficient in terms of hydrogen production compared to photofermentation. Photofermentation, on the other hand, is more efficient in terms of hydrogen yield but is generally more sensitive to environmental conditions. Key steps in the biohydrogen production process are described below.

7.7.2.1 Feedstock Preparation

Biomass feedstocks are prepared and converted into suitable substrates for the microorganisms involved in biohydrogen production.

7.7.2.2 Inoculation

The selected microorganisms, typically hydrogen-producing bacteria, are introduced into the reactor containing the biomass substrate [61].

7.7.2.3 Fermentation/Photofermentation

The bacteria break down the organic compounds in the biomass substrate, producing hydrogen gas as a by-product. This process occurs either in the absence of light (dark fermentation) or in the presence of light (photofermentation).

7.7.2.4 Harvesting

The produced hydrogen gas needs to be separated and collected for further use. Various methods, such as membrane separation or gas stripping, can be employed for this purpose [61]. Biohydrogen production from biomass is considered a promising and environmentally friendly way to generate hydrogen, as it utilizes renewable resources and produces a clean fuel. However, challenges such as low yields, reactor design, and substrate availability still must be taken care of in order to make biohydrogen production financially feasible on a large scale. Researchers are actively working on improving the efficiency and scalability of biohydrogen production processes.

7.8 Biorefinery Approaches

Biorefinery approaches in biomass for energy production represent a pivotal strategy in the quest for sustainable and renewable energy sources. By integrating various conversion processes, biorefineries strive to maximize the utilization of biomass feedstocks, including agricultural residues, energy crops, and forestry waste, to produce a spectrum of bio-based products and energy. Biochemical, thermochemical, and hybrid processes are key categories in biorefinery approaches [62]. In biochemical biorefineries, microorganisms play a central role in converting biomass into biofuels or chemicals through fermentation and enzymatic processes. This approach is exemplified by the production of bioethanol from sugars derived through enzymatic hydrolysis of cellulose and hemicellulose. Thermochemical biorefineries, on the other hand, leverage high-temperature processes like gasification and pyrolysis to convert biomass into syngas or bio-oil, which can be used for power generation or further refined into biofuels [63]. Hybrid biorefineries combine both biochemical and thermochemical processes, aiming to capitalize on the synergies between them. In addition to increasing overall efficiency, this integration makes it possible to produce a wide range of goods, such as minerals, chemicals, and biofuels. Such hybridization increases the biorefinery's economic viability and permits a more thorough exploitation of the biomass feedstock. These biorefinery approaches not only address energy production needs but also align with broader goals of sustainability. Through a focus on circular resource utilization, biorefineries reduce waste and encourage a greener, more resilient energy industry. Technology must continue to progress in order to optimize biorefinery operations and enhance the integration of biomass into our energy landscape, which will promote the shift to a more diverse and sustainable energy future. Policies and incentives must also be in place to assist this process [63].

7.9 Economical and Environmental Aspects

The development and operation of biorefineries involve various economic and environmental aspects that are crucial for evaluating their sustainability and overall impact. Biomass for energy production is characterized by a dual impact on both economic and environmental aspects, making it a key player in the transition towards sustainable and renewable energy sources. From an economic perspective, biomass presents opportunities for rural development and job creation, particularly in regions where feedstocks like agricultural residues and energy crops are abundant. The establishment of biomass supply chains can stimulate local economies, providing income for farmers and supporting businesses involved in the biomass sector. Additionally, biomass is often considered a cost-competitive option for energy production, especially in comparison to certain renewable technologies [64].

On the environmental front, biomass for energy production offers an assuring possible replacement to fossil fuels, contributing to the reduction of greenhouse gas (GHG) emissions and mitigating climate change. Biomass is considered carbon-neutral, as the carbon dioxide released during combustion is offset by the carbon dioxide absorbed during the growth of the biomass feedstock. This cycle creates a closed loop, making biomass an environmentally sustainable energy source. Moreover, the utilization of biomass helps in the management of organic waste, reducing the environmental impact of landfill disposal. However, it is crucial to acknowledge the complexities and trade-offs associated with biomass for energy production. Sustainable sourcing and management of biomass feedstocks are imperative to prevent negative impacts on biodiversity, land use, and food production. The energy conversion processes, such as combustion or gasification, need to be optimized to minimize air pollutants and maximize energy efficiency. In conclusion, the economic and environmental aspects of biomass for energy production are intertwined, and a holistic approach is essential to ensure its long-term sustainability. Advancements in technology, coupled with supportive policies and efficient supply chain management, can enhance the positive economic and environmental contributions of biomass, making it a vital component of a diversified and sustainable energy portfolio [65]. Here's an overview of these aspects.

7.9.1 Economic Aspects

7.9.1.1 Cost of Feedstock

The cost of feedstock in biorefinery processes is impacted by different components, counting biomass type and accessibility, transportation costs, regularity, and market demand. Biomass feedstocks sourced from rural residues, forestry buildups, and organic waste streams may change in cost depending on variables such as crop yields, land accessibility, and harvesting strategies. Moreover, transportation costs play a critical part, particularly for biomass sources found far from biorefinery facilities [66]. Biorefineries utilize a few procedures to optimize feedstock costs and progress economic productivity. One approach is to expand feedstock sources to moderate supply chain dangers and decrease reliance on a single biomass sort. By utilizing locally accessible biomass assets and building up strategic organizations with biomass suppliers, biorefineries can minimize transportation costs and ensure a steady supply of feedstock. The availability and cost of biomass feedstocks significantly impact the economic viability of a biorefinery or biomass processing operation. Access to affordable and sustainable feedstocks is essential for maintaining competitiveness [67].

7.9.1.2 Technological Costs

The initial capital investment and ongoing operational costs of biorefinery technologies influence their economic feasibility. Advances in technology and process optimization can help reduce these costs over time.

7.9.1.3 *Product Market Value*

The market value of bio-based products, such as biofuels, chemicals, and materials, plays a crucial role. The demand and pricing of these products can affect the overall profitability of a biorefinery [64].

7.9.1.4 *Policy Support and Incentives*

Government policies, subsidies, and incentives can significantly impact the economic feasibility of biorefineries. Supportive policies, such as renewable energy mandates and tax credits, can enhance the attractiveness of bio-based products [68].

7.9.1.5 *Scale of Operation*

The scale of a biorefinery operation can influence its economics. Larger-scale facilities may benefit from economies of scale, potentially reducing production costs [69].

7.9.1.6 *Market Competitiveness*

Biorefineries must be competitive with traditional fossil-based industries. Continuous improvement in efficiency, technology, and cost-effectiveness is necessary for long-term success [65].

7.9.2 Environmental Aspects

7.9.2.1 *Greenhouse Gas Emissions*

Biorefineries are often considered environmentally friendly due to the potential to bring down greenhouse gas (GHG) emissions compared to traditional fossil-based processes. When biomass wastes are utilized for energy generation, the carbon dioxide (CO_2) radiated amid combustion is considered constituent of the natural carbon cycle, because it is reabsorbed by plants through photosynthesis. In contrast, fossil fuels discharge CO_2 (Carbon dioxide) that has been sequestered for many years (likely millions of years), playing a key role to the accumulation of greenhouse gases within the nature and compounding climate change. By uprooting fossil fuels with biomass-derived vitality, biomass-to-energy projects can offer assistance relieve climate alter by diminishing net CO_2 emissions [70]. However, in general, the environmental impact relies on some other factors too, such as how and where it's made — feedstock production, processing and transportation [70].

7.9.2.2 *Waste Diversion from Landfills*

Although landfilling causes some environmental impacts, counting greenhouse gas (GHG) emissions and leachate generation, it gives a basic and ethical means of waste transfer to energy. By introducing the gas recuperation frameworks, landfills give a possible and potential origin of methane (CH_4) as a fuel source. Progressively revegetation is executed on conventionally overseen landfill plots to moderate degradation of environment, which too gives a sufficient source of biomass wastes for generation of energy [71].

The combustion or burning of landfill gas for energy generation contributes to the decrement of greenhouse gases (GHG) emission primarily by anticipating the discharge of methane (CH_4) into the environment. Biomass from landfill areas can be changed over to bioenergy through different forms counting liquefaction, pyrolysis and gasification. It impacts the environmental sustainability of biorefineries significantly [72].

7.9.2.3 *Land Use and Biodiversity*

The choice of biomass feedstock and the scale of cultivation can impact land use and biodiversity. Land-use alter could be an essential cause of biodiversity loss, influencing key biological system forms such as the generation of living organisms, supplement and hydrobiological cycling as well as soil arrangement and maintenance. The loss of a fundamental species can disturb biological systems for a long time, and interchanges within the interactions among species can, moreover, lead to dismissive impacts on biological system forms. Changes in land utilization and land cover that influence biodiversity can in turn influence nature/climate, and a few components of biological systems/biodiversity influence carbon sequestration and are critical in fighting climate alteration. Sustainable practices that minimize land-use change and preserve biodiversity are essential for the environmental sustainability of biorefineries [73].

7.9.2.4 *Water Usage and Quality*

The water footprint of biorefinery processes, including cultivation and processing, should be considered. The renewable biofuel production depends on many factors, including feedstock availability, refinery and shipment infrastructure, and water but the availability of water, its utilization and waste water management play an important role. A normal biorefining supply chain includes significant impact on the regional water assets. Undoubtedly, a complex relationship exists between the generation of biomass, the preparing of biomass into fuels and chemicals, the utilization of new water, and the release of waste water. Efficient water use and the prevention of water pollution are critical for environmental stewardship [74].

7.9.2.5 *Waste Management*

Biorefineries are essential to the shift towards a sustainable bio-based economy by converting biomass feedstocks into a variety of valuable products such as biofuels, biochemicals, and bioplastics through incorporation of relevant conversion technologies. However, the production processes in biorefineries also generate various by-products and waste streams, presenting environmental challenges that need to be addressed [75]. It is additionally found that the environmental effect of bioconversion techniques that yield valued bioproducts is much lower than that of traditional techniques. Including bioconversion procedures increases process efficiency and recovers resources in a sustainable manner [76]. This can be achieved through process optimization, efficiency improvements, and recycling of process intermediates. By adopting cleaner production techniques and maximizing the utilization of biomass feedstocks, biorefineries can reduce the volume and toxicity of waste generated, thereby minimizing their environmental footprint. Many of the by-products generated in biorefinery processes have the potential to be valorized into valuable products or feedstocks for other industrial processes. For example, lignin that is a by-product of biomass pretreatment can be converted into high-value chemicals, materials, or bio-based adhesives. For residual waste streams that cannot be valorized or recycled, sustainable disposal options must be considered to minimize environmental impacts.

7.9.2.6 *Energy Efficiency*

One key aspect of biorefinery operations that significantly impacts the environment is energy efficiency. Energy efficiency measures in biorefineries contribute to the reduction of greenhouse gas emissions by minimizing the energy intensity of production processes. By optimizing energy use, improving process efficiency and implementing energy-saving technologies such as cogeneration and waste heat recovery, biorefineries can significantly decrease their dependence on fossil fuels and lower their carbon footprint.

This results in lower emissions of greenhouse gases such as carbon dioxide (CO_2), methane (CH_4), and nitrous oxide (N_2O), thus mitigating climate change and its associated environmental impacts. Energy efficiency in biorefineries also promotes the conservation of natural resources by reducing the consumption of fossil fuels and other non-renewable energy sources. By utilizing renewable biomass feedstocks and maximizing the efficiency of energy conversion processes, biorefineries minimize the depletion of finite resources and promote the sustainable use of natural resources. Improved energy efficiency in biorefineries leads to a reduction in the overall environmental footprint of bio-based production processes. By optimizing resource use, minimizing waste generation, and reducing energy-related emissions, biorefineries can lower their environmental impact across various environmental indicators, including water consumption, land use, air pollution, and waste generation. In conclusion, energy efficiency in biorefineries plays a critical role in mitigating environmental impacts and promoting sustainable development. By reducing greenhouse gas emissions, conserving natural resources, and minimizing environmental footprint, energy-efficient biorefineries contribute to the transition towards a low-carbon, resource-efficient economy.

7.9.2.7 Life Cycle Assessment (LCA)

LCA (Life Cycle Assessment) is a system that determine the comprehensive environmental/climate impacts of a product produced from biorefineries, framework, or benefit over its lifetime. LCA can be connected to a wide extend of products and advances, from the apparently straightforward framework of creating a paper pack to the large-scale generation of electricity. LCA also evaluates the utilization of resources such as water, land, and nutrients in biorefinery processes. By assessing resource inputs and outputs throughout the life cycle of bio-based products, LCA identifies opportunities to optimize resource utilization and minimize environmental impacts. For example, water consumption in biomass cultivation, land use for feedstock production, and nutrient inputs for fermentation processes are important considerations in LCA studies of biorefineries. In conclusion, Life Cycle Assessment (LCA) plays an important role in assessing the environmental perspectives of biorefinery processes. By evaluating energy proficiency, asset utilization, greenhouse gas emissions, and other environmental indicators, LCA makes a difference in recognizing opportunities for enhancement and directing decision-making towards more economical biorefinery operations.

7.9.2.8 Sustainability Certifications

Sustainability certifications in biorefinery forms regularly emphasize asset productivity by advancing the proficient utilization of biomass feedstocks and minimizing asset utilization all through the generation chain. Certified biorefineries are required to execute measures to optimize asset utilization, diminish energy and water utilization, and minimize waste generation. Another key perspective of sustainability certifications in biorefinery forms is waste management. Certified biorefineries are required to execute successful waste management practices to play down the generation of waste by-products and confirm appropriate treatment and transfer of remaining waste streams. This may incorporate reusing, valorization or feasible transfer strategies such as composting or anaerobic absorption. By effectively overseeing waste, certified biorefineries moderate environmental contamination and decrease the burden on landfills and biological systems. In conclusion, sustainability certifications play an important role in ensuring that biorefinery forms follow the environmental guidelines and promote sustainable practices. By emphasizing resource proficiency, waste management, and the diminishment of greenhouse gas emissions, certified biorefineries contribute to environmental preservation and supportability.

Balancing economic and environmental considerations is essential for the long-term success of biorefineries. Sustainable practices, technological innovation, and supportive policies contribute to achieving both economic viability and positive environmental outcomes.

7.10 Conclusion

The valorization of biomass waste for vitality generation represents a pivotal strategy in the pursuit of sustainable and renewable energy sources. As the world grapples with the challenges of climate change and diminishing fossil fuel reserves, biomass emerges as a versatile and environmentally friendly alternative. Through processes like biomass gasification and biohydrogen production, we can harness the energy stored in organic materials to generate power, heat, and fuels. The economic and environmental benefits of biomass valorization are evident, as it offers a cleaner and more sustainable energy solution along with the contribution to the effective management of organic waste and the reduction of greenhouse gas emissions. As technology continues to advance and synergies between various biomass valorization approaches are explored, the potential for biomass to keep a central role in the adaptation to a greener and more resilient energy landscape becomes increasingly promising. By embracing biomass valorization, we pave the way for a future where renewable resources power our societies, fostering energy security and environmental well-being. The literature emphasizes the importance of integrating biomass valorization into a broader framework of renewable energy strategies, considering factors such as feedstock sustainability, economic viability, and the reduction of greenhouse gas (GHG) emissions. Additionally, there is a growing recognition of the need for holistic assessments, including life cycle analyses and environmental impact assessments, to ensure that biomass energy production aligns with the principles of sustainability. The current chapter collectively highlights the potential of biomass valorization as a key contributor to a more sustainable and resilient energy future, paving the way for a diversified, environmentally friendly, and economically viable energy landscape.

References

1. Clauser, N.M., González, G., Mendieta, C.M., Kruyeniski, J., Area, M.C., Vallejos, M.E., Biomass waste as sustainable raw material for energy and fuels. *Sustainability*, 13, 2, 794, 2021 Jan 15.
2. Agbor, V.B., Cicek, N., Sparling, R., Berlin, A., Levin, D.B., Biomass pretreatment: Fundamentals toward application. *Biotechnol. Adv.*, 29, 675–685, 2011.
3. Pishvaee, M.S., Mohseni, S., Bairamzadeh, S., An overview of biomass feedstocks for biofuel production, in: *Biomass to Biofuel Supply Chain Design and Planning under Uncertainty*, pp. 1–20, Elsevier, Amsterdam, The Netherlands, 2021.
4. Konstantinavičienė, J. and Vitunskienė, V., Definition and Classification of Potential of Forest Wood Biomass in Terms of Sustainable Development: A Review. *Sustainability*, 15, 12, 9311, 2023.
5. https://www.energy.gov/eere/bioenergy/biomass-resources.
6. Kumar, J.A., Sathish, S., Prabu, D., Renita, A.A., Saravanan, A., Deivayanai, V.C., Anish, M., Jayaprabakar, J., Baigenzhenov, O., Hosseini-Bandegharaei, A., Agricultural waste biomass for sustainable bioenergy production: Feedstock, characterization and pre-treatment methodologies. *Chemosphere*, 331, 138680, 2023 Aug 1.
7. Balaman, SY., Introduction to biomass—resources, production, harvesting, collection, and storage, in: *Decision-Making for Biomass-based Production Chains*, pp. 1–23, Elsevier, San Diego, 2019.
8. Fernando, A.L., Costa, J., Barbosa, B., Monti, A., Rettenmaier, N., Environmental impact assessment of perennial crops cultivation on marginal

soils in the Mediterranean Region. *Biomass Bioenergy*, 111, 174–86, 2018 Apr 1.

9. Hauk, S., Gandorfer, M., Wittkopf, S., Müller, U.K., Knoke, T., Ecological diversification is risk reducing and economically profitable–The case of biomass production with short rotation woody crops in south German land-use portfolios. *Biomass Bioenergy*, 98, 142–52, 2017 Mar 1.
10. Baruya, P., World forest and agricultural crop residue resources for cofiring, IEA Clean Coal Centre, London, UK, 2015 Apr.
11. Enters, T., *Asia-Pacific Forestry Commission*, vol. 16, Rap Publication, 2001.
12. Lindkvist, E., Resource-efficient treatment of organic industrial waste: Optimization of different treatment options using reMIND. *Resour. Conserv. Recycl.*, 197, 107065, 2023 Oct 1.
13. Kumar, A., Srivastava, N.K., Gera, P., Removal of color from pulp and paper mill wastewater-methods and techniques-A review. *J. Environ. Manage.*, 298, 113527, 2021 Nov 15.
14. https://www.eia.gov/energyexplained/biomass/waste-to-energy.php.
15. Lee, S.Y., Sankaran, R., Chew, K.W., Tan, C.H., Krishnamoorthy, R., Chu, D.T., Show, P.L., Waste to bioenergy: a review on the recent conversion technologies. *BMC Energy*, 1, 1, 1–22, 2019 Dec.
16. Shi, L., Han, L., Yang, F., Gao, L., The evolution of sustainable development theory: Types, goals, and research prospects. *Sustainability*, 11, 7158, 2019.
17. Sajid, M.J., Yu, Z., Rehman, S.A., The coal, petroleum, and gas embedded in the sectoral demand-and-supply Chain: Evidence from China. *Sustainability*, 14, 3, 1888, 2022 Feb 7.
18. Niu, X., Zhan, Z., Li, B., Chen, Z., Environmental governance and cleaner energy transition: Evaluating the role of environment friendly technologies. *Sustain. Energy Technol. Assess.*, 53, 102669, 2022 Oct 1.
19. Megia, P.J., Vizcaíno, A.J., Calles, J.A., Carrero, A., Hydrogen production technologies: from fossil fuels toward renewable sources. A mini review. *Energy Fuels*, 35, 20, 16403–15, 2021 Oct 7.
20. Wang, Y. and Wu, J.J., Thermochemical conversion of biomass: Potential future prospects. *Renew. Sustain. Energy Rev.*, 187, 113754, 2023 Nov 1.
21. Zhang, L., Xu, C.C., Champagne, P., Overview of recent advances in thermochemical conversion of biomass. *Energy Convers. Manage.*, 51, 5, 969–82, 2010 May 1.
22. Naqvi, S.R., Ullah, Z., Taqvi, S.A., Khan, M.N., Farooq, W., Mehran, M.T., Juchelková, D., Štěpanec, L., Applications of machine learning in thermochemical conversion of biomass-A review. *Fuel*, 332, 126055, 2023 Jan 15.
23. Wang, C., Wang, Z., Wang, X., Li, N., Tao, J., Zheng, W., Yan, B., Cui, X., Cheng, Z., Chen, G., A review on the hydrothermal treatment of food waste: processing and applications. *Processes*, 10, 11, 2439, 2022 Nov 17.

24. Chen, W.H., Huang, M.Y., Chang, J.S., Chen, C.Y., Torrefaction operation and optimization of microalga residue for energy densification and utilization. *Appl. Energy*, 154, 622–30, 2015 Sep 15.
25. Dai, L., Wang, Y., Liu, Y., Ruan, R., He, C., Yu, Z., Jiang, L., Zeng, Z., Tian, X., Integrated process of lignocellulosic biomass torrefaction and pyrolysis for upgrading bio-oil production: A state-of-the-art review. *Renew. Sustain. Energy Rev.*, 107, 20–36, 2019 Jun 1.
26. Chen, D., Gao, A., Ma, Z., Fei, D., Chang, Y., Shen, C., In-depth study of rice husk torrefaction: Characterization of solid, liquid and gaseous products, oxygen migration and energy yield. *Bioresour. Technol.*, 253, 148–53, 2018 Apr 1.
27. Ong, H.C., Chen, W.H., Farooq, A., Gan, Y.Y., Lee, K.T., Ashokkumar, V., Catalytic thermochemical conversion of biomass for biofuel production: A comprehensive review. *Renew. Sustain. Energy Rev.*, 113, 109266, 2019 Oct 1.
28. Lam, M.K., Loy, A.C., Yusup, S., Lee, K.T., Biohydrogen production from algae, in: *Biohydrogen*, pp. 219–245, Elsevier, 2019 Jan 1.
29. Alvira, P., Tomás-Pejó, E., Ballesteros, M., Negro, M.J., Pretreatment technologies for an efficient bioethanol production process based on enzymatic hydrolysis. *Bioresour. Technol.*, 101, 13, 4851–61, 2010.
30. Hatfield, J.L., Prueger, J.H., Sauer, T.J., Dold, C., O'Brien, P., Wacha, K., Applications of vegetative indices from remote sensing to agriculture: Past and future. *Inventions*, 4, 4, 71, 2019 Dec 6.
31. Haque, S., Singh, R., Pal, D.B., Harakeh, S., Alghanmi, M., Teklemariam, A.D., Abujamel, T.S., Srivastava, N., Gupta, V.K., Recent Update on anaerobic digestion of paddy straw for biogas production: Advancement, limitation and recommendations. *Environ. Res.*, 215, 114292, 2022 Dec 1.
32. Encinar, J.M., Pardal, A., Sánchez, N., Nogales, S., Biodiesel by transesterification of rapeseed oil using ultrasound: a kinetic study of base-catalysed reactions. *Energies*, 11, 9, 2229, 2018 Aug 25.
33. Tran, D.T., Chang, J.S., Lee, D.J., Recent insights into continuous-flow biodiesel production *via* catalytic and non-catalytic transesterification processes. *Appl. Energy*, 185, 376–409, 2017 Jan 1.
34. Chen, W.H., Lo, H.J., Aniza, R., Lin, B.J., Park, Y.K., Kwon, E.E., Sheen, H.K., Grafilo, L.A., Forecast of glucose production from biomass wet torrefaction using statistical approach along with multivariate adaptive regression splines, neural network and decision tree. *Appl. Energy*, 324, 119775, 2022 Oct 15.
35. Yao, R., Qi, B., Deng, S., Liu, N., Peng, S., Cui, Q., Use of surfactants in enzymatic hydrolysis of rice straw and lactic acid production from rice straw by simultaneous saccharification and fermentation. *BioResources*, 2, 3, 389–98, 2007 Aug 25.
36. Wang, L., Hanna, M.A., Weller, C.L., Jones, D.D., Technical and economical analyses of combined heat and power generation from distillers grains and corn stover in ethanol plants. *Energy Convers. Manage.*, 50, 7, 1704–13, 2009 Jul 1.

37. Keshwani, D.R. and Cheng, J.J., Switchgrass for bioethanol and other value-added applications: a review. *Bioresour. Technol.*, 100, 4, 1515–23, 2009 Feb 1.
38. Carvalho, L., Gomes, A., Aranda, D., Pereira, N., Ethanol from lignocellulosic residues of palm oil industry, in: *11th International Conference on Advanced Materials*, Rio de Janeiro, Brazil, 2009 Sep 20.
39. Zheng, Y., Pan, Z., Zhang, R., Overview of biomass pretreatment for cellulosic ethanol production. *Int. J. Agric. Biol. Eng.*, 2, 3, 51–68, 2009 Sep 30.
40. Lindstedt, J., Alcohol production from lignocellulosic feedstock.
41. Mousdale, D.M., *Biofuels: Biotechnology, Chemistry, and Sustainable Development*, CRC Press, 2008 Jun 4.
42. Plant, P., The DOE Bioethanol Pilot Plant.
43. Demirbas, A., The importance of bioethanol and biodiesel from biomass. *Energy Sources Part B*, 3, 2, 177–85, 2008 Mar 26.
44. Shelare, S.D., Belkhode, P.N., Nikam, K.C., Jathar, L.D., Shahapurkar, K., Soudagar, M.E., Veza, I., Khan, T.Y., Kalam, M.A., Nizami, A.S., Rehan, M., Biofuels for a sustainable future: Examining the role of nano-additives, economics, policy, internet of things, artificial intelligence and machine learning technology in biodiesel production. *Energy*, 128874, 2023 Aug 29.
45. Maroušek, J. and Maroušková, A., Economic considerations on nutrient utilization in wastewater management. *Energies*, 14, 12, 3468, 2021 Jun 11.
46. Aro, E.M., From first generation biofuels to advanced solar biofuels. *Ambio*, 45, Suppl 1, 24–31, 2016 Jan.
47. Mishra, B., Mohanta, Y.K., Reddy, C.N., Reddy, S.D., Mandal, S.K., Yadavalli, R., Sarma, H., Valorization of agro-industrial biowaste to biomaterials: An innovative circular bioeconomy approach. *Circ. Econ.*, 100050, 2023 Aug 19.
48. Balat, M., Bio-diesel from vegetable oils *via* transesterification in supercritical ethanol. *Energy Educ. Sci. Technol.*, 16, 1/2, 45, 2006.
49. Fukuda, H., Kondo, A., Noda, H., Biodiesel fuel production by transesterification of oils. *J. Biosci. Bioeng.*, 92, 5, 405–16, 2001 Jan 1.
50. Couto, N., Rouboa, A., Silva, V., Monteiro, E., Bouziane, K., Influence of the Biomass Gasification Processes on the Final Composition of Syngas. *Energy Procedia*, 36, 596–606, 2013, doi: 10.1016/j.egypro.2013.07.068.
51. Demirbaş, A., Hydrogen Production from Biomass by the Gasification Process. *J. Energy Sources*, 24, 59–68, 2002.
52. Skoulou, V.K. and Zabaniotou, A.A., Co-gasification of crude glycerol with lignocellulosic biomass for enhanced syngas production. *J. Anal. Appl. Pyrolysis*, 99, 110–116, 2013.
53. Madankar, C.S., Borde, P., Meshram, P.D., Studies on Extraction Techniques of Bio-Hydrogen, in: *Biofuel Extraction Techniques*, pp. 291–306, 2023 May 2.
54. Chong, M.L., Sabaratnam, V., Shirai, Y., Hassan, M.A., Biohydrogen production from biomass and industrial wastes by dark fermentation. *Int. J. Hydrogen Energy*, 34, 8, 3277–87, 2009 May 1.
55. Kumar, S., Sharma, S., Thakur, S., Mishra, T., Negi, P., Mishra, S., Hesham, A.E., Rastegari, A.A., Yadav, N., Yadav, A.N., Bioprospecting of microbes

for biohydrogen production: current status and future challenges, in: *Bioprocessing for Biomolecules Production*, pp. 443–71, 2019 Nov 18.
56. Gopalakrishnan, B., Khanna, N., Das, D., Dark-fermentative biohydrogen production, in: *Biohydrogen*, pp. 79–122, Elsevier, 2019 Jan 1.
57. Veeravalli, S.S., Shanmugam, S.R., Ray, S., Lalman, J.A., Biswas, N., Biohydrogen production from renewable resources, in: *Advanced Bioprocessing for Alternative Fuels, Biobased Chemicals, and Bioproducts*, pp. 289–312, Woodhead Publishing, 2019 Jan 1.
58. Pandit, S. and Das, D., Biohydrogen, January 2013, https://www.researchgate.net/publication/312977475_Biohydrogen_chapter_15, 2017.
59. Usman, T.M., Banu, J.R., Gunasekaran, M., Kumar, G., Biohydrogen production from industrial wastewater: an overview. *Bioresour. Technol. Rep.*, 7, 100287, 2019 Sep 1.
60. Dincer, I. (Ed.), *Comprehensive Energy Systems*, Elsevier, 2018 Feb 7.
61. Sinha, P. and Pandey, A., An evaluative report and challenges for fermentative biohydrogen production. *Int. J. Hydrogen Energy*, 36, 13, 7460–78, 2011 Jul 1.
62. Patel, A. and Shah, A.R., Integrated lignocellulosic biorefinery: Gateway for production of second generation ethanol and value added products. *J. Bioresour. Bioprod.*, 6, 2, 108–28, 2021 May 1.
63. Singh, Y.D., Mahanta, P., Bora, U., Comprehensive characterization of lignocellulosic biomass through proximate, ultimate and compositional analysis for bioenergy production. *Renew. Energy*, 103, 490–500, 2017 Apr 1.
64. Abbasi, T. and Abbasi, S.A., Biomass energy and the environmental impacts associated with its production and utilization. *Renew. Sustain. Energy Rev.*, 14, 3, 919–37, 2010 Apr 1.
65. Faaij, A.P. and Domac, J., Emerging international bio-energy markets and opportunities for socio-economic development. *Energy Sustain. Dev.*, 10, 1, 7–19, 2006 Mar 1.
66. Sadhukhan, J., Martinez-Hernandez, E., Amezcua-Allieri, M.A., Aburto, J., Economic and environmental impact evaluation of various biomass feedstock for bioethanol production and correlations to lignocellulosic composition. *Bioresour. Technol. Rep.*, 7, 100230, 2019 Sep 1.
67. Hughes, S.R. and Qureshi, N., Biomass for biorefining: Resources, allocation, utilization, and policies, in: *Biorefineries*, pp. 37–58, Elsevier, 2014 Jan 1.
68. Pourhashem, G., Hung, S.Y., Medlock, K.B., Masiello, C.A., Policy support for biochar: Review and recommendations. *GCB Bioenergy*, 11, 2, 364–80, 2019 Feb.
69. Zetterholm, J., Bryngemark, E., Ahlström, J., Söderholm, P., Harvey, S., Wetterlund, E., Economic evaluation of large-scale biorefinery deployment: A framework integrating dynamic biomass market and techno-economic models. *Sustainability*, 12, 17, 7126, 2020 Sep 1.
70. https://www.energy.gov/eere/bioenergy/articles/ethanol-vs-petroleum-based-fuel-carbon-emissions.

71. Seshadri, B., Bolan, N.S., Thangarajan, R., Jena, U., Das, K.C., Wang, H., Naidu, R., Biomass energy from revegetation of landfill sites, in: *Bioremediation and Bioeconomy*, pp. 99–109, Elsevier, 2016 Jan 1.
72. Bolan, N.S., Thangarajan, R., Seshadri, B., Jena, U., Das, K.C., Wang, H., Naidu, R., Landfills as a biorefinery to produce biomass and capture biogas. *Bioresour. Technol.*, 135, 578–87, 2013 May 1.
73. Abidin, I.S.Z., Haseeb, M., Azam, M., Islam, R., Foreign direct investment, financial development, international trade and energy consumption. *Int. J. Energy Econ. Policy*, 5, 3, 841–850, 2015b.
74. López-Díaz, D.C., Lira-Barragán, L.F., Rubio-Castro, E., Ponce-Ortega, J.M., El-Halwagi, M.M., Optimal location of biorefineries considering sustainable integration with the environment. *Renew. Energy*, 100, 65–77, 2017 Jan 1.
75. Ubando, A.T., Felix, C.B., Chen, W.H., Biorefineries in circular bioeconomy: A comprehensive review. *Bioresour. Technol.*, 299, 122585, 2020 Mar 1.
76. Ahmed, S.F., Kabir, M., Mehjabin, A., Oishi, F.T., Ahmed, S., Mannan, S., Mofijur, M., Almomani, F., Badruddin, I.A., Kamangar, S., Waste biorefinery to produce renewable energy: Bioconversion process and circular bioeconomy. *Energy Rep.*, 10, 3073–91, 2023 Nov 1.

8

Valorization of Biowastes for Energy Production Using Halophiles

Saloni Singh, Ayushi Goyal and Kakoli Dutt*

Dept. of Life Sciences, J.C. Bose University of Science and Technology, YMCA, Faridabad, Haryana, India

Abstract

Waste generation from various human activities, spanning domestic, industrial, agricultural, and livestock sectors poses a significant challenge. Both non-degradable elements like plastics, metals, and hazardous chemicals, as well as biodegradable waste, have exceeded natural recycling capacities. While incineration seems to be a straightforward solution, it leads to air quality deterioration. To address this, exploring alternate routes to convert biowaste into valuable resources is imperative. Leveraging the inherent energy in organic waste through biofuel production stands out as a promising solution. Among different microorganisms, halophiles, due to their diverse metabolic abilities encompassing fermentation, photolysis, and anaerobic digestion, offer a unique opportunity to convert lignocellulosic waste into bioethanol, biobutanol, biodiesel, biohydrogen, and biogas. However, their potential in biowaste transformation remains largely untapped. While biotransformation of waste to energy offers economic benefits by cutting waste disposal costs and generating valuable by-products, challenges do persist. Economic viability hinges on feedstock availability, technological efficiency, and market demand for biofuels and by-products. Overcoming these challenges necessitates meticulous planning, extensive research, and collaborative efforts to unleash the full economic potential of halophile-driven biowaste-to-energy projects.

***Keywords*:** Halophiles, lignocellulosic wastes, microbial intervention, energy storage, physico-chemical processes, biofuel, economic benefits, limitations

**Corresponding author*: duttkakoli79@gmail.com; kakolidutt@jcboseust.ac.in

Priyanka Singh, Pooja Agarwal and V. Vivekanand (eds.) Solid Waste Management: Challenges, Sustainability and Advancements, (251–276) © 2026 Scrivener Publishing LLC

8.1 Introduction

The evolution of organic life form on this planet was the beginning of generation of biowastes which through natural degradation mechanisms were eliminated, resulting in a balance between generation and degradation. This scenario continued till the start of the Renaissance period followed by the industrialization era characterized by massive and rapid expansion of human activities, growth, lifestyle, influences and impacts. These led to the generation of many by-products and waste products of variable chemical composition like metal, paper, plastic, organic, etc., leading to categorization of wastes into degradable and non-degradable, hazardous and non-hazardous, persistent and non-persistent, organic, metallic and radioactive (Figure 8.1). Currently, waste generation is significantly higher than waste removal by natural or artificial means. Although organic-based wastes have been considered as non-problematic, the sheer volume generated has started impacting the ecological balance by causing eutrophication in water bodies and soil deterioration, along with the release of methane, a strong greenhouse gas contributing to climate change [63].

The growing concerns for food security have led to massive agricultural developments and reshaping of the food processing sector. Along with it, horticultural sectors along with dairy and livestock for sourcing milk, meat, eggs and leather are equally developing worldwide. These developments have helped in reducing food concerns, generated jobs, enhanced the economy and improved quality of life, but on the flip side, the quantity of waste being generated as an outcome of these activities has slowly started becoming a major concern. Incidentally, organic biomass is considered as being biodegradable but the accumulating lignin-based biomass is among the most recalcitrant substrate on this planet with very few natural degraders being present [53]. This is the major reason that wood-based products are able to survive for decades if not centuries. It is extremely concerning that biowaste creation, which is sped up by urbanization, industry, and agriculture is expected to further expand considerably each year. These concerns make it imperative to devise strategies to tackle these wastes in an ecologically sustainable method as mere incineration of such wastes can lead to massive generation of greenhouse gases and decrease air quality. In the greener alternatives available, many microbial techniques are currently being promoted which convert these wastes to value. In this chapter, these techniques will be compared and analysed with special focus towards the use of halophiles in waste treatment [64].

8.2 Biowaste and Its Types

The biowaste generation is unavoidable and can be divided into rural, urban and natural with the large-scale production from large agricultural farms, urban centres and processing units (Figure 8.1). It is further fuelled by population growth and population migration and negatively impacts air, water and soil [65].

a) **Urban biowastes:** This category covers mostly biodegradable organic wastes produced by urban households, institutions, commercial entities, and processing sectors. Food scraps, yard clippings, paper, cardboard, and other biodegradable materials make up the majority of these wastes [48]. Because of their organic composition, urban biowastes are highly biodegradable under optimal conditions, distinguishing them from other waste categories. Among these, food waste and certain green waste variants (Table 8.1) present optimal biowaste options synergistic with halophiles [26]. This typically contains alcohols (like ethanol) and organic acids (e.g., acetic acid, lactic acid) which impart distinctive flavours to fermented foods and consequently are present in the resulting waste. Food waste, especially processed or fermented food waste (Kimchi factory, soy sauce, sauerkraut, etc.) with elevated salt content aligns well with the salt-tolerant nature of halophiles and encompasses a diverse range of components influenced by the specific food types and fermentation techniques employed. Similarly, specific green waste, like yard trimmings or plant residues from saline environments (such as coastal regions), may contain higher salt levels,

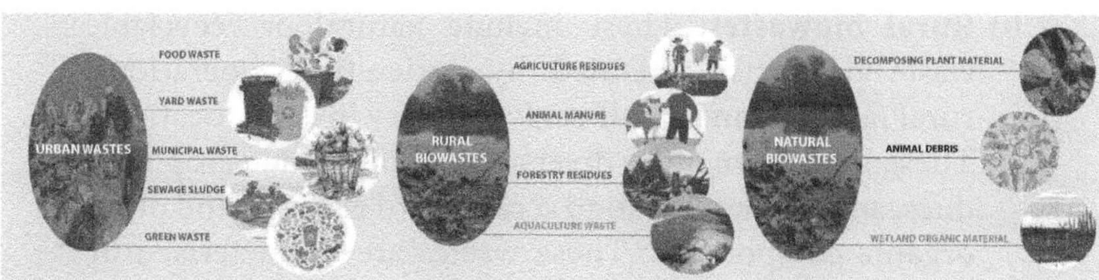

Figure 8.1 Categories of biowaste generated worldwide (sourced and modified from [48]).

Table 8.1 Biowaste variants synergistic with halophilic valorization.

Biowaste	Composition	Technique	Halophiles	Value-added product	References
Fermented food waste	Carbohydrates, proteins, and fats	Anaerobic Digestion, NaOH treatment	*Nesterenkonia* sp., *Clostridia sp.*	CH_4, CO_2 and VFAs, Protein, phenolic compounds, sugars	[66]
Agricultural residues	Cellulose, hemicellulose, lignin, protein	Anaerobic digestion, hydrolysis	*Pseudolateromonas phenolica, Micrococcus luteus, Pseudoalteromonas peptidolytica, Halomonas socia, Marinobacter maritimus, Exiguobacterium aurantiacum*	Bioethanol, enzymes, biofertilizers, biogas	[67, 68]
Sea food waste	Chitin, protein, lipids, chitosin, Carotenoid pigments, Collagen, gelatin	Enzymatic hydrolysis	*Lactobacillus plantarum*	Pharmaceuticals, Fertilizers	[69]

making them suitable for halophile-driven biofuel production. Halophiles can utilize these optimal biowaste variants by employing their diverse metabolic pathways and specialized enzymes in saline conditions. These microorganisms break down organic matter within the biowaste, and halophilic enzymes play a crucial role in converting these materials into biofuels through direct or indirect strategies.

b) **Rural biowastes:** These include natural or recyclable waste products generated by agricultural operations, rural families, and small-scale industry in rural locations. These wastes are often diverse and can include crop residues, animal manure, food scraps, yard garbage, and other organic components. These wastes are characterised by a higher number of agricultural leftovers such as straw, husks, and stalks [46]. Due to overuse or continuous use of fertilizers and growth promoters, there is generation of salinity in the soil which may impact the final agriculture residue too [56]. Cyanobacteria and microalgae can offer specific remediation of such soil. Such soil often reduces plant growth [4].

c) **Natural biowastes:** These are organic waste materials originating from unexploited environments and includes forests, oceans and other wildlife habitats. Biowastes here include fallen leaves, dead plant material, animal carcasses, and other organic debris found in ecosystems [1]. These materials undergo natural decomposition forming a part of the detritivore's food chain. While these natural biowastes contribute to nutrient cycling and the health of ecosystems, their excessive accumulation can sometimes affect local environments [59].

8.3 Conversion to Value-Added Products

According to a recent analysis by Huang *et al.* [70], several businesses such as food, agricultural, urban and municipal, and starch sectors collectively generate an estimated 1.3 billion tonnes of organic waste each year. It can turn into an economic opportunity with ecological and energy benefits when they are transformed into value-added products [73]. Approximately 10–14% of the world's energy production is now expected to come from biomass, which should be increased [71]. Utilised as a source of carbon and energy by various microbes, organic compounds found in industrial by-products, municipal pollutants, and agricultural waste can be converted into valuable goods [72, 73]. This makes microbial biotransformation a very potent biotechnological tool for waste conversion.

The three steps of the bio-transformation approach are as follows: the first is the composition-based selection of the organic waste biomass. Applying suitable pretreatments determined by an accurate analysis of the organic waste's content can improve the efficiency of biomass conversion. Scientists can employ metabolic engineering to gain a better understanding of microbe metabolic processes, which in turn allows them to use gene editing to increase end productivity [74]. The second factor that increases bioprocess productivity is midstream regulation. Optimisation of the fermentation process is a tried-and-true technique for boosting the concentration of products obtained from the conversion of organic waste biomass. To increase the rate of organic waste conversion, it is essential to consider the fermenter configuration and transformation parameters. Separating products from the conversion system is best accomplished using an efficient downstream technique. Recovering resources from these biological streams is one way to alleviate resource constraints through the implementation of a circular economy approach [75].

8.4 Biowastes as Energy Reservoir

The biomass or organic waste generated contains energy in the form of various chemical compounds. The chemical composition of these biowastes makes them fall predominantly in the category of carbohydrates, proteins or lipids. These can be utilized by various microorganisms for various catabolic and anabolic processes. Upon utilizing these substrates through fermentation for energy production, the balancing of reductive and oxidative processes generally lead to production of various alcoholic or acidic by-products. Additionally, biogas production can also be observed. As an outcome of fermentation, aerobic or anaerobic respiration, these wastes are converted into myriad of products. Various enzymes break down the polymers and the monomeric into glycolytic, sugar and sugar molecules.

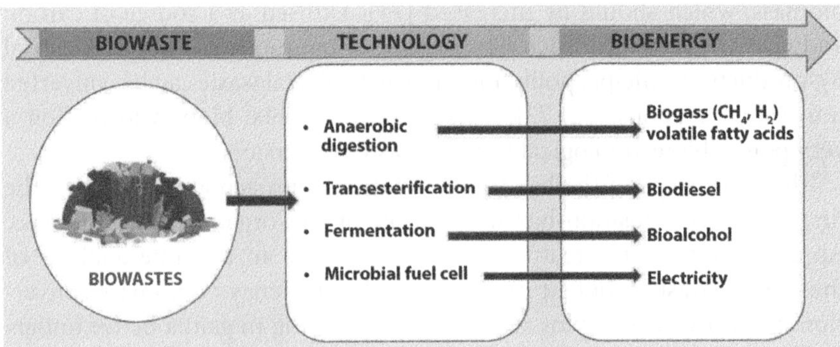

Figure 8.2 Bioenergy production from biowastes (sourced and modified from [76]).

Through substrate level and oxidative level phosphorylation, there is generation of ATP or formation of energy-rich compounds. Several researchers [36, 77–79] have revealed a variety of fuels generated from biowastes, both directly and *via* enzyme-mediated methods (Figure 8.2).

8.4.1 Direct Conversion Strategies

Organic wastes can be converted into biofuels using myriad strategies which include anaerobic digestion, fermentation and microbial fuel cell.

a) Anaerobic digestion

In anaerobic digestion (AD), a community of microbes breaks down organic matter without the presence of oxygen [5, 36]. This method is employed for the maintenance of biological waste and the production of biogas, which mostly consists of methane, carbon dioxide, and small quantities of other gases [36]. Multiple microbial groups carry out the sequential reactions that constitute AD, which include hydrolysis, acidogenesis, acetogenesis, and methanation (Figure 8.3). Anaerobic digestion begins with the production of hydrolases by halophiles such as *Pseudoalteromonas peptidolytica, Halomonas socia, Micrococcus luteus, Exiguobacterium aurantiacum* strain 2, *Marinobacter maritimus*, and *Pseudoalteromonas phenolica* [80]. Essa *et al.* [81] found that acidogenic marine microalgal species, such as *Nannochloropsis oculata* and *Tetraselmis chuii*, ferment these molecules even further, resulting in a variety of volatile fatty acids and gaseous components, such as CO_2. The acetogenic halophiles *Marinobacter hydrocarbonoclasticus, Marinobacter Pelagius*, and *Marinobacter daepoensis* further convert these components into acetic acid [41, 44]. The intermediate products of the various stages are fermented into methane, CO_2, and water by methanogenic halophiles (*Methanohalophilus* and *Methanohalobium spp.*) [82]. The AD method is applicable to any kind of biological waste, including that from the food, agro-industry, animals, and municipal wastes [22, 49, 62].

b) Transesterification

The process of transesterification involves the conversion of oils derived from organic materials into biodiesel. Sun *et al.* [84] examined the progress in technology for producing microbial biodiesel from volatile fatty acids (VFAs) produced from industrial organic wastes. Their summary provided

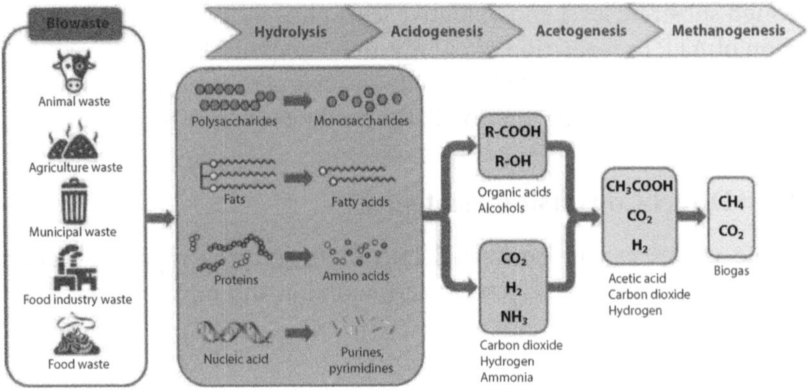

Figure 8.3 Different phases of anaerobic digestion for biogas production (sourced and modified from [83]).

an overview of the present state of various critical technological processes involved in the production of microbial biodiesel. These processes include acidogenic fermentation of biowastes to collect volatile fatty acids, lipid accumulation in oleaginous microbes, microbial lipid extraction, transesterification of microbial lipids into crude biodiesel, and crude biodiesel purification as shown in Figure 8.4. To create bio-oil, Hariram *et al.* [85] used the hypersaline, unicellular greenish-orange micro-alga *Dunaliella salina*. The microalgae were grown in an environment high in carbon dioxide using F/2 nutrition media, as well as a trace metal and vitamin solution. This bio-oil was extracted using an ultrasonic technique operating at 60 Hz for a duration of 90 minutes. The yield was 650 mL. An extraction efficiency of 87% was achieved in a single-stage based-catalyzed transesterification process using methanol and sodium hydroxide, which resulted in 380 mL of pure *Dunaliella salina* biodiesel. In order to explore the potential of cultivating *Dunaliella sp.* for use as internal combustion (IC) engines [86] phytochemical screening was carried out on the organisms. A sustainable approach was used to synthesise novel carbon-based solid acid catalysts for biodiesel generation from lipid-extracted microalgal residue of *Dunaliella tertiolecta.*

According to Salam *et al.* [54], Skorupskaite *et al.* [57] and Akubude *et al.* [2], microbial lipids can be further transformed into biodiesel through direct or indirect transesterification. Biodiesel manufacturing is an established technology, but producing biodiesel from microbial lipids is an emerging area of study that has grown rapidly in the past decade.

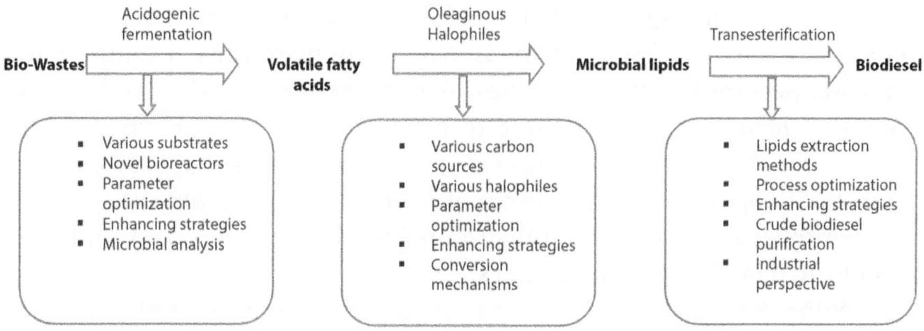

Figure 8.4 Flow chart of biodiesel production based on microbial lipids extracted from oleaginous halophiles grown with biowastes [84].

c) Fermentation

Alcohols are chemically important solvents used in various processes and fuel blends [10, 32, 34, 58]. According to Bhatia *et al.* [11], most of the alcohol is made from food crops rich in starch, which could compromise food security. According to the World Ethanol Research and Forecasting Association [87], the United States accounts for 54% of global ethanol output, while Brazil accounts for 26%. The sugarcane in Brazil and the maize in the United States are the primary sources of glucose, which is the substrate used to support large-scale ethanol biofuel production. Corn, starch, and sugar are all ingredients in ethanol, but their feed and food value make them unappealing for current consumption. The cost is also a major consideration for expanding bioethanol production on a wide scale. The current food-versus-fuel dilemma, which arises from bio-ethanol production that is dependent on grain, can be sidestepped by using green gold petroleum derived from lignocellulosic waste [88]. Lignocellulosic biomass has the potential to produce 442 billion litres of bio-ethanol per year, according to Kim and Dale [38]. The combined yield of crop residues and wasted crops might reach 491 billion litres of bio-ethanol per annum, which is approximately sixteen times greater than the current global bio-ethanol production. Cellulosic materials are abundant, inexpensive, and a renewable resource including agro-waste, grass, sawdust, wood chips, and crop residues. For the last 20 years, a lot of researchers have been trying to figure out cost-efficient production of ethanol from lignocellulosics with significant achievements [12, 14, 15, 23, 25, 47, 60, 89]. Therefore, making bioethanol could be the key to finding a good use for all that leftover food and yard trash. When it comes to biomass, the most abundant agro-residues include rice straw, wheat straw, corn straw, cotton seed hair, paper, banana stem, pineapple leaf, oil palm fronds, sugarcane bagasse and Jatropha waste [38]. This enzyme is involved in hydrolysis of cellulose to glucose. Several halophilic species, *Aspergillus terreus, Thalassobacillus sp., Marinobacter sp.* are capable of producing cellulase enzyme. After this, fermentation process is required for conversion of glucose to ethanol by halophiles, e.g., *Haloferax volcanii, Aspergillus terreus* [3, 28]. These processes are required for sustainable bioethenol production from cellulosic material of agro-residues.

Another potential biofuel is butanol, which, at 85% concentration, can be used directly to power an unmodified petrol engine [6]. *Nesterenkonia sp.* strain F, a moderately halophilic bacterium isolated from Aran-Bidgol Lake (Iran), can also produce ethanol, butanol, and acetone [90]. However, after ethanol, butanol is the most important product of fermentation. Butanol can be made from a variety of biowastes [30]. Agricultural waste requires pretreatment with a variety of physical and chemical processes due to its recalcitrance. Two agricultural by-products, sugarcane bagasse and rice straw, were alkaline-treated before cellulase hydrolysis [20]. The hydrolysate was fermented with *Clostridia sp.* to produce butanol at concentrations of 2.93 g/L from rice straw and 1.95 g/L from bagasse, respectively. According to Joshi *et al.* [35], 19.5 g/L of butanol was produced from orange peel hydrolysed by steam explosion. The hydrolysate was then fermented using *Clostridium acetobutylicum* NCIM 2877. There is currently no commercially viable bioprocess for producing isobutanol, 1,3 propandiol, or 2,3 butanediol from waste by fermentation; however, this could change in the near future due to filling up of the existing technological gap [11].

Priya *et al.* [91] explored the development of a novel fed-batch fermentation process for PHBV (poly(3-hydroxybutyrate-*co*-3-hydroxyvalerate) production by *H. mediterranei* from food waste with pulse-fed levulinic acid as an inducer. The use of food waste as a cheap and renewable feedstock in the production of PHBV has economic and environmental advantages that could assist the biopolymer and waste management industry in converting waste to wealth. According to Priya *et al.* [91], *H. mediterranei* was cultivated for PHBV production in fed-batch bioreactor with pulse fed levulinic acid. The maximum biomass of 3.19 ± 0.66 g/L was achieved after 140 h of cultivation with 3 g/L of levulinic acid. A decrease in *H. mediterranei* growth was noticed with the increase in levulinic acid concentration in the range of 3–10 g/L. The overall yields of PHBV at 3, 5, 7 and 10 g/L of levulinic acid were 18.23%, 56.70%, 31.54%, 21.29%, respectively. The optimum concentration of 5 g/L of levulinic acid was found to produce the maximum yield of 56.70% PHBV with 18.55 mol% 3HV content. A correlation between levulinic acid concentrations and PHBV production established in this study can serve as an important reference for future large-scale production.

d) Microbial fuel cells

Microbial fuel cells (MFC) are a developing environmental-friendly technology that has numerous advantages in comparison to conventional methods of using biomass. This technique is designed to process wet organic waste by purifying it and converting its chemical energy into electricity. The generated electricity is then utilised to connected peripherals and specific applications [33]. Electricity production from biowaste using microbial fuel cells (MFCs) starts with introducing the biowaste into the system. Microorganisms within the anodic chamber of the MFC metabolise the organic matters in the biowaste, resulting in the release of electrons. The liberated electrons are collected by the anode, producing an electrical current that flows through an external circuit. Meanwhile, protons are generated through microbial metabolism and traverse an electrolyte or membrane to reach the cathodic chamber. Within the cathodic chamber, an oxidising substance, often oxygen, reacts with electrons and protons, leading to the generation of water or other non-hazardous by-products. The movement of electrons from the anode to the cathode in the external circuit produces electrical energy, which can be utilised as a power source.

The utilisation of halophilic bacteria for the biological treatment of high saline wastewater is a highly effective approach for addressing the issue of salty waste treatment [92, 93]. Exoelectrogenic properties of halophilic bacteria (*Bacillus sp., Staphylococcus sp., Bacillus sp., Virgibacillus sp., and Staphylococcus sp.*) and saline water treatment have been successfully demonstrated in studies [51, 94–98]. At a concentration of 1200 mM NaCl, Paul *et al.* [51] utilised a haloalkaliphilic bacteria (*Halanaerobium hydrogeniformans*) from soda lake to reach a power density of 1.6 mW/m2. Grattieri *et al.* [98] reported that halotolerant bacteria from the Great Salt Lake, USA, were used to reach 36 mW/m2 of power density at 35 g/l NaCl (599 mM). To date, the maximum salt concentration measured in the MFC is 250 g/l (4277.89 mM). The bacteria recovered from the saline pond was used to produce power densities of 70.37 mW/m2 and 5.47 mW/m2 at NaCl concentrations of 175 g/l (2994.52 mM) and 250 g/l, respectively.

According to Vijay *et al.* [97], halophilic bacteria that break down starch (*Bacillus sp.* strain SLA-01, *Staphylococcus sp.* strain SLA-02, *Bacillus sp.* strain SLA-03, *Virgibacillus sp.* strain SLA-04, *Staphylococcus sp.* strain SLA-05) were used to develop MFC that can treat saline starch water. At the anode, amounts of sodium chloride (NaCl) ranging from 500 mM to 3000 mM were measured. The researchers utilised nitrate as an electron acceptor at the biocathode, and their findings demonstrated an efficient removal of starch and nitrate, resulting in a simultaneous power production of 207.05 mW/m2 at a NaCl concentration of 1000 mM. An enhancement in power density, from 71.06 mW/m2 to 207.05 mW/m2 (a 2.92-fold increase), was noted when elevating the NaCl concentration from 500 mM to 1000 mM. A decrease in power density was noted when salt concentrations higher than 1000 mM were employed. A concentration of 3000 mM facilitated optimal power output, starch breakdown, and amylase activity.

Vijay *et al.* [99] conducted a recent study focusing on the development of a dual-chambered microbial fuel cell (MFC) with an anode that effectively treated saline wastewater. The anodic chamber of the MFC was filled with halophilic exo-electrogenic bacteria that were enhanced from saltwater collected from the Arabian Sea in Mumbai, India. The study of the halophilic community revealed that in the anodic chamber, denitrification was mostly carried out by denitrifying bacteria belonging to the genera *Clostridium, Shewanella,* and *Bacillus*. Their investigation validated the system's ability to effectively remove organic compounds from saline wastewater and eliminate nitrates from polluted wastewater. Additionally, the system demonstrated its capability to generate power using a microbial fuel cell (MFC).

Jamal *et al.* [100] conducted research on the use of potential halophiles from desalination plants for the treatment of wastewater from seafood and the production of power in conditions with high salinity (8%). An air cathode microbial fuel cell (ACMFC) reactor was employed to study the wastewater treatment efficacy of halophiles under high saline condition. Their study revealed potential treatment of seafood industrial wastewater and energy production. Halophilic species like *Ochrobactrum, Marinobacter, Martelella, Rhodococcus, Bacillus, Stenotrophomonas, Alicyclobacillus, Microbacterium, Xanthobacter, Pseudomonas,* and *Sedimentibacter* were found in the bacterial community analysis carried out with samples from the anode and sludge region.

8.4.2 Indirect Conversion Strategies

a) Enzymatic conversion

Passarin *et al.* [50] state that halophilic fungi and their salt-tolerant enzymes, which primarily break down lignin, appear to have significant potential for bioremediation processes. Several fungi are known to break down persistent pollutants such as PHAs and poisonous dyes [29].

The biodegradation of organic pollutants by fungi mainly occurs through the catalytic action of extracellular enzymes released by the fungi as shown in Table 8.2. The primary extracellular enzymes that fungi purify and are accountable for PAH breakdown are ligin peroxidase and manganese peroxidase [13]. Fungal laccase is an enzyme that has found widespread application in environmental remediation processes [9]. Laccases are copper-containing oxidases, which can transform various compounds, such as some toxic chemical wastes (e.g., polycyclic aromatic hydrocarbons, chlorinated aromatic compounds, nitroaromatics, and pesticides) and dyes, into less harmful products [61]. According to Boucherit *et al.* [101], the laccase activities produced by halophilic fungi *P. chrysogenum* on raw olive wastes were reported. *P. chrysogenum* produced 183.0 and 203.0 U/L of laccase activity in 10% and 20% unsupplemented olive mill wastewaters, respectively. These strong enzymatic activities could be connected to the great ability of *P. chrysogenum* to decolorize industrial wastewater from olive oil extraction.

Starch: Starch is the reserved form of carbohydrate in plants and therefore is one of the most frequent biomasses on Earth as it is highly produced by plants annually. In industrial applications, starch is hydrolyzed to produce glucose, glucose syrups, and high-fructose corn syrups. The resulting glucose could be fermented further to bioethanol [102]. Among all the industries which proceed by using enzymes, the starch industry consumes 15–20% of the total amount, and α-amylase, β-amylase, glucoamylase, and glucose isomerase are the main enzymes in this industry [103].

α-Amylase: α-Amylase (also known as endo-1, 4-α-D-glucan glucanohydrolase EC 3.2.1.1) cleaves α-1,4 linkages between adjacent glucose units of starch and produces glucose, maltose, and maltotriose to form linear amylose chains (Sivaramakrishnan *et al.*, 2006). As mentioned above, the produced glucose could be a substrate for bioethanol production. The presence of this extracellular endoenzyme has been reported in several halophile microorganisms. This enzyme has been found in various groups of halophilic microorganisms, including mainly bacteria [40, 55] (Onishi and Hidaka, 1978; Onishi and Sonoda, 1979; Coronado *et al.*, 2000; Deutch, 2002; Mijts and Patel, 2002; Amoozegar *et al.*, 2003; Aygan *et al.*, 2008; Kiran and Chandra, 2008; Prakash B. *et al.*, 2009; Shafiei *et al.*, 2011, 2012; Yamaguchi *et al.*, 2011; Li and Yu, 2012a, c; Uzyol *et al.*, 2012), archaea (Kobayashi *et al.*, 1992; Pérez-Pomares *et al.*, 2003; Fukushima *et al.*, 2005; Hutcheon *et al.*, 2005; Moshfegh *et al.*, 2013), fungi (Ali *et al.*, 2014, 2015), marine bacteria (Qin *et al.*, 2014), and Actinobacteria [17] (Chakraborty *et al.*, 2009). The molecular weight of these α-amylase enzymes varies from 30 to 140 kDa. Most of them can work properly in the presence of high concentrations of salt while some are active in a broad range of temperature and pH values (Table 8.1). Among them, the α-amylase from *Nesterenkonia sp.* strain F was

Table 8.2 Enzymatic conversion into biofuels using halophiles.

Halophiles	Enzymes	Substrate	Biofuels	Mechanism	References
Haloferax volcanii	Alcohol dehydrogenase	Fermented food wastes	Bio alcohols	Fermentation	[3]
Halomonas sp.	Alkaline phosphatases	Biowaste	Bioethanol, Biobutenol, Biohydrogen	Anaerobic digestion, fermentation	[8]
Amphibacillus sp., Exiguobacterium sp., Marinobacter sp., Saccharopolyspora sp., Nesterenkonia sp.	α-Amylases	Biowaste	Bioethanol	Fermentation	[17, 18, 40, 45, 55]
Aspergillus terreus, Thalassobacillus sp., Marinobacter sp.	Cellulases	Biowaste	Bioethanol, Biobutenol	Fermentation	[28, 42]
Planococcus rifitoensis, Halobacterium sp., Virgibacillus sp.	Chitinases	Biowaste	Bioethanol	Fermentation	[25]
Haloarcula marismortui	Esterases	Biowaste	Biodiesel	Transesterification	[16]
Halorubrum lacusprofundi, Haloferax alicantei	β-Galactosidase	Biowaste	Bioethanol	Fermentation	[31, 37]
Haloferax volcanii	Laccase	Biowaste	Biogas Bioethanol	Anaerobic digestion, Fermentation	[39]

(*Continued*)

Table 8.2 Enzymatic conversion into biofuels using halophiles. (*Continued*)

Halophiles	Enzymes	Substrate	Biofuels	Mechanism	References
Haloarcula sp., Marinobacter lipolyticus, Fusarium solani	Lipases	Lignocellulosic biowastes	Biodiesel	Transesterification	[27, 43, 52]
Bacillus sp., Micrococcus varians	Nucleases	Lignocellulosic biowastes	Biodiesel	Transesterification	[7]
Bacillus sp., Geomicrobium sp., Halobacterium sp.	Proteases	Biowaste	Bioethanol	Fermentation	[21]
Gracilibacillus sp., Thermoanaerobacterium saccharolyticum,	Xylanases	Biowaste	Bioethanol, Biobutenol	Fermentation	[19]

interesting. Three of the amylase enzymes produced by this strain have been purified with molecular masses of 57, 100 and 110 kDa. These enzymes had a maximum activity at pH 6.5–7.5 and 40°C and in a broad range of NaCl concentrations (0–4 M), with optimal activity at 0.25 M NaCl. One of these amylases had the ability to metabolize starch, which makes it very important in biotechnology. The activity of the enzymes was not hindered by Ca^{2+}, Rb^{+}, Li^{+}, Cs^{+}, Mg^{2+}, and Hg^{2+}, whereas Fe^{3+}, Cu^{2+}, Zn^{2+}, and Al^{3+} strongly inhibited the enzyme activity. This α-amylase was inhibited by EDTA, though PMFS and β-mercaptoethanol had no inhibitory effect. This enzyme was also the first and only reported to increase the microbial α-amylase activity in the presence of organic solvents [55] (Shafiei *et al.*, 2011; 2012). Furthermore, as mentioned above, *Nesterenkonia sp.* strain F also can produce ethanol and butanol directly from glucose [90]. This suggests that with some modifications ethanol and butanol could be produced from starch only by utilizing this strain.

β-Amylase: β-Amylase is an exoenzyme that hydrolyzes starch by removing maltose from the no reducing end of the starch. Unlike α-amylase, β-amylases are very rare. This enzyme is secreted by several species of the genus Bacillus, including *B. polymyxa, B. cereus, and B. megaterium*, and by *Clostridium thermosulfurogenes* (Kawazu *et al.*, 1987). Up to now, the presence of β-amylase has been reported only in two halophile bacteria. These two β-amylases are from two moderately halophilic bacteria, *Halobacillus sp.* strain LY9 (Li and Yu, 2011) and *Salimicrobium halophilum* strain LY20 (Li and Yu, 2012a). These enzymes showed activity under high temperatures and pH values and the optimal activity of both enzymes was at 1.7 M NaCl which revealed the potential of these β-amylases in industrial processes.

There are several reports of biofuel production from cellulolytic products using halophilic microorganisms. The haloarchaeon *Haloarcula sp.* strain LLSG7 with cellulolytic activity was isolated from the saline soil of Yuncheng Salt Lake, China. Cellulase production by this strain was strongly influenced by the salinity of the culture medium, where the maximum amount was obtained in the presence of 25% NaCl. Substrate specificity tests showed that the crude cellulase was a multicomponent enzyme system, and zymogram analysis revealed that strain LLSG7 secreted five different endoglucanases. The optimal cellulase activity of *Haloarcula sp.* strain LLSG7 was at 50°C, pH 8.0, and 20% NaCl. In addition, this cellulase was highly active and stable in broad ranges of temperature (40–80°C), pH (7.0–11.0), and NaCl concentration (17.5–30%). It displayed remarkable stability in the presence of non-polar organic solvents with log Pow ≥ 1.97. The crude cellulase secreted by *Haloarcula* sp. strain LLSG7 was further utilized to hydrolyze alkali-pretreated rice straw. Then, the enzymatic hydrolysate was used as the substrate for bioethanol fermentation by *S. cerevisiae*. The production of bioethanol increased along the fermentation period and maximal yield of 10.7 g/L was observed after 30 h. No increase in ethanol production was obtained after 30 h. However, the concentration of total reduced sugars decreased along the incubation time and remained at about 1.4 g/L by the end of incubation. As reported, the obtained ethanol yield was about 0.177 g/g of dry substrate, with a conversion efficiency of 41.9%; these results were much higher than those reported from other fungal cellulases, which later were used in bioethanol fermentation with *S. cerevisiae* (Li and Yu, 2013). In another study, a novel halotolerant cellulolytic bacterium, *Bacillus methylotrophicus* RYC01101, was isolated from ruminant faeces in Thailand. This strain could grow at 10% (w/v) NaCl. The cellulase activity of the bacterium on CMCase activity and FPase activity was reported at 0.230 ± 0.004 and 0.080 ± 0.007 U/mL, respectively. As for ethanol production, a pretreated cassava stalk (source of carbon with high cellulose content) was saccharified by the extracellular cellulase of *B. methylotrophicus* RYC01101 under incubation conditions for 72 h, yielding a cassava stalk hydrolysate containing 0.41 ± 0.01 mg/mL of glucose. This hydrolysate was subsequently co-cultured with *S. cerevisiae* TISTR 5111. At the end of the fermentation process, 1.38 ± 0.40 g/L of bioethanol was produced. However, further optimizations seem to be necessary (Chantarasiri, 2014). Another study was based on a halophilic bacterium *Gracilibacillus sp.* strain SK1 showing cellulolytic activity; that strain was isolated from Yuncheng Salt Lake. The salinity of the culture medium had a great influence on cellulase production of this strain with maximal levels of cellulase in the presence of 10% NaCl. The substrate specificity tests indicated that the crude cellulase is a multi-component enzyme system with a combined activity of endoglucanase, exoglucanase, and β-glucosidase. As zymogram analysis indicated, six different endoglucanases were secreted by this strain. They reported that the crude enzyme was highly active and stable over broad ranges of temperature (40–70°C), pH (6.0–10.0), and NaCl concentration (7.5–17.5%), with an optimum at 60°C, pH of 8.0, and 12.5% NaCl. These observations showed that this enzyme is splendidly thermostable, alkali-stable and halostable. Moreover, the crude cellulases of the *Gracilibacillus sp.* strain SK1 displayed high stability in the presence of hydrophobic organic solvents. For bioethanol production, first, the corn stover and rice straw were selected as the feedstocks because the isolated cellulase was more effective against them. Maximal production of reduced sugar was obtained as 27.1 g/L in 48 h and 20.4 g/L in 64 h from corn stover and rice straw, respectively. Total amounts of reduced sugars released from 1 g of dry feedstock were 0.678 g/g (dry substrate) for corn stover and 0.502 g/g (dry substrate) for rice straw. Interestingly, it was observed that the considerable saccharification yield using the crude cellulase from *Gracilibacillus sp.* strain SK1 was close to some commercial enzyme, such as Celluclast, and Novozyme 188. *S. cerevisiae* was used for ethanol production from enzymatic hydrolysates of corn stover. The maximal yield of ethanol (13.5 g/L) was observed after 56 h of fermentation. Meanwhile, at the end of fermentation, the content of reduced sugars reached 0.9 g/L. The obtained ethanol yield was about 0.186 g/g (dry substrate), and the conversion efficiency of reduced sugars to ethanol was about 52.8%. These gained values were much higher than those reported for other fungal cellulases, which were usually used for bio-ethanol production (Yu and Li, 2015). Several halophiles with cellulolytic activity have been reported. Furthermore, several cellulases have been isolated and purified from halophilic microorganisms. Although no bioethanol production from these cellulases has yet been reported, their existence bolds the potential of halophiles in bioethanol production. Cellulolytic enzymes from *Salinivibrio sp.* NTU-05 (Wang *et al.*, 2009), *Bacillus sp.* BG-CS10 (Zhang *et al.*, 2012), *Thalassobacillus sp.* LY18 3 [42], *Bacillus sp.* L1 4 (Li and Yu, 2012c), and *Alkalilimnicola sp.* NM-DCM1 (Mesbah and Wiegel, 2017) have been purified. These enzymes showed interesting biochemical properties and have activity in the presence of organic solvents, high temperature, alkaline pH, and NaCl.

8.5 Niche of Halophiles in Biowaste Valorisation

Halophiles present a promising avenue for biowaste valorization due to their adeptness in extreme conditions. Their potential contribution spans various facets of waste management. They exhibit promise in treating saline-rich biowastes from industries like food processing or agricultural runoff, efficiently degrading such high-salt content where other organisms might falter. Additionally, their enzymatic and metabolic adaptations tailored to high-salt environments position them as candidates for bioremediation, aiding in the breakdown of organic pollutants within saline waste, thereby detoxifying these environments. Moreover, certain halophiles harbour metabolic pathways capable of producing biofuels and valuable compounds from organic substrates present in biowaste. This includes the generation of hydrogen gas, biodiesel-suitable lipids, and other valuable compounds during their metabolic processes, offering prospects for bioenergy production from saline biowaste. Their propensity to produce stable enzymes adaptable to high-salt conditions further accentuates their significance in various biowaste conversion processes, facilitating the breakdown of complex organic molecules for biofuel synthesis or other industrial applications. Furthermore, halophiles contribute to nutrient recycling by efficiently decomposing organic matter in saline biowaste, converting it into simpler compounds usable by other organisms in the ecosystem, thereby fostering nutrient cycling and environmental equilibrium.

8.6 Conclusion

Halophiles are still unexplored with respect to both their diversity and utilization in different applications. The investigations carried out so far prove the potential of halophiles in effective remediation processes which can give the added benefit of converting waste to value. Among the important products obtained, different biofuels hold a great relevance due to the unmitigated energy demand of a technologically developing society. The diversity of metabolism and adaptation tailor make them for certain specific conditions where salt concentration does not allow other microbes to even grow. However, for effective commercialization of these strategies, still lot of concerted efforts are needed for viable, scalable and sustainable waste to fuel processes.

References

1. Agarwal, S., Darbar, S., Saha, S., Singh, R.P., Microbiological digestion of agricultural biomass: prospects and challenges in generating clean and green energy, in: *Agriculture Waste Management and Bioresource: The Circular Economy Perspective*, pp. 105–130, 2023.
2. Akubude, V.C., Nwaigwe, K.N., Dintwa, E., Production of biodiesel from microalgae *via* nanocatalyzed transesterification process: a review. *Mater. Sci. Energy Technol.*, 2, 216–225, 2019.
3. Alsafadi, D. and Paradisi, F., Effect of organic solvents on the activity and stability of halophilic alcohol dehydrogenase (ADH2) from *Haloferax volcanii*. *Extremophiles*, 17, 115–122, 2013.
4. Alvarez, A.L., Weyers, S.L., Goemann, H.M., Peyton, B.M., Gardner, R.D., Microalgae, soil and plants: A critical review of microalgae as renewable resources for agriculture. *Algal Res.*, *54*, 102200, 2021.
5. Angelidaki, I., Treu, L., Tsapekos, P., Luo, G., Campanaro, S., Wenzel, H., *et al.*, Biogas upgrading and utilization: current status and perspectives. *Biotechnol. Adv.*, 36, 452–66, 2018.
6. Antoni, D., Zverlov, V.V., Schwarz, W.H., Biofuels from microbes. *Appl. Microbiol. Biotechnol.*, 77, 23–35, 2007.

7. Aouida, M., Li, L., Mahjoub, A., Alshareef, S., Ali, Z., Piatek, A., Mahfouz, M.M., Transcription activator-like effector nucleases mediated metabolic engineering for enhanced fatty acids production in Saccharomyces cerevisiae. *J. Biosci. Bioeng.*, *120*, 4, 364–371, 2015.
8. Arai, S., Yonezawa, Y., Ishibashi, M., *et al.*, Structural characteristics of alkaline phosphatase from the moderately halophilic bacterium *Halomonas sp.* 593. *Acta Crystallogr. D Biol. Crystallogr.*, 70, 811–820, 2014.
9. Arun, A., Raja, P.P., Arthi, R., Ananthi, M., Kumar, K.S., Eyini, M., Polycyclic aromatic hydrocarbons (PAHs) biodegradation by basidiomycetes fungi, Pseudomonas isolate, and their cocultures: Comparative *in vivo* and *in silico* approach. *Appl. Biochem. Biotechnol.*, 151, 132–142, 2008.
10. Bhatia, S.K., Bhatia, R.K., Yang, Y.-H., Biosynthesis of polyesters and polyamide building blocks using microbial fermentation and biotransformation. *Rev. Environ. Sci. Bio/Technol.*, 15, 639–63, 2016.
11. Bhatia, S.K., Kim, S.-H., Yoon, J.-J., Yang, Y.-H., Current status and strategies for second generation biofuel production using microbial systems. *Energy Convers. Manage.*, 148, 1142–56, 2017.
12. Binod, P., Sindhu, R., Singhania, R.R., Vikram, S., Devi, L., Nagalakshmi, S., *et al.*, Bioethanol production from rice straw: an overview. *Bioresour. Technol.*, 101, 4767–74, 2010.
13. Bonugli-Santos, R.C., Durrant, L.R., da Silva, M., Sette, L.D., Production of laccase, manganese peroxidase and lignin peroxidase by Brazilian marine-derived fungi. *Enzyme Microb. Technol.*, 46, 32–37, 2010.
14. Buaban, B., Inoue, H., Yano, S., Tanapongpipat, S., Ruanglek, V., Champreda, V., *et al.*, Bioethanol production from ball milled bagasse using an on-site produced fungal enzyme cocktail and xylose-fermenting *Pichia stipitis*. *J. Biosci. Bioeng.*, 110, 1, 18–25, 2010.
15. Cadoche, L. and Lopez, G.D., Assessment of size reduction as a preliminary step in the production of ethanol from lignocellulosic wastes. *Biol. Wastes*, 30, 153–157, 1989.
16. Camacho, R.M., Mateos, J.C., Gonzalez-Reynoso, O., *et al.*, Production and characterization of esterase and lipase from *Haloarcula marismortui*. *J. Ind. Microbiol. Biotechnol.*, 36, 901–909, 2009.
17. Chakraborty, S., Khopade, A., Biao, R., *et al.*, Characterization and stability studies on surfactant, detergent and oxidant stable α-amylase from marine haloalkaliphilic Saccharopolyspora sp. A9. *J. Mol. Catal. B: Enzym.*, 68, 52–58, 2011.
18. Chang, J., Lee, Y.S., Fang, S.J., *et al.*, Recombinant expression and characterization of an organic-solvent-tolerant alpha-amylase from *Exiguobacterium* sp. DAU5. *Appl. Biochem. Biotechnol.*, 169, 1870–1883, 2013.
19. Chaudhary, R., Kuthiala, T., Singh, G., Rarotra, S., Kaur, A., Arya, S.K., Kumar, P., Current status of xylanase for biofuel production: a review on classification and characterization. *Biomass Convers. Biorefin.*, 1–19, 2021.

20. Cheng, C.-L., Che, P.-Y., Chen, B.-Y., Lee, W.-J., Lin, C.-Y., Chang, J.-S., Biobutanol production from agricultural waste by an acclimated mixed bacterial microflora. *Appl. Energy*, 100, 3–9, 2012.
21. Das, A., Anuradha, Sampath, M.K., Techno-economic analysis of protease enzyme production and its biofuel application, in: *National Statistics Forum (FNE) and Latin-American Congress of Statistical Societies (CLATSE)*, Springer Nature Singapore, Singapore, pp. 149–156, 2022.
22. De la Rubia, M.A., Villamil, J.A., Rodriguez, J.J., Borja, R., Mohedano, A.F., Mesophilic anaerobic co-digestion of the organic fraction of municipal solid waste with the liquid fraction from hydrothermal carbonization of sewage sludge. *Waste Manage. (Oxford)*, 76, 315–322, 2018.
23. Duff, S.J.B. and Murray, W.D., Bioconversion of forest products industry waste cellulosics to fuel ethanol: a review. *Bioresour. Technol.*, 55, 1–33, 1996.
24. Erdei, B., Barta, Z., Sipos, B., Réczey, K., Galbe, M., Zacchi, G., Ethanol production from mixtures of wheat straw and wheat meal. *Biotechnol. Biofuels*, 3, 16, 2010.
25. Essghaier, B., Rouaissi, M., Boudabous, A., *et al.*, Production and partial characterization of chitinase from a halotolerant Planococcus rifitoensis strain M2-26. *World J. Microbiol. Biotechnol.*, 26, 977–984, 2010.
26. Faisal, S., Zaky, A., Wang, Q., Huang, J., Abomohra, A., Integrated marine biogas: a promising approach towards sustainability. *Fermentation*, *8*, 10, 520, 2022.
27. Geoffry, K. and Achur, R.N., A novel halophilic extracellular lipase with both hydrolytic and synthetic activities. *Biocatal. Agric. Biotechnol.*, *12*, 125–130, 2017.
28. Gunny, A.A., Arbain, D., Edwin Gumba, R., *et al.*, Potential halophilic cellulases for *in situ* enzymatic saccharification of ionic liquids pretreated lignocelluloses. *Bioresour. Technol.*, 155, 177–181, 2014.
29. Haritash, A.K. and Kaushik, C.P., Biodegradation aspects of Polycyclic Aromatic Hydrocarbons (PAHs): *A review. J. Hazard. Mater.*, 169, 1–15, 2009.
30. Heidari, F., Asadollahi, M.A., Jeihanipour, A., Kheyrandish, M., Rismani-Yazdi, H., Karimi, K., Biobutanol production using unhydrolyzed waste acorn as a novel substrate. *RSC Adv.*, 6, 9254–60, 2016.
31. Holmes, M.L., Scopes, R.K., Moritz, R.L., *et al.*, Purification and analysis of an extremely halophilic beta-galactosidase from *Haloferax alicantei*. *Biochim. Biophys. Acta*, 1337, 276–286, 1997.
32. Hun-Suk, S., Hyung-Min, S., Jong-Min, J., Yu-Mi, M., Won, H.J., Gi, H.Y., *et al.*, Enhanced isobutanol production from acetate by combinatorial overexpression of acetyl-CoA synthetase and anaplerotic enzymes in engineered *Escherichia coli*. *Biotechnol. Bioeng.*, 115, 1971–8, 2018.
33. Ieropoulos, I. and Greenman, J., The future role of MFCs in biomass energy. *Front. Energy Res.*, *11*, 1108389, 2023.

34. Jeon, J.M., Song, H.S., Lee, D.G., Hong, J.W., Hong, Y.G., Moon, Y.M., *et al.*, Butyrate-based nbutanol production from an engineered *Shewanella oneidensis* MR-1. *Bioprocess. Biosyst. Eng.*, 41, 1195–1204, 2018.
35. Joshi, S.M., Waghmare, J.S., Sonawane, K.D., Waghmare, S.R., Bio-ethanol and bio-butanol production from orange peel waste. *Biofuels*, 6, 55–61, 2015.
36. Kadam, R. and Panwar, N.L., Recent advancement in biogas enrichment and its applications. *Renew. Sustain. Energy Rev.*, 73, 892–903, 2017.
37. Karan, R., Capes, M.D., DasSarma, P., *et al.*, Cloning, overexpression, purification, and characterization of a polyextremophilic beta-galactosidase from the Antarctic haloarchaeon *Halorubrum lacusprofundi. BMC Biotech.*, 13, 3, 2013.
38. Kim, S. and Dale, B.E., Global potential bioethanol production from wasted crops and crop residues. *Biomass Bioenergy*, 26, 361–75, 2004.
39. Kudanga, T. and Le Roes-Hill, M., Laccase applications in biofuels production: current status and future prospects. *Appl. Microbiol. Biotechnol., 98*, 6525–6542, 2014.
40. Kumar, S. and Khare, S.K., Purification and characterization of malto oligosaccharide-forming alpha-amylase from moderately halophilic *Marinobacter sp.* EMB8. *Bioresour. Technol.*, 116, 247–251, 2012.
41. Lan, N.N., Van Dung, V., Lien, N.T.K., Thoa, N.K., Nghi, D.H., Hoang, N.H., Isolation and identification of indole acetic acid producing bacteria from the coasts of Ben Tre and Tra Vinh Provinces. *Acad. J. Biol., 41*, 4, 55–67, 2019.
42. Li, X., Wang, H.L., Li, T., *et al.*, Purification and characterization of an organic solvent tolerant alkaline cellulase from a halophilic isolate of *Thalassobacillus. Biotechnol. Lett.*, 34, 1531–1536, 2012.
43. Li, X. and Yu, H.Y., Characterization of an organic solvent-tolerant lipase from *Haloarcula* sp. G41 and its application for biodiesel production. *Folia Microbiol. (Praha)*, 59, 455–463, 2014.
44. Li, R., Song, M., Yin, D., Ye, X., Yu, J., Chen, X., Indole-3-acetic acid mediated removal of sludge toxicity by microalgae: Focus on the role of extracellular polymeric substances. *Bioresour. Technol., 387*, 129700, 2023.
45. Mesbah, N.M. and Wiegel, J., Halophilic alkali- and thermostable amylase from a novel polyextremophilic Amphibacillus sp. NM-Ra2. *Int. J. Biol. Macromol.*, 70, 222–229, 2014.
46. Nath, A., Das, K., Dhal, G.C., Global Status of Agricultural Waste-Based Industries, Challenges, and Future Prospects, in: *Agricultural Waste to Value-Added Products: Technical, Economic and Sustainable Aspects*, pp. 21–45, 2023.
47. Nguyen, T.A.D., Kim, K.R., Han, S.J., Cho, H.Y., Kim, J.W., Park, S.M., *et al.*, Pretreatment of rice straw with ammonia and ionic liquid for lignocellulose conversion to fermentable sugars. *Bioresour. Technol.*, 101, 19, 7432–8, 2010.
48. Noor, T., Javid, A., Hussain, A., Bukhari, S.M., Ali, W., Akmal, M., Hussain, S.M., Types, sources and management of urban wastes, in: *Urban Ecology*, pp. 239–263, 2020.

49. Parthiba Karthikeyan, O., Trably, E., Mehariya, S., Bernet, N., Wong, J.W.C., Carrere, H., Pretreatment of food waste for methane and hydrogen recovery: a review. *Bioresour. Technol.*, 249, 1025–39, 2018.
50. Passarini, M.R.Z., Rodrigues, M.V.N., da Silva, M., Sette, L.D., Marine-derived filamentous fungi and their potential application for polycyclic aromatic hydrocarbon bioremediation. *Mar. Pollut. Bull.*, 62, 364–370, 2011.
51. Paul, V.G., Minteer, S.D., Treu, B.L., Mormile, M.R., Ability of a haloalkaliphilic bacterium isolated from Soap Lake, Washington to generate electricity at pH 11.0 and 7% salinity. *Environ. Technol.*, 35, 8, 1003–1011, 2014.
52. Pérez, D., Martín, S., Fernández-Lorente, G., *et al.*, A novel halophilic lipase, LipBL, showing high efficiency in the production of eicosapentaenoic acid (EPA). *PLoS One*, 6, e23325, 2011.
53. Saini, R., Kaur, A., Saini, J.K., Patel, A.K., Varjani, S., Chen, C.W., Singhania, R.R., Dong, C.D., Trends in lignin biotransformations for bio-based products and energy applications. *Bioenergy Res.*, *16*, 1, 88–104, 2023.
54. Salam, K.A., Velasquez-Orta, S.B., Harvey, A.P., A sustainable integrated *in situ* transesterification of microalgae for biodiesel production and associated co-producta review. *Renew. Sustain. Energy Rev.*, 65, 1179–1198, 2016.
55. Shafiei, M., Ziaee, A.A., Amoozegar, M.A., Purification and biochemical characterization of a novel SDS and surfactant stable, raw starch digesting, and halophilic α-amylase from a moderately halophilic bacterium, *Nesterenkonia sp. strain F. Process Biochem.*, 45, 694–699, 2010.
56. Sharma, P.C. and Singh, A., Reviving the productivity of salt-affected lands: technological options, constraints and research needs, in: *Research Developments in Saline Agriculture*, pp. 591–627, 2019.
57. Skorupskaite, V., Makareviciene, V., Gumbyte, M., Opportunities for simultaneous oil extraction and transesterification during biodiesel fuel production from microalgae: a review. *Fuel Process. Technol.*, 150, 78–87, 2016.
58. Song, H.-S., Jeon, J.-M., Kim, H.-J., Bhatia, S.K., Sathiyanarayanan, G., Kim, J., *et al.*, Increase in furfural tolerance by combinatorial overexpression of NAD salvage pathway enzymes in engineered isobutanol-producing *E. coli*. *Bioresour. Technol.*, 245, 1430–5, 2017.
59. Srivastava, R.K., Shetti, N.P., Reddy, K.R., Nadagouda, M.N., Badawi, M., Bonilla-Petriciolet, A., Aminabhavi, T.M., Valorization of biowastes for clean energy production, environmental depollution and soil fertility. *J. Environ. Manage.*, *332*, 117410, 2023.
60. Talebnia, F., Karakashev, D., Angelidaki, I., Production of bioethanol from wheat straw: an overview on pretreatment, hydrolysis and fermentation. *Bioresour. Technol.*, 101, 13, 4744–53, 2010.
61. Whiteley, C.G. and Lee, D.J., Enzyme technology and biological remediation. *Enzyme Microb. Technol.*, 38, 291–316, 2006.
62. Yong, Z., Dong, Y., Zhang, X., Tan, T., Anaerobic co-digestion of food waste and straw for biogas production. *Renew. Energy*, 78, 527–30, 2015.
63. Smith, K.R., Jerrett, M., Anderson, H.R., Burnett, R.T., Stone, V., Derwent, R., Atkinson, R.W., Cohen, A., Shonkoff, S.B., Krewski, D., Pope, C.A., Public

health benefits of strategies to reduce greenhouse-gas emissions: health implications of short-lived greenhouse pollutants. *Lancet*, *374*, 9707, 2091–2103, 2009.
64. Zhao, J., Wu, Q., Tang, Y., Zhou, J., Guo, H., Tannery wastewater treatment: conventional and promising processes, an updated 20-year review. *J. Leather Sci. Eng.*, *4*, 1, 10, 2022.
65. Shah, S.W.A., Khalid, T., Fayyaz, S., Impact Of Foreign Direct Investment, Population Density, Consumption Of Oil And Gas On Environmental Degradation In Pakistan. *Bull. Bus. Econ. (BBE)*, *11*, 1, 163–171, 2022.
66. Dutta, B. and Bandopadhyay, R., Biotechnological potentials of halophilic microorganisms and their impact on mankind. *Beni-Suef Univ. J. Basic Appl. Sci.*, *11*, 1, 75, 2022.
67. Adhikari, S., Nam, H., Chakraborty, J.P., Conversion of solid wastes to fuels and chemicals through pyrolysis, in: *Waste biorefinery*, pp. 239–263, 2018.
68. Bala, S., Garg, D., Sridhar, K., Inbaraj, B.S., Singh, R., Kamma, S., Tripathi, M., Sharma, M., Transformation of agro-waste into value-added bioproducts and bioactive compounds: Micro/nano formulations and application in the agri-food-pharma sector. *Bioengineering*, *10*, 2, 152, 2023.
69. Tan, Y.N., Zhang, J.H., Chen, W.N., GC-MS-based metabolomics analysis of prawn shell waste co-fermentation by Lactobacillus plantarum and Bacillus subtilis. *Polysaccharides*, *1*, 1, 31–50, 2020.
70. Huang, Y., Zhao, C., Gao, B., Ma, S., Zhong, Q., Wang, L., Cui, S., Life cycle assessment and society willingness to pay indexes of food waste-to-energy strategies. *J. Environ. Manage.*, *305*, 114364, 2022.
71. Abo-State, M.A., Ragab, A.M., EL-Gendy, N.S., Farahat, L.A., Madian, H.R., Effect of different pretreatments on egyptian sugar-cane bagasse saccharification and bioethanol production. *Egypt. J. Pet.*, *22*, 1, 161–167, 2013.
72. Kasirajan, L. and Maupin-Furlow, J.A., Halophilic archaea and their potential to generate renewable fuels and chemicals. *Biotechnol. Bioeng.*, *118*, 3, 1066–1090, 2021.
73. Awasthi, M.K., Sarsaiya, S., Patel, A., Juneja, A., Singh, R.P., Yan, B., Awasthi, S.K., Jain, A., Liu, T., Duan, Y., Pandey, A., Refining biomass residues for sustainable energy and bio-products: An assessment of technology, its importance, and strategic applications in circular bio-economy. *Renewable Sustainable Energy Rev.*, *127*, 109876, 2020.
74. Mussatto, S.I., Dragone, G., Guimarães, P.M., Silva, J.P.A., Carneiro, L.M., Roberto, I.C., Vicente, A., Domingues, L., Teixeira, J.A., Technological trends, global market, and challenges of bio-ethanol production. *Biotechnol. Adv.*, *28*, 6, 817–830, 2010.
75. Kim, M.H., Song, H.B., Song, Y., Jeong, I.T., Kim, J.W., Evaluation of food waste disposal options in terms of global warming and energy recovery: Korea. *Int. J. Energy Environ. Eng.*, *4*, 1, 1, 2013.
76. Munir, N., Safi, S.Z., Sarwar, Z., Arshad, M., Hasnain, M., Haq, R., Techniques and strategies for bioenergy production from manure, in: *Climate Changes Mitigation and Sustainable Bioenergy Harvest Through Animal Waste:*

Sustainable Environmental Implications of Animal Waste, pp. 125–148, Springer Nature Switzerland, Cham, 2023.

77. Srivastava, N., Shrivastav, A., Singh, R., Abohashrh, M., Srivastava, K.R., Irfan, S., Srivastava, M., Mishra, P.K., Gupta, V.K., Thakur, V.K., Advances in the structural composition of biomass: fundamental and bioenergy applications. *J. Renewable Mater.*, *9*, 4, 615–636, 2021.
78. Hosseinzadeh-Bandbafha, H., Nazemi, F., Khounani, Z., Ghanavati, H., Shafiei, M., Karimi, K., Lam, S.S., Aghbashlo, M., Tabatabaei, M., Safflower-based biorefinery producing a broad spectrum of biofuels and biochemicals: A life cycle assessment perspective. *Sci. Total Environ.*, *802*, 149842, 2022.
79. Millati, R., Wikandari, R., Ariyanto, T., Hasniah, N., Taherzadeh, M.J., Anaerobic digestion biorefinery for circular bioeconomy development. *Bioresour. Technol. Rep.*, *21*, 101315, 2023.
80. Johnson, J., Sudheer, P.D., Yang, Y.H., Kim, Y.G., Choi, K.Y., Hydrolytic activities of hydrolase enzymes from halophilic microorganisms. *Biotechnol. Bioprocess Eng.*, *22*, 4, 450–461, 2017.
81. Essa, D., Abo-Shady, A., Khairy, H., Abomohra, A.E.F., Elshobary, M., Potential cultivation of halophilic oleaginous microalgae on industrial wastewater. *Egypt. J. Bot.*, *58*, 2, 205–216, 2018.
82. McGenity, T.J. and Sorokin, D.Y., Methanogens and methanogenesis in hypersaline environments, in: *Biogenesis of hydrocarbons*, pp. 283–309, Springer, Cham, 2019.
83. Bhatia, S.K., Joo, H.S., Yang, Y.H., Biowaste-to-bioenergy using biological methods–a mini-review. *Energy Convers. Manage.*, *177*, 640–660, 2018.
84. Sun, J., Zhang, L., Loh, K.C., Review and perspectives of enhanced volatile fatty acids production from acidogenic fermentation of lignocellulosic biomass wastes. *Bioresour. Bioprocess.*, *8*, 1, 68, 2021.
85. Hariram, V., Janarthanan, M., Paul, R.C., Sivasankar, A., Akram, M.W., Sangeethkumar, E., Ramanathan, V., Khan, P.S., Reddy, S.M., Biodiesel from Dunaliella salina Microalgae Using Base Catalyzed Transesterification–An Assessment through GC/MS, FTIR and NMR Studies. *Nat. Environ. Pollut. Technol.*, *22*, 4, 2023.
86. Ryu, Y.J., Kim, Z.H., Lee, S.G., Yang, J.H., Shin, H.Y., Lee, C.G., Development of carbon-based solid acid catalysts using a lipid-extracted alga, Dunaliella tertiolecta, for esterification, 2018.
87. www.ethanolrfa.org/pages/annual-industry-outlook, 2022.
88. Bjerre, A.B., Olesen, A.B., Fernqvist, T., Plöger, A., Schmidt, A.S., Pretreatment of wheat straw using combined wet oxidation and alkaline hydrolysis resulting in convertible cellulose and hemicellulose. *Biotechnol. Bioeng.*, *49*, 5, 568–577, 1996.
89. Sarkar, N., Ghosh, S.K., Bannerjee, S., Aikat, K., Bioethanol production from agricultural wastes: an overview. *Renewable Energy*, *37*, 1, 19–27, 2012.
90. Amiri, H., Azarbaijani, R., Parsa Yeganeh, L., Shahzadeh Fazeli, A., Tabatabaei, M., Hosseini Salekdeh, G., Karimi, K., Nesterenkonia sp. strain F,

a halophilic bacterium producing acetone, butanol and ethanol under aerobic conditions. *Sci. Rep.*, *6*, 1, 18408, 2016.
91. Priya, A., Hathi, Z., Haque, M.A., Kumar, S., Kumar, A., Singh, E., Lin, C.S., Effect of levulinic acid on production of polyhydroxyalkanoates from food waste by Haloferax mediterranei. *Environ. Res.*, *214*, 114001, 2022.
92. Woolard, C.R. and Irvine, R.L., Biological treatment of hypersaline wastewater by a biofilm of halophilic bacteria. *Water Environ. Res.*, *66*, 3, 230–235, 1994.
93. Abou-Elela, S.I., Kamel, M.M., Fawzy, M.E., Biological treatment of saline wastewater using a salt-tolerant microorganism. *Desalination*, *250*, 1, 1–5, 2010.
94. Miller, L.G. and Oremland, R.S., Electricity generation by anaerobic bacteria and anoxic sediments from hypersaline soda lakes. *Extremophiles*, *12*, 6, 837–848, 2008.
95. Abrevaya, X.C., Sacco, N., Mauas, P.J., Cortón, E., Archaea-based microbial fuel cell operating at high ionic strength conditions. *Extremophiles*, *15*, 6, 633, 2011.
96. Rousseau, R., Dominguez-Benetton, X., Délia, M.L., Bergel, A., Microbial bioanodes with high salinity tolerance for microbial fuel cells and microbial electrolysis cells. *Electrochem. Commun.*, *33*, 1–4, 2013.
97. Vijay, A., Arora, S., Gupta, S., Chhabra, M., Halophilic starch degrading bacteria isolated from Sambhar Lake, India, as potential anode catalyst in microbial fuel cell: a promising process for saline water treatment. *Bioresour. Technol.*, *256*, 391–398, 2018.
98. Grattieri, M., Suvira, M., Hasan, K., Minteer, S.D., Halotolerant extremophile bacteria from the Great Salt Lake for recycling pollutants in microbial fuel cells. *J. Power Sources*, *356*, 310–318, 2017.
99. Vijay, A., Ghosh, P.C., Mukherji, S., Power generation by halophilic Bacteria and assessment of the effect of salinity on performance of a denitrifying microbial fuel cell. *Energies*, *16*, 2, 877, 2023.
100. Jamal, M.T., Pugazhendi, A., Jeyakumar, R.B., Application of halophiles in air cathode MFC for seafood industrial wastewater treatment and energy production under high saline condition. *Environ. Technol. Innovation*, *20*, 101119, 2020.
101. Boucherit, Z., Flahaut, S., Djoudi, B., Mouas, T.N., Mechakra, A., Ameddah, S., Potential of halophilic Penicillium chrysogenum isolated from Algerian saline soil to produce laccase on olive oil wastes. *Curr. Microbiol.*, *79*, 7, 178, 2022.
102. Bai, F.W., Anderson, W.A., Moo-Young, M., Ethanol fermentation technologies from sugar and starch feedstocks. *Biotechnol. Adv.*, *26*, 1, 89–105, 2008.
103. Bertoldo, C. and Antranikian, G., Starch-hydrolyzing enzymes from thermophilic archaea and bacteria. *Curr. Opin. Chem. Biol.*, *6*, 2, 151–160, 2002.

9

Valorization of Biological Solid Waste for Agricultural Sectors

Indu Sharma[1*], Komal Yadav[1], Khushboo Tamboli[1], Rajeev Kumar[2], Tanvi Taneja[3] and Raj Singh[3]

[1]Nims Institute of Allied Medical Science and Technology, NIMS University Rajasthan, Jaipur, Rajasthan, India
[2]Department of Bioscience and Bioengineering, Lovely Professional University, Phagwara, Punjab, India
[3]Department of Bio-Sciences and Technology, Maharishi Markandeshwar (Deemed to be University), Mullana, Ambala, Haryana, India

Abstract
The efficient management of biological solid waste has become a pressing global concern due to its significant environmental impacts and the burgeoning demand for sustainable agricultural practices. This chapter brings together comprehensive information to assert that biological waste can be converted into a resource or a raw material for value addition. These wastes, if not treated properly, may harm human health and our nature. Therefore, these wastes need to be treated or utilized in various applications. Waste valorization encompasses the natural recycling principles of zero waste, loop closing, and underlines the importance of sustainable and environmentally friendly alternatives of reduce, reuse, and recycle the waste into biofuel, biofertilizer, manure, compost, vermicomposting, biopesticides and bioplastics. These methods not only facilitate the conversion of waste into nutrient-rich organic fertilizers and value-added products but also contribute to the reduction of greenhouse gas emissions and the conservation of natural resources. Worldwide, a huge amount of biological solid wastes and by-products containing valuable bioactive compounds are generated, especially throughout the entire supply chain. Although many countries implement policies to reduce biological solid wastes and by-products, and different management methods are available to utilize biological solid wastes, they are still produced annually. In conclusion, the valorization of biological solid waste holds immense promise for

**Corresponding author*: endusharma@gmail.com

Priyanka Singh, Pooja Agarwal and V. Vivekanand (eds.) Solid Waste Management: Challenges, Sustainability and Advancements, (277–334) © 2026 Scrivener Publishing LLC

addressing the dual challenges of waste management and sustainable agriculture. By harnessing the potential of biological solid waste as a valuable resource, we can create a more resilient and environmentally sustainable agricultural sector, thereby contributing to food security and ecosystem health on a global scale.

Keywords: Valorization, biological solid waste, biofertilizer, compost, biopesticides, sustainable agriculture

9.1 Introduction

Increased global waste production has led to a heightened strain on the environment, resulting in adverse impacts on soil, air, and water resources. These effects pose a threat to both human health and the long-term sustainability of ecosystems. Biological solid waste disposal stands out as a significant environmental concern, with detrimental consequences on ecosystems stemming from the negligent dumping of such waste. According to various studies [7, 21], the disposal of untreated and underutilized biological solid waste often involves practices like burning, dumping, or landfilling.

Hence, significant interventions are imperative for the sustainable utilization of biological solid agricultural waste. This may manifest in the form of advancing sustainable energy technologies. Despite the potential environmental hazards posed by composite wastes, they have the potential to serve as fundamental constituents in various agricultural, culinary, and medicinal products [23]. The European Commission states that significant waste can be derived from urban and animal waste sources, which are repurposed into various value-added products within the agricultural sector.

Biomass resources, including residues from forestry biomass, by-products from the agricultural residues, wood industry, municipal solid waste (MSW), and energy crops, also contribute significantly to waste generated by agriculture [43]. Agricultural waste products such as straws, husks, shells, and stalks can be transformed into value-added such as biofuels, biochar, bioplastics, natural fibers, animal feed, and organic fertilizers. Additionally, these materials can be used as raw materials for the production of biocomposites, activated carbon, and industrial chemicals, contributing to sustainable agriculture and the circular economy.

Bio-fuels production uses various techniques, including thermochemical and biological processes. Thermochemical techniques involve processes such as pyrolysis and gasification. Pyrolysis refers to the thermal decomposition of organic materials in the absence of oxygen, resulting in solid residue, liquid fractions, and gases such as carbon monoxide, carbon dioxide, hydrogen and methane. Gasification involves the conversion of carbonaceous fuel into gaseous energy in the presence of heat, carbon monoxide, carbon dioxide, hydrogen, methane, and some hydrocarbons [54]. Bagasse, another example of agricultural waste, is generated primarily by the sugar and alcohol industries, particularly in countries like Brazil [103].

Bagasse, the residual lignocellulosic material left after extracting sugarcane juice, is currently utilized by heating it in boilers to generate electrical energy, resulting in the production of a considerable amount of ash, including bottom ashes and fly ash. This ash contains a high proportion of organic material, including bagasse debris and approximately 35% charcoal by weight. Previous studies indicated that both the organic and inorganic components of this ash hold industrial utility [102, 104]. In another investigation, onion skin waste was explored for the production of medicinal products by industries [22]. Onions possess both nutritional and medicinal value, and with the global increase in onion production, the volume of onion skin waste has surged, exceeding 500,000 tons annually in the European Union alone, posing environmental challenges [12]. Onion skin waste comprises brown skin, outer layers, roots, and stalks, which are typically discarded as they are not suitable as cattle feed. However, the brown skin and outer layers contain significant amounts of phenolic compounds and fiber, such as quercetin and flavonoids. Quercetin, a natural compound found in these wastes, is widely utilized in the treatment of various diseases, including breast cancer, colorectal cancer, ovarian cancer, and kidney cancer and prostate cancer [52]. Therefore, it is worthwhile to investigate the extraction of these valuable components from onion skin waste, as it holds potential benefits for both the environment and various industries. The valorization of agricultural waste from rice husks represents a significant waste product of the agricultural industry, constituting approximately 22% of the 700 million tons of rice produced globally [45, 107]. Proper disposal or treatment of these husks presents a challenge in environmental conservation efforts. The primary constituents of rice husks include cellulose, hemicellulose, lignin and other organic matter (comprising 80%-83% of the weight), along with a minor silica content (17 to 20% by weight) and trace impurities such as aluminum chloride, alkali oxides, and alkali earth metals. Utilizing rice husks offers numerous advantages, including biodegradability, low cost, low density, and high deformability. Due to its enhanced mechanical properties, rice husk has been utilized to produce both carbon-silica-based black rice husk ash and silica-based white rice husk ash by various researchers. These materials find applications as reinforcing fillers in rubber and polymers, insulating refractories, thermal dissipation materials, silicon carbide production, and absorbents for water purification, among others. However, rice husk ashes contain ionic impurities such as potassium chlorides and sodium, which can lead to electrical failures and corrosion within metal connections present in integrated circuits when exposed to moisture, thus adversely affecting the performance of electronic packaging [45]. To address these issues, the development of biodegradable carbon-silica materials from rice husks is being pursued for electronic packaging. The main reasons for fabricating carbon-silica-based materials from rice husks include their favorable filler-matrix interactions and low ionic content achieved through pre-acid and post-hydrothermal methods. Thus, this chapter underscores the valorization of various agricultural waste products for diverse value-added applications.

Microbial biotechnology and nanotechnology assume a crucial role in the enzymatic and bioactive compound conversion of agricultural waste within the pharmaceutical industry, the production of vermi-compost, organic fertilizers, and biofuels in agriculture, as well as the development of nutraceuticals and food items [9, 10, 41, 82, 108].

Nanostructured materials possess the capacity to transform the food industry due to their distinctive characteristics and expansive surface area [60]. Nano-encapsulation, employing a matrix or inert substance, represents a technique for encapsulating substances (such as food or flavor molecules/ingredients) in liquid, solid, or gaseous states. The process of stabilizing bioactive compounds through nano-encapsulation enables precise regulation of their release at physiologically active sites [81].

9.2 Biomass Sources

Biomass sources represent renewable reservoirs that can be harnessed in their original or altered states. Biomass exhibits versatility, as it can be transformed into liquid transportation biofuels comparable to those derived from fossil fuels, such as gasoline, aviation fuel, and diesel. Bioenergy technologies alleviate the conversion of carbon sourced from biomass and waste materials into low-emission fuels for automobiles, trucks, aircraft and vessels, as well as into bioproducts and renewable energy [95].

Renewable biomass resources, whether utilized directly as fuel or converted into alternative energy products, are commonly referred to as feedstock. This feedstock encompasses energy crops, residues from agricultural and forestry activities, algae cultivation, and municipal solid waste.

9.2.1 Energy Crops

Energy non-food crops include bamboo, switchgrass, wheatgrass, hybrid willow, hybrid poplar, and sycamore. A single bamboo pole can be used to power a rural household for a month. These are non-food crops that can grow on marginal land specifically to provide biomass. Energy crops are divided into herbaceous and woody types [77]. Energy crops include some plant biomass, which is planted exclusively use for the biomass as raw materials for bioenergy generation. Energy crops includes as switchgrass (*Panicum virgatum*), miscanthus (*Miscanthus giganteus*), kenaf (*Hibiscus cannabinus*), reed canary grass (*Phalaris arundinacea*), giant reed (*Arundo donax*), and eucalyptus (*Eucalyptus globulus*) are regarded as appropriate plants for the formation of bio-hydrogen renewable bioenergy. Short rotation woody crops are harvested years after planting. These are fast-growing hardwood trees. They include silver maple, eastern cottonwood, black walnut, green ash, sweet gum, sycamore, hybrid poplar and hybrid willow. These plant species help to improve water (capillary water, electrical conductivity), and soil quality (organic matter, microbial load, moisture, water holding capacity), or wildlife habitat related to annual crops, diversify sources of income and improve overall farm productivity. Mixed or single microbial culture can be employed to transform energy crops into bio-hydrogen. Utilization of microorganisms, microalgae, plant biomass is utilized as energy crops for biohydrogen production.

9.2.2 Biological Solid Waste

Refers to any organic material generated from living organisms or their metabolic processes. This includes waste composition based on agricultural practices and waste sources; composition and variability of biological solid waste; and factors influencing waste generation and characteristics. These wastes can come from various sources and may include those shown in Figure 9.1 and described below.

Figure 9.1 Different types of biological solid waste. (Self made)

9.2.2.1 Bio-Agro Wastes

Agricultural waste denotes the remnants acquired from agricultural activities, including the processing and manufacturing of agricultural commodities such as vegetables, crops, meat, poultry, fruits, and dairy products [83]. Broadly, agricultural wastes can be grouped into four categories: livestock waste, waste from agricultural industries, and waste from fruits, vegetables, and crop residues (FVWs) [97].

9.2.2.2 Agricultural Crop Residues

Crop residues such as stalks, leaves, husks and straw left over after harvesting, as well as animal manure and bedding materials from livestock operations. Globally, there is an abundance of agricultural waste suitable for biomass production, including stalks, leaves, husks, cobs, and various straw types like wheat, oat, barley, and rice straw. Farmers have the opportunity to sell this waste for biomass generation, providing them with additional income without disrupting feed, food, or fiber production. Stalks and leaves are prevalent and diverse across the United States, with examples including wheat straw, corn straw, and corn stover (comprising stalks, leaves, husks and cobs). Selling residue to local bio-refineries presents farmers with an income-generating opportunity.

9.2.2.3 Agro-Industrial Residues

Waste generated from food processing industries, breweries, distilleries, and other agricultural processing facilities, including fruit place, spent grains, and sugar beet pulp. Agro-industrial residues refer to the by-products generated from agricultural and industrial processes that involve the production, processing, or transformation of agricultural raw materials. These residues are often considered waste but can be repurposed into value-added products through sustainable practices. Common examples of agro-industrial residues include fruit and vegetable waste, sugarcane bagasse, animal waste, dairy waste.

9.2.3 Biomass Residues

Organic waste from forestry, wood processing, and bioenergy production, including wood chips, sawdust, and agricultural residues used for bio-fuel production. Biomass residues refer to organic materials that are the by-products of agricultural, forestry, or industrial activities, and are not directly consumed or utilized in their original form. These residues, often considered waste, can be repurposed as a renewable resource for energy production, biofuels, or other value-added products. They are a critical component in sustainable energy systems and the circular economy. Biomass residues can be categorized into several types: agricultural biomass residues, forestry biomass residues, industrial biomass residues, animal biomass residues.

9.2.4 Forestry Residue

Forest-related biomass sources include dead, poorly formed, unmerchantable or whole trees specifically planted for biomass generation. This biomass could be harvested without having a negative impact on the environment [42]. Forest biomass includes forest residue left after logging timber (limbs, tops, and culled trees and tree components) and whole-tree biomass harvested explicitly for biomass. Dead, diseased, poorly formed trees are often left in the woods after timber harvest. There are also opportunities to use excess biomass in millions of acres of forests. Harvesting excessive woody biomass can reduce the risk for fire and pests as well as aid in forest restoration, productivity, vitality and resilience. This biomass could be harvested for bioenergy without negatively affecting the health and stability of forest ecological structures and functions.

9.2.5 Aquatic-Biomass

Aquatic biomass is a promising form of biothermal energy, and there are many types of aquatic biomass that can be used for this purpose. Macroalgae, microalgae, aquatic plants, and aquatic animals are some of the most promising forms of aquatic biomass. Organic matter derived from aquatic environments, such as algae, seaweed, and aquatic plants.

9.2.5.1 Algae

Algae, including microalgae and cyanobacteria, have the capability to utilize sunlight and nutrients to produce biomass suitable for biofuel production. They can be cultivated in various nutrient-rich environments, including agricultural and municipal wastewater.

9.2.5.2 Food Waste

Organic waste generated from food production, processing, distribution, and consumption. Food waste refers to any edible food that is discarded, lost, or uneaten. It includes both the waste generated at the household level and in the food supply chain. This includes food scraps, spoiled produce, expired products, and food processing by-products.

9.2.6 Yard Waste

Organic materials from residential and commercial landscaping activities, such as grass clippings, tree trimmings, leaves, and branches. Yard waste is typically collected separately for composting, mulching, or other eco-friendly disposal methods. Many municipalities offer curbside yard waste collection or have designated composting sites. Managing yard waste properly helps reduce landfill use and supports natural recycling through composting.

9.2.7 Green Waste

Green waste, also known as "green" or "organic" waste, refers to biodegradable waste that comes primarily from plants. Organic waste generated from parks, gardens, and green spaces, including plant trimmings, prunings, and landscape maintenance debris. Green waste can be processed into compost, mulch, or even biofuel, making it a key element in sustainable waste management. Composting green waste helps return nutrients to the soil and reduce the volume of waste sent to landfills.

9.2.8 Sewage Sludge

Sewage sludge is the semi-solid solid residue generated from wastewater treatment processes, containing organic matter, nutrients, and microorganisms. It consists primarily of organic materials, nutrients, chemicals, and pathogens that are removed during the wastewater treatment process. Properly treated sewage sludge, known as biosolids, can be beneficial when applied to land as fertilizer, as it is rich in nutrients like nitrogen and phosphorus that improve soil health. Regulations often govern the application of biosolids to protect public health and the environment.

9.2.9 Sorted Municipal Waste

This includes yard clippings, vegetable waste, leather, rubber, and textile waste [13]. Resources in municipal solid waste include residential and commercial garbage such as paper and paperboard, rubber, textiles, and food waste. Bioenergy municipal solid waste represents an opportunity to reduce commercial and residential waste by diverting volume from landfills.

9.2.10 Organic Municipal Solid Waste (MSW)

This refers to the biodegradable portion of waste that comes from households, businesses, and other municipal sources. It primarily consists of materials that can decompose naturally. Organic MSW is a significant part of the overall waste stream and includes household waste containing organic materials, such as food waste, kitchen scraps, paper products, and yard trimmings.

9.2.11 Biomedical Waste

Organic waste generated from healthcare facilities, including expired pharmaceuticals, biomedical samples, and biological tissues. These are just some examples of biological solid waste types, and the composition and characteristics of these wastes can vary depending on factors such as geographic location, industrial activities, and waste management practices.

Proper management and valorization of biological solid waste are essential for minimizing environmental pollution and maximizing resource recovery in various sectors, including agriculture, energy production, and waste management.

9.3 Sources and Availability of Biological Solid Waste in Agricultural Sectors

In the environment, a diverse array of agricultural waste exists, characterized by their origins and abundance. This waste primarily comprises cellulose, hemicellulose and lignin. Lignocelluloses, owing to their intricate structure, pose challenges in degradation [32, 110]. The agricultural and food processing industries generate vast quantities of waste, primarily in the form of crop residue and agro-industrial waste, constituting the two largest categories of available waste [4]. Cattle manure, a significant contributor to noxious gases, harmful microorganisms, and unpleasant odors, raises concerns regarding public health and environmental impact [101]. Improper disposal of agro-residue, undiluted chemicals, or pesticide

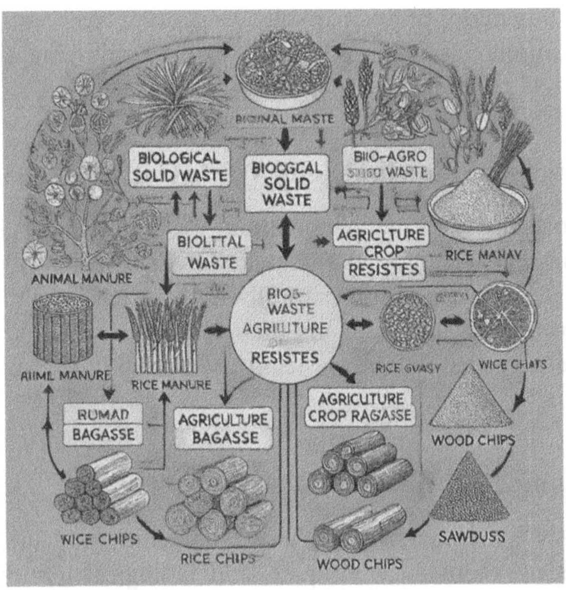

Figure 9.2 Sources and availability of biological solid waste in agricultural sectors. (Self made)

containers could lead to contamination of surface water and groundwater. Managed utilization of farm waste has the potential to enhance irrigation and erosion control practices [112] (Figure 9.2).

9.3.1 Agricultural Industry Wastes

Agricultural wastes include residues generated from agricultural industrial processes. This includes by-products and leftover derived from food processing industries, such as sugarcane bagasse, fruit peels, remaining after juice or oil extraction, eggs, meat, chicken skin, and animal fat from meat processing facilities and slaughterhouses [88]. It also comprises starch residue from starch manufacturing plants, de-oiled seed cakes from edible oil production facilities, and molasses/black treacle from sugar manufacturing facilities.

Among the various agricultural industrial wastes, sugarcane bagasse, this fibrous residue is obtained after the extraction of juice in industries, with a projected global availability of around 180 million tons [87]. Additionally, residues like orange peel, apple pomace, and fruit wastes resulting from fruit juice extraction, cider production, and other food processing units are considered agricultural industrial wastes. Moreover, certain non-food-based agricultural industries, such as de-oiled seed cakes obtained from non-edible oil plants like *Pongamia pinnata* and *Jatropha curcas* are agricultural industrial wastes [68].

9.3.2 Crop Residues

The waste residues directly resulting from agricultural production mainly consist of crop residues like stovers, seed pods, leaves, and straws found at the field level. These crop residues are considered the most abundant and cost-effective organic waste, capable of being transformed into various value-added products [96]. Globally, wheat straws, corn straws and rice straws are the primary crop residues utilized in bio-ethanol production. These crops are available year-round, with a small portion used for biofuel production or as fodder, while the remainder is often burned, leading to significant environmental challenges [6]. Among these three major crop residues, rice straw stands out as one of the most promising and valuable biomasses worldwide, with a substantial annual production of about 731 million tons, primarily in Asia [1]. Additionally, crop residues from sorghum, barley and oats are also agricultural waste [49].

9.3.3 Livestock Wastes

Crop cultivation and livestock rearing support many farming families globally. They utilize livestock manure as an energy source and are widely practiced in some communities for biogas production or by forming manure cakes for combustion [14]. Examples include using chicken litter as fuel, digesting animal slurry, and utilizing significant quantities of straw for combined heat and power generation [50]. However, the economic viability of such projects often depends on finding markets for the generated heat (industrial use in nearby factories) and by-products (fertilizers for agricultural use). Generated power can be utilized on the farm or distributed to nearby communities through piping, and surplus waste heat can be employed on the farm itself [50].

9.3.4 Fruit and Vegetable Wastes

Fruit and vegetable wastes include unprocessed fruits and vegetables like oranges, mangoes, pineapples, bananas, tomatoes, potatoes, jackfruit and others, constituting a significant portion of agricultural waste (Table 9.1). Fruits and vegetable wastes generated from food processing industries and wholesale markets are substantial [94]. Additionally, these organic wastes pose significant environmental concerns due to their highly biodegradable nature [83]. Statistics indicate that the waste generated from the processing, packaging, distribution and consumption of fruits and vegetables in countries like the Europe, United States, China, the Philippines, and India amounts to approximately 15, 32, 6.5, and 1.8 million tons, respectively [83]. Currently, in India, approximately 5.6 million tons of fruits and vegetable wastes are produced annually and are disposed of by dumping on the outskirts of cities [85] (Figure 9.2).

Table 9.1 Optimum condition for extraction of agricultural food waste and by-products [94].

Sr. no.	Agricultural food waste/ byproduct	Extraction method	Optimal conditions
1.	Kiwi juice pomace	Microwave-assisted extraction with different conditions	Extraction at a microwave power of 400 W and pressure of 350 psi→ solvent composition: 50% ethanol: water, solid-to-solvent ratio: 1:15 at 75 °C for 15 min

(Continued)

Table 9.1 Optimum condition for extraction of agricultural food waste and by-products [94]. (*Continued*)

Sr. no.	Agricultural food waste/ byproduct	Extraction method	Optimal conditions
2.	Pistachio hard shells	Extraction with different solvents/microwave assisted extraction	Microwave-assisted extraction→ using EtOH at 1000 W for 270 s
3.	Tomato processing waste	Ultrasound-assisted extraction/conventional organic solvent extraction	Ultrasound-assisted extraction→ solvent: hexane: acetone:ethanol (2:1:1 v/v/v) including 0.05% (w/v) butylated hydroxy toluene (BHT), solid liquid ratio (1:35 w/v) at 15°C, 90 W for 30 or 15 min
4.	Pulp of hot pepper paste	Ultrasound-assisted extraction/ maceration extraction	Ultrasound-assisted extraction→60% amplitude and 60°C for 5 min optimal condition for maceration extraction for 8 h → 50 °C
5.	Beetroot waste	Pressurized liquid extraction (at 40°C and 7.5, 10, and 12.5 MPa for 90 min, flow rate of 3 mL min^{-1})	Pressurized liquid extraction at 40°C for 90 min→ the mixture of ethanol:water 70/30 at 10 MPa
6.	Grape pomace	Ultrasound-assisted extraction/ microwave-assisted extraction	Microwave assisted extraction→ with 2% citric acid at 1000 W for 10 min
7.	Sesame bran	Enzyme-assisted extraction/ Ultrasound-assisted extraction/ultrasound-assisted enzymatic extraction	Ultrasound-assisted enzymatic extraction→ pH (9.8), and 1.248 AU (Anson unit)/100 g enzyme concentration at 836 W, 43°C for 98 min
8.	Coffee husk	Supercritical CO_2 extraction	Supercritical CO_2 extraction→ solvent to raw material mass ratio: 197 kg CO_2/kg husks at 373 K and 300 bar
9.	Pumpkin seeds	Ultrasound-assisted three phase partitioning extraction	Ultrasound-assisted three phase partitioning extraction→ $(NH_4)_2SO_4$ addition:30 g/100 mL, a *t*-butanol to slurry ratio of 1.0:1.0 (mL:mL), pH 5, and a duty cycle of 60% at 118W for 20 min irradiation time
10.	Rice bran	Conventional organic solvent-based Soxhlet extraction/subcritical CO_2 Soxhlet extraction	Subcritical CO_2 Soxhlet extraction→ solvent to feed ratio: 24:1 at 68–70 bar and 27–29 °C

(*Continued*)

Table 9.1 Optimum condition for extraction of agricultural food waste and by-products [94]. *(Continued)*

Sr. no.	Agricultural food waste/ byproduct	Extraction method	Optimal conditions
11.	Wheat bran	Ultrasound-assisted enzymatic extraction	Ultrasound-assisted enzymatic extraction→ raw material concentration: 50 g/L, enzyme dose: 4.5 g/L, at 180 W, 50 °C for 70 min
12.	Green plantain peels	Wet extraction	Wet extraction→ 5% w/v ascorbic acid concentration
13.	Broccoli stalk	Extraction with 0.1 M nitric acid	Extraction with 0.1M nitric acid→ liquid to solid ratio of 25 (v/w), 30 min
14.	Watermelon rind	Microwave-assisted extraction with different acid solutions(for15min at 39.9 W)	Microwave-assisted extraction→ 1N sulfuric acid for 15 min at 39.9
15.	Banana peels	Extraction with 0.5 N hydrochloric acid/citric acid (at 90 °C for 1, 2, 3, 4 h)	Extraction → 0.5 N HCl at 90°C for 4 h
16.	Orange peels	Hot water extraction/rapid solid liquid dynamic (RSLD) extraction/microwave-assisted extraction	Acidic hot water extraction→ liquid-to-solid ratio (20), pH (1.5) at 70 °C for 60 min
17.	Kiwi seeds	Conventional Soxhlet extraction/microwave-assisted extraction/supercritical CO_2 extraction/ultrasound-assisted extraction/microwave integrated Soxhlet extraction	Ultrasound-assisted extraction→ 10 mg seed powder in 400 mL *n*-hexane at 50°C, 80 W for 30 min
18.	Lemon peels	sequential extraction with a combination of microwave-assisted extraction and hydrodistillation and microwave assisted extraction	Combination of microwave-assisted extraction and hydrodistillation → water to solid ratio: 0.3 mL/g, 1st step (irradiation power:1.2 W/g for 5 min) and 2nd step (irradiation power: 0.7 W for 15 min
19.	Lemon peels	Conventional Soxhlet extraction/high pressure high temperature extraction	High pressure high temperature extraction→ at 150°C for 30 min and matrix solvent ratio: 1:15
20.	Citrus peels	Conventional hydrodistillation/ microwave-assisted hydrodistillation	Sure (300 mbar), solid to liquid ratio (1:1.5)→ 1st step (irradiated with 785 W for 5 min) and 2nd step (irradiated with 250 W for 15 min)

9.4 Methods of Biological Solid Waste Management

9.4.1 Bioethanol Production

The raw materials, including wheat straw, rice straw, corn straw and sugarcane bagasse, have the potential for the production of bioethanol. These material are rice in cellulose, hemicellulose and lignin (Figure 9.3). There are three processes for bioethanol production: (1) physical pretreatment: this technique is used to free cellulose and hemicellulose from lignocellulose; (2) hydrolysis of cellulose and hemicellulose so that fermentable

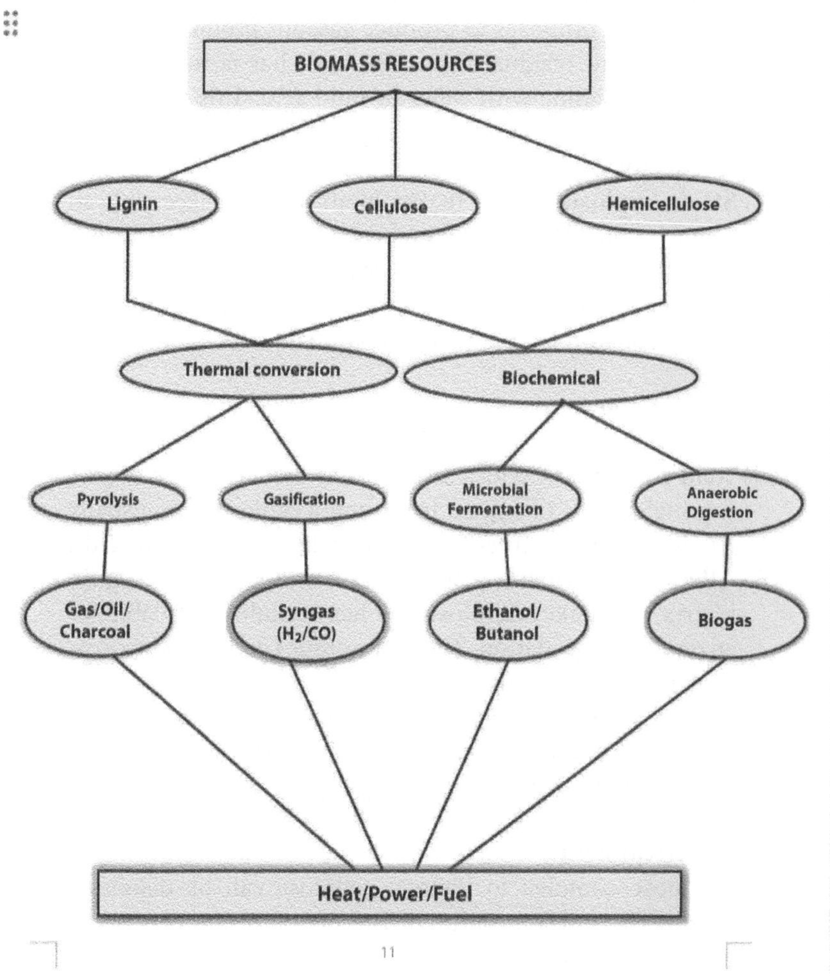

Figure 9.3 Biomass sources and their valorization [119].

sugars are produced, such as glucose, xylose, arabinose, galactose, and mannose; and (3) fermentation of reducing sugars.

Physical pretreatment: This is a technique in which raw material is treated physically.

(A) **Mechanical size reduction:** This method includes mechanical size reduction through milling, comminution, grinding or chipping. This pretreatment can improve the quality of downstream processing. The method employed depends on the type of raw material used.
(B) **Pyrolysis:** In this method, materials are treated at 300°C and cellulose decomposes to produce gaseous products such as H_2, CO_2, and residual char. Residual char is further subjected to leaching with water or mild acid. This water contains enough carbon source to support microbial growth for bioethanol.
(C) **Steam explosion:** This makes biomass more accessible to cellulose attack. It involves heating using high-pressure steam for a few minutes. The reaction is stopped by sudden decompression to atmospheric pressure. The purpose of this reaction is to separate individual fibers.

Chemical pretreatment:

(A) **Acid treatment:** Acid treatment has the advantage of giving high yields of sugars from lignocellulose. It is usually carried out by treatment with concentrated or diluted acids.
(B) **Alkaline treatment:** This treatment digests the lignin matrix and makes cellulose and hemicellulose available for enzymatic degradation. The end residue can be used to produce paper or cellulose derivatives.
(C) **Wet oxidation:** This treatment feedstock material is treated with water and by air or oxygen above 120°C. The product of hemicellulose hydrolysis is sugar polymers.

Biological pretreatment:
The lignocellulose complex to liberate cellulose can be degraded with brown rot, white rot, and soft rot fungi. The aim of this process is to release sugars from the lignocellulose matrix [5].

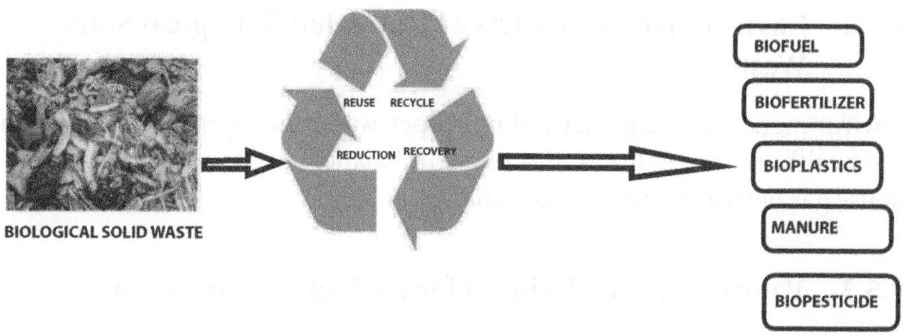

Figure 9.4 Biological solid waste valorization is a process that aims to reduce, reuse, and recycle the waste into usable, value-added product development. (Self made)

9.5 Valorization

The valorization of biological solid waste presents a significant opportunity to address both environmental challenges and agricultural sustainability. Waste valorization is a process that aims to reduce, reuse, and recycle the waste into usable, value-added, and environmental benign raw materials which can be a source of energy. It is an alternative to traditional waste disposal and has become an important concept for businesses and cities (Figure 9.4).

Waste valorization involves converting waste materials into useful products or resources, thereby adding value to what would otherwise be considered waste. It encompasses various processes such as recycling, upcycling, biorefining and energy recovery. Organic waste materials, such as food waste, agricultural residues, and wastewater sludge, can be valorized into products such as biofuels, biochemicals, fertilizers, bioplastics, and other high-value materials. Valorization can help reduce the environmental impact of waste disposal, conserve resources, and contribute to the circular economy by promoting the reuse and recycling of materials (Figure 9.4).

9.5.1 Novel Approaches for the Valorization of Biological Solid Waste

This valorization technique explores various strategies for effectively utilizing biological solid waste to enhance agricultural productivity while minimizing environmental impact. Overview of the significance of biological solid waste in agriculture. Importance of sustainable waste management practices. Overview of the importance of sustainable waste management in agriculture. Introduction to the concept of valorization and its potential benefits for agricultural sectors.

9.5.2 Environmental Impacts of Untreated Biological Solid Waste

Environmental consequences of improper waste management. Impact on soil health, water quality, and ecosystem balance. Contributions to greenhouse gas emissions and climate change.

9.5.3 Valorization Techniques for Biological Solid Waste

Composting: Principles, methods, and applications in agriculture.
Anaerobic digestion: Process, biogas production, and nutrient recovery.
Vermicomposting: Mechanisms, benefits, and considerations for large-scale implementation. All are innovative valorization methods and emerging technologies.

9.6 Value-Added Products

Utilizing food waste as a substrate enables the recovery of value-added products such as fine chemicals, nutraceuticals, antioxidants, bioactives, biopolymers, biopeptides, antibiotics, industrial enzymes, bio-nanocomposites, single-cell proteins, polysaccharides, activated carbon adsorbent, chitosan, corrosion inhibitors, organic acids, pigments, sugars, wax esters, and xanthan gum [58, 113].

Conventional methods of treating food waste, such as landfilling and incineration, result in environmental, economic, and social challenges. Therefore, various valorization methods have emerged as sustainable and profitable alternatives to manage food waste and obtain the aforementioned value-added products. Moreover, specialized chemicals derived from food waste, ranging from solvents to antioxidant materials, play crucial roles in nutraceutical and biomaterial applications [113]. To selectively extract and modify preferred components from food wastes and by-products, combined methods incorporating biochemical, chemical, and physical steps should be employed. These methods must be implemented cautiously to mitigate microbiological risks and ensure that the final products meet consumer preferences and adhere to food regulations [8].

Efficient utilization of renewable resources and the harnessing of renewable carbon are essential for substituting fossil resources in the production of chemicals, materials, polymers, fuels, and energy. Industrial sustainability can be achieved through various means, including the extraction of specialty products, conversion *via* green chemical or biotechnological processes, integrated biorefining, industrial symbiosis, cascade processing, and on-site processing of seasonal waste streams. These strategies involve effectively exploiting agricultural and forestry residues, aquatic biomass, and diverse waste streams [66].

Co-products unsuitable for food applications should be repurposed as energy sources through fermentation, biogas production, and composting. This highlights the potential of an integrated biorefinery approach for valorizing food waste and producing bioactive molecules for pharmaceutical, cosmetic, food, and non-food applications [8].

9.7 Valuable Biomaterials

Biopolymers, bioplastics, biofertilizers, enzymes, organic acids, single-cell protein (microbial biomass) are also obtained from agricultural food wastes/byproducts with the application of different treatments like fermentation and composting. These valuable biomaterials are used in the cosmetics, pharmaceutical, chemical, food, and beverage industries [75].

Single-cell protein, the extracted protein from microbial biomass like bacteria, yeast, algae, and fungi, can be used as a supplement protein source instead of conventional high-cost protein sources in the staple human diet to alleviate problems related to protein scarcity [3, 74]. Besides the nutritional benefit of using single-cell proteins in human or animal diet, another advantage is reducing the costs of fish products during the formulation of food and fodder stocks, rich in protein, by using bioconversion products from wastes of agriculture and industry.

Biopolymers, bioplastics, biofertilizers, enzymes, organic acids, and single-cell protein (microbial biomass) can be derived from agricultural food wastes or by-products through various treatments such as fermentation and composting. These valuable biomaterials find applications in industries including cosmetics, pharmaceuticals, chemicals, food, and beverages [75].

Single-cell protein, which is extracted protein from microbial biomass like bacteria, yeast, algae, and fungi, offers an alternative protein source to conventional, high-cost protein sources in the human diet, addressing issues related to protein scarcity [3, 74]. Apart from the nutritional benefits of incorporating single-cell proteins into human or animal diets, another advantage is the cost reduction in the final products by utilizing bioconversion products from agricultural and industrial waste in protein-rich food and fodder formulations.

9.8 Biorefinery for Agricultural Food Wastes/By-Products

The efficient utilization of by-products obtained during biomass production, such as agricultural residues, food processing waste, and food leftovers, has the potential to enhance the global bioeconomy [61]. The concept of biorefinery, widely embraced as a sustainable alternative within the scientific community, revolves around energy and commercial production through the recycling of food waste. Biotechnological methods, including anaerobic digestion, fermentation, and composting, facilitate the conversion of abundant and low-cost waste biomass into biorefinery products such as biofuels, biomass biofertilizers, bioplastics, and secondary chemicals [61, 75, 78].

Furthermore, these biotechnological methods have the capability to transform agricultural food wastes into effective biobased adsorbents utilized in the bioremediation of various pollutants present in wastewater [75]. Utilizing food wastes more extensively in chemical synthesis aims to reshape public perceptions of food waste and establish a closed-loop economy by establishing a renewable supply chain. For instance, the EU recently reached a provisional agreement on renewable energy policy, with a focus on mandating the development of biofuels derived from waste. This agreement outlines the significance of a biorefinery based on food wastes in contributing to a more sustainable and environmentally friendly society in the future [113]. As a result, managing food waste within the framework of biorefinery concepts yields positive environmental outcomes, including reduced greenhouse gas emissions, decreased environmental burden associated with waste disposal, and reduced dependence on fossil-based sources for fuel generation [75, 78].

9.9 Biofuel

Biofuel, derived from biomass and its refined products, encompasses solid, liquid, and gaseous forms, including bioethanol, biodiesel, biokerosene, and natural gas [47]. Throughout human history, biofuel has been extensively utilized for activities such as cooking, lighting, and heating. The global support for biofuel production as an alternative to petroleum and coal-based fuels is increasing due to concerns about production issues and sustainability [55].

Efforts are underway worldwide to utilize food waste as a source of fuel. For instance, Nordic countries have implemented policies to boost the utilization of food waste and explore new technologies to convert waste into transport fuels. Denmark, Finland, and Sweden are currently utilizing food waste from various sources, including fruit and vegetable waste, animal-based waste, bakery waste, and household, agricultural, industrial, and commercial biowaste, in the production of biodiesel, bioethanol, and biogas.

Biodiesel, derived from plant oils like soybean, rapeseed, and canola, is produced through direct or indirect transesterification [55]. Researchers have demonstrated the use of low-cost food waste in biodiesel production by extracting lipids from food waste through fungal hydrolysis, followed by transesterification.

Biogas, another biofuel and renewable energy source, is obtained by anaerobic digestion of agro-food biomass residuals, aligning with the renewable energy legislation of the EU [16]. However, certain components in biogas, such as hydrogen sulfide and carbon dioxide, reduce its calorific value compared to natural gas. Various physicochemical and biological techniques are employed to upgrade raw biogas quality. Innovative technologies, such as solid-state micro-anaerobic digestion, have been developed to valorize food waste by enhancing anaerobic digestion efficiency [53]. Bio-alcohol, particularly ethanol, is a promising alternative liquid fuel produced through microbial fermentation of feedstocks like potatoes, molasses, corn, stover, wheat, sugar cane, and various carbohydrate-rich sources [55, 75]. Biochar, a carbon-rich biofuel, is produced through charring or pyrolysis of biomass above 250°C under limited or no air conditions [63]. It finds applications in soil amendment and environmental management due to its ability to mediate soil acidity, water holding capacity, nutrient retention, and cation exchange capacity. Additionally, biochar serves as a suitable electrode in supercapacitors, facilitating green energy storage.

9.10 Valorization of Biowastes for Production of Biofertilizer

Biowaste, stemming from agricultural residues, food scraps, and organic municipal waste, presents a pressing challenge in waste management. However, through innovative approaches, biowaste can be transformed into valuable biofertilizers, offering a sustainable solution to enhance soil fertility and promote agricultural productivity. This abstract explores the process, benefits, challenges, and applications of converting biowaste into biofertilizer.

The valorization of biowaste into biofertilizer involves several key steps. Initially, biowaste undergoes composting or microbial fermentation processes, where organic matter is broken down by microorganisms to release essential nutrients. Concurrently, specific strains of beneficial microorganisms, such as nitrogen-fixing bacteria and phosphate-solubilizing fungi, may be introduced to enrich the compost with plant-available nutrients. This nutrient-rich compost serves as the basis for producing biofertilizer formulations.

9.11 Agricultural Waste-Based Organic Fertilizers

Organic materials encompass various applications such as the production of biofertilizers and soil amendments, energy recovery, manufacturing of chemicals like alcohols and volatile organic acids, and utilization in farm animal nutrition. Organic fertilizers consist of a diverse range of plant, animal, and mineral-based components. Plant-based fertilizers include cotton seed meal, molasses, compost tea, legume cover crops, green manure cover crops, and kelp seaweed. Similarly, animal-based fertilizers include fish emulsion, milk, urea, and manure tea. Manure-based organic fertilizers are further categorized as calcium and Epsom salt (magnesium and sulfur). Compared to chemical fertilizers, these organic fertilizers generally have lower phosphorus and nitrogen content. Additionally, the moisture content in these organic-based fertilizers dilutes or reduces the concentration of phosphorus and nitrogen. Selected agricultural by-products and animal manures are commonly utilized as organic fertilizers. The application of organic fertilizers promotes an increase in heterotrophic bacterial biomass, facilitating nutrient mineralization to stimulate primary productivity and support other secondary productivity [31].

Traditional methods for treating agricultural waste include composting and anaerobic digestion. Common agricultural wastes used for producing biofertilizers include residues from vegetable processing (such as aubergines and peppers), livestock farming (such as cattle manure), and olive oil production (including wet olive pomace and olive pruning). Olive pomace, in particular, is a significant organic waste in Mediterranean countries, containing a high organic matter content of around 90%. Therefore, recycling this agricultural waste through composting can yield profitable biofertilizers for agricultural use [70]. Digestates, on the other hand, are by-products derived from biogas plants that produce methane gas through anaerobic digestion of various organic materials.

Livestock manure, derived from farm animals, can be separated into liquid and solid fractions. The solid fraction of the digestate typically contains 60-80% phosphorus and organic matter, 20-25% nitrogen, and 10-15% potassium [30].

Anaerobic digestion is a process involving the decomposition of organic matter through microbial action in the absence of oxygen. It can be applied to various feedstocks, including plant residues, industrial and municipal wastewaters, and agricultural, municipal, and food industry wastes. Anaerobic digestion produces biogas, which offers several advantages over other waste treatment methods. These benefits include reduced production of biomass sludge compared to aerobic processes, more efficient removal of pathogens, minimized odor emissions due to the decomposition of volatile compounds during combustion (e.g., H_2S forms SO_2), and the production of digestate (slurry), which serves as an enhanced fertilizer for increased plant yields. The stepwise process of anaerobic digestion from collected wastes (agricultural wastes and manures), as well as the co-products (primarily solid fertilizers and soil amendments, along with liquid) obtained from the anaerobic digester (Figure 9.5).

However, digestates can emit phytotoxicity and odor, limiting their direct application to agricultural soils. Therefore, pre-treatments are necessary to produce high-quality bio-based fertilizers that support sustainable crop cultivation without posing environmental risks. For improving the quality of such pre-processed solid fraction co-composting is done, which can provide degradable materials to substitute the activity caused by microbes during the waste treatment [105]. The co-composting method is biological degradation under aerobic conditions in a controlled manner, where the organic compounds are converted into shorter molecular chains and more stable humus. This enables enhanced agricultural-related production and ways to recycle and reuse the organic matters [79]. The obtained anaerobic digestate (co-composted) can easily be obtained from the farms;

including the waste products available in the farm provides better prospects as a partial substitute or alternative for the inorganic-based fertilizers. Also, the use of organic-based fertilizers improves the actions caused by the microorganisms. In addition, it has been observed that the application of organic-based fertilizers improves water- and nutrient-holding capacity of the soil, soil organic matter, crop yield, and growth vigorously [17]. But these organic-based fertilizers offer different disadvantages, one of which is that the nutrients required are not available immediately to plants and need to be broken down by microorganisms present in soil into simple and inorganic ions and molecules. Organic fertilizers can lead to pollution of the groundwater after the process of decomposition, when applied in huge volumes. Organic-based fertilizers contain lower amounts of macronutrients and micronutrients, and a large amount of these materials is needed to provide the plant requirements for essential nutrients. Also, use of these organic-based fertilizers requires many skilled laborers and is an expensive process. Therefore, using agricultural wastes to produce organic-based fertilizers is both an economically and environmentally viable process.

9.12 Bio-Waste-Derived Fertilizers: Properties and Benefits

Bio-Waste-Derived Fertilizers: Properties and Agricultural Benefits:

- Nutrient content and quality of bio-waste-derived fertilizers
- Effects on soil fertility, microbial communities, and crop growth

Comparative Analysis with Conventional Fertilizers:

- Examination of the nutrient content and agronomic benefits of bio-waste-derived fertilizers.
- Comparison of bio-waste-derived fertilizers with conventional chemical fertilizers in terms of effectiveness and environmental impact.

Biological waste valorization involves the conversion of organic waste materials into valuable products using biological processes. This can include composting, anaerobic digestion, fermentation, and other biochemical processes. Biofertilizers and nanofertilizers are two examples of valuable products that can be derived from biological waste valorization processes.

9.12.1 Biofertilizers

- Biofertilizers are microbial inoculants that contain beneficial microorganisms such as bacteria, fungi, or algae, which enhance nutrient availability, soil fertility, and plant growth.
- They can be produced from organic waste materials through processes such as composting or fermentation, where specific microorganisms are isolated, cultured, and applied to soil or plant surfaces.
- Biofertilizers provide nutrients to plants in a form that is readily available and easily absorbed, promoting sustainable agriculture practices by reducing reliance on chemical fertilizers.
- Examples of biofertilizers include nitrogen-fixing bacteria (*Rhizobium, Azotobacter*), phosphate-solubilizing bacteria (*Bacillus, Pseudomonas*), and mycorrhizal fungi.

9.12.2 Biofertilizers Derived from Biowaste Offer Numerous Benefits

- **Soil Health Enhancement:** Biofertilizers replenish soil nutrients and improve soil structure, promoting a healthy soil microbiome and enhancing nutrient cycling processes.
- **Sustainable Agriculture:** By reducing reliance on chemical fertilizers, biofertilizers mitigate environmental pollution, minimize soil degradation, and support sustainable agricultural practices.
- **Increased Crop Yield and Quality:** Biofertilizers enhance nutrient uptake by plants, leading to increased crop yield, improved crop quality, and enhanced resistance to environmental stressors.
- **Carbon Sequestration:** The composting process involved in biofertilizer production contributes to carbon sequestration, aiding in climate change mitigation efforts.
- **Circular Economy:** Valorizing biowaste into biofertilizer closes the loop on organic waste streams, transforming waste into a valuable resource and promoting circular economy principles.

Despite these benefits, challenges remain in scaling up biofertilizer production, ensuring product quality and consistency, and navigating regulatory frameworks. Collaboration among researchers, policymakers, waste managers, and agricultural stakeholders is essential to address these challenges and promote the widespread adoption of biofertilizers derived from biowaste. The quality of soil is predominantly degraded by both climate change and human activities, leading to significant declines in productivity and ecosystem functioning. Therefore, it's crucial to develop concepts and strategies aimed at restoring soil fertility to support the rapid growth of the human population and meet increasing demands. Utilizing agricultural wastes as organic amendments represents a valuable strategy for mitigating greenhouse gas emissions such as carbon dioxide and carbon monoxide while sustaining productivity. Agricultural wastes from various sources, when recycled, can be effectively utilized in the production of fertilizers. Incorporating such organic wastes into agriculture contributes to soil fertility enhancement and crop productivity by improving nutrient levels, soil porosity, aggregate stability, cations, and exchange capacity [39]. Additionally, organic amendments foster microbial activity in the soil, resulting in changes to bacterial communities [36]. Therefore, fertilizers are employed to enhance soil fertility, serving as key components to achieve optimal agricultural yields. Before agricultural waste can be utilized, it typically undergoes various treatment processes (Figure 9.5). Agro-waste residues primarily consist of lignocellulosic materials, which necessitate chemical, physical, and biological treatments to facilitate their breakdown [29]. Pretreatments such as mechanical, thermal, and sonic methods enhance the biodegradation kinetics of wheat straw but require significant energy input [89] (Table 9.2). These pretreatments reduce biomass size, increasing its surface area and facilitating hydrolysis through improved heat and mass transfer [86].

Chemical pretreatments, involving acids, alkalis, ozonolysis, organosolv, and wet oxidation, break internal linkages of lignin and hemicellulose [117]. For instance, pretreatments with H_2SO_4 significantly reduced cassava bagasse weight and increased cellulose yield compared to untreated bagasse [20]. Biological processes, employing microbes, involve hydrolysis and saccharification mediated by exoenzymes, followed by fermentation using whole-cell systems or enzymatic digestion of solid or liquid substrates.

Green pretreatments methods, such as ionic liquids (ILs), deep eutectic solvents, and natural deep eutectic solvents, have been explored. ILs pretreatments, for example, improves glucose production from enzymatic hydrolysis by altering cellulose crystallinity and reducing particle size and lignin content [72].

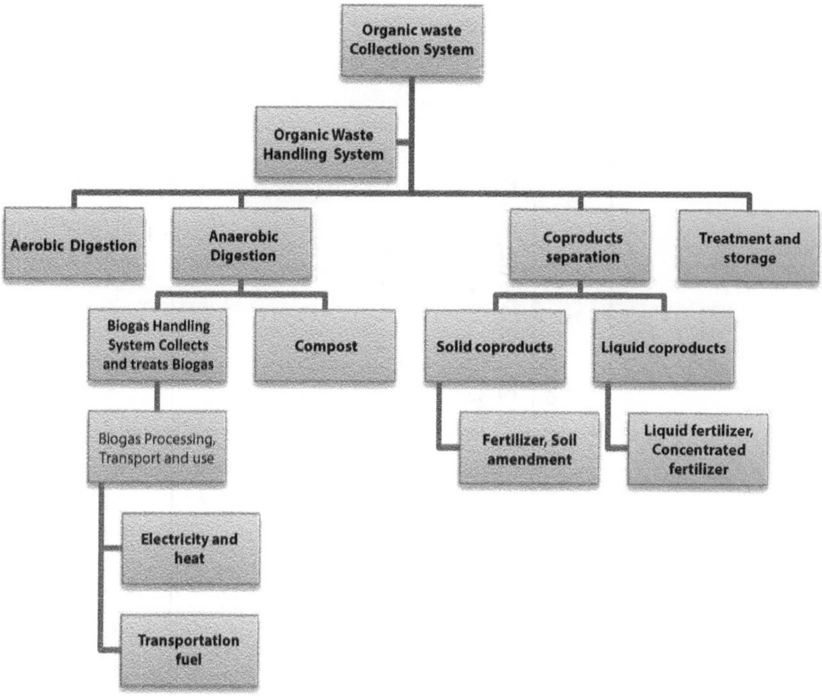

Figure 9.5 Anaerobic digestion process for obtaining organic fertilizers. (Modified from source: https://www.epa.gov/agstar/anaerobic-system-design-and-technologys)

Enhanced breakdown of lignin and cellulose occurs by disrupting intramolecular and intermolecular hydrogen bonds [69]. These pretreatment techniques enhance enzyme accessibility for hydrolysis, increasing surface area while reducing operational costs [82]. Also discussed were strategies for valorizing fruit wastes into value-added compounds and potential approaches for converting agro-waste into useful bioproducts and biochemicals.

9.12.3 Bioactive Compounds from Agro-Waste

Functional foods contain higher levels of bioactive compounds compared to regular foods, which are believed to offer superior health benefits. These substances contribute to improved health outcomes and are considered more advantageous than conventional foods (Table 9.3). Food additives are incorporated during food processing to enhance food quality and prolong shelf life [11]. In recent years, increased consumer interest in health and wellness has driven the growth of the dietary supplement and

Table 9.2 Application-based pretreatment methods and related pros and cons [89].

Application	Pretreatment methods	Pros	Cons	References
Agricultural Sector Biofuels and manure Enzymatic digestibility Ethanol production Bio-oil and biochar formation	Grinding Ultrasonic Steaming explosion Microwave Pyrolysis Irradiations	Biomass, a fine powder with a crystallinityof up to 0.2 mm Is produced. Easing the process of breaking down a variety of lignocellulosic materials.	Lack of long-term viability in technique calls for a lot of energy	[2, 7, 28, 46, 59, 62, 73, 76, 84, 93, 108, 109, 112, 114–116]

Table 9.3 Agricultural waste fermentation by microorganisms [11].

Agricultural food waste/ byproduct	**Fermentation method**	**Microorganism**	**Effects of fermentation**
Mango seeds	Solid-state fermentation	*Aspergillus niger*	Mobilization of polyphenolic compounds improvement of nutraceutical properties practical method for releasing the bound phenolics
Plum fruit (*Prunus domestica* L.) byproducts	Solid-state fermentation	*Aspergillus niger* and *Rhizopus oligosporus*	Increase of total phenolic and favonoid contents (over 70% for total phenolics and 38% for total favonoids with *R. oligosporus*, and more than 30% for total phenolics and 12% for total favonoids with *A. niger*
Spent coffee grounds	Solid-state fermentation	*Bacillus clausi*	Significant increase of total phenolic and antioxidant levels (for total phenolic content: higher than 30% with *R. oligosporus* and higher than 21% with *A. niger*)
Spent coffee grounds	Solid-state fermentation	*Bacillus clausi*	Increase of total phenolic and *flavonoid* contents and antioxidant capacity (36%, 13%, and 15%, respectively) improvement of antimicrobial activity against gram (+) and gram (–) bacteria
Tomato seed meal extract	Solid-state fermentation	*Lactobacillus plantarum*	Increase of total proteins, soluble proteins, and protein hydrolysates amounts (2.7-, 2.2-, and 1.2-fold, respectively)
Peanut press cake	Solid-state fermentation	*Aspergillus awamori*	Reduction of crude and soluble proteins contents (18.44% and 68.99%, respectively) due to *L. plantarum* growth on the substrate after 24 h increase of radical scavenging activity after 24 h (87%) due to different bioactive peptides production significant reduction of total amino acids (specifically glutamic acid and aspartic acid) concentration improvement of new amides and aromatic compounds formation
Jussara pulp	*Lactobacillus* fermentation	*Lactobacillus* and *Bifidobacterium* strains	Improvement of phenolic and antioxidant properties more extensive changing of jussara anthocyanins by *Lactobacillus deubruekii* the main bioconversion products from anthocyanins → protocatechuic acid
Rice bran	solid-state fermentation	*Rizhopus oryzae*	Increase of phenolic compound content (more than 2-fold) the highest increase of phenolic acid → gallic and ferulic acid phenolic extract from fermented rice bran → inhibition of the peroxidase enzyme

nutraceutical markets. Nutraceuticals and dietary supplements encompass a variety of products with health-promoting properties [25]. However, nutraceuticals constitute a distinct category of dietary supplements as they contain pharmaceutical-grade substances but are not subject to the same testing requirements as pharmaceuticals. These products, often derived from plants, vegetables, and microorganisms, were initially developed for medicinal purposes due to their pharmacological properties [111].

9.12.4 Composting

Composting is a biological process that involves the decomposition of organic waste materials by microorganisms under controlled conditions. It typically involves mixing organic waste materials, such as food scraps, yard waste, and manure, with bulking agents such as wood chips or shredded paper to create a compost pile. Microorganisms break down the organic matter into compost, a nutrient-rich soil amendment that can be used to improve soil fertility and structure in agriculture, landscaping, and gardening. Composting reduces the volume of organic waste, stabilizes nutrients, and suppresses pathogens and weed seeds, resulting in a valuable end product that can substitute for chemical fertilizers and reduce the need for landfilling or incineration of organic waste.

Relationship between Waste Valorization and Composting

- Composting can be considered a form of waste valorization since it converts organic waste materials into a valuable end product (compost) that can be used to enrich soil fertility.
- However, waste valorization encompasses a broader range of processes and products beyond composting, including the production of biofuels, biochemicals, and other high-value materials from organic waste.
- Composting is a sustainable waste management practice that aligns with the principles of waste valorization by converting organic waste into a useful resource while minimizing environmental impacts.
- Waste valorization strategies may include composting as one of the methods for converting organic waste into valuable products, depending on factors such as the composition of the waste stream, available resources, and end-use applications.

9.13 Vermicomposting

Vermicomposting presents a cost-effective and environmentally friendly approach to managing agricultural waste. *Eisenia foetida*, one of the most common earthworm species utilized in vermicomposting, aids in the breakdown and stabilization of materials such as straw, leaves, twigs, stubbles, and various grasses and weeds generated during crop and agricultural production. Through the collaborative efforts of earthworms and aerobic microorganisms, vermicomposting leads to the stabilization of these organic materials [91]. Physicochemical analyses have shown that vermicomposting results in a reduction in total organic carbon (TOC) and the carbon-nitrogen (C/N) ratio, while simultaneously increasing nitrogen-phosphorus-potassium (NPK) content compared to traditional composting methods and other agricultural waste management techniques [27, 71]. Studies have also demonstrated that vermicomposting, particularly using *E. fetida* with cow dung, results in significantly higher concentrations of nitrogen, phosphorus, and potassium in the compost compared to compost made from other waste sources. Additionally, certain earthworm species have been found to secrete phosphatase enzymes that facilitate decomposition [71]. Consequently, vermicomposting serves as a valuable tool for promoting environmental conservation efforts.

9.14 Biofertilizers

Soil biofertilizers, derived from microorganisms like bacteria, cyanobacteria, fungi, and algae, are substances that enhance soil quality by introducing nutrients and carbon substrates. Common examples of soil biofertilizers include cyanobacterial supplements and formulations containing bacteria such as *Azotobacter* sp., *Azospirillum* sp., *Trichoderma* sp., and arbuscular mycorrhizal fungus (AMF), which are widely utilized in agriculture [30, 99]. Research has been directed towards creating biofertilizers from agricultural waste, including discarded fruits like melons, pineapples, oranges, bananas, and papayas. Notably, banana biofertilizer exhibited the highest potassium concentration among these sources [65]. Insects have the ability to assimilate the nutrients present in organic wastes, thereby reducing the volume of waste material and generating a more valuable and homogeneous biomass. For instance, the biotreatment of food waste using larvae of the black soldier fly (*Hermetia illucens*) results in the reduction of waste volume and the production of high-quality animal feed. This approach enables the recovery, recycling, and valorization of food waste materials as components of animal feed and grass fertilizers [64, 67].

9.15 Bioenergy Biofuels

Biofuels, derived from renewable organic biomass, are recognized as a means to reduce dependence on fossil fuels, mitigate greenhouse gas emissions—particularly in the transportation sector and enhance fuel supply security [9, 41, 80, 82, 108]. Among biofuels, the utilization of sugars derived from lignocellulose for ethanol production has garnered significant attention. Ethanol can be efficiently synthesized from glucose and other hexoses using the yeast *Saccharomyces cerevisiae*, which is widely employed in ethanol production [19]. Transitioning from ethanol to biogas production through anaerobic digestion leads to improvements in energy efficiency, greenhouse gas emissions, and biomass conversion. Large-scale production of acetone, butanol, and ethanol from starchy materials has been made possible through the utilization of solventogenic Clostridium [56].

9.15.1 Food Sector

A broad range of bioactive compounds derived from agricultural waste presents a rich source of natural antioxidants and antimicrobials with extensive potential applications in the food and pharmaceutical sectors. Additionally, various micro- and nano-encapsulation technologies are employed to encapsulate these bioactive compounds, thereby expanding their utility in the food industry. These bioactive compounds, sourced from agricultural waste or by-products, encompass polyphenolic compounds, vitamins, minerals, fatty acids, volatiles, anthocyanins, and pigments, all of which offer significant health benefits. For instance, the extracted bioactive compounds can be utilized in food fortification initiatives, including the development of innovative functional and health-promoting food products [57].

9.15.2 Pharma Sector

9.15.2.1 Antioxidant Properties

Agro-waste has also been utilized as a substrate for the development of antioxidant and anticancer drugs. For instance, the discarded parts of various fruits and vegetables, such as peels, which are commonly considered waste, have been studied for their antioxidant properties. Duda-Chodak and Tarko investigated the antioxidant capabilities of fruit seeds and peels, revealing that selected fruit peels exhibit notably high antioxidant activity and polyphenol concentration. Pomegranate by products, for instance, yield significant amounts of waste, containing potent antioxidants like ellagitannins, punicalagin, and punicalin [57].

9.15.2.2 Antibacterial and Anticancer Properties

Agricultural and food industry waste holds promise for enhancing the body's absorption of various pharmaceuticals. Phytochemical substances found in such waste are rich sources of both organic and inorganic compounds, including sugars and numerous phenolic compounds. These phenolic compounds possess therapeutic properties attributed to their anti-inflammatory, antibacterial, immunomodulatory, and antifungal effects. Moreover, agricultural waste fosters the production of antibiotics by microbes, further enhancing its antibacterial potential [40, 111]. Waste products such as orange and lemon juice residues contain abundant hydroxycinnamic acids and flavonoids, particularly flavanone glycosides, flavanones, and flavone aglycons [34].

9.16 Nanofertilizers

Nanofertilizers are fertilizers that incorporate nanomaterials, typically nanoparticles, to improve nutrient delivery, uptake efficiency, and targeted release. They can be synthesized from biological waste materials or conventional fertilizers through nanotechnology-based approaches, such as nanoparticle synthesis, encapsulation, or surface modification. Nanofertilizers offer several potential benefits, including controlled release of nutrients, reduced nutrient leaching and runoff, enhanced plant uptake efficiency, and improved soil fertility and crop productivity. Examples of nanofertilizers include nanostructured materials such as nanoclays, nanocomposites, nanoemulsions, and nanoparticles of essential nutrients like nitrogen, phosphorus, and potassium.

9.16.1 Relationship Between Biological Waste Valorization, Biofertilizers, and Nanofertilizers

- Biological waste valorization processes, such as composting or fermentation, can be used to produce organic materials and microbial biomass that serve as raw materials for biofertilizer production.
- Biofertilizers derived from biological waste valorization processes contribute to sustainable agriculture by recycling nutrients from organic waste materials and improving soil health and fertility.
- Nanotechnology can be applied to enhance the effectiveness of biofertilizers by developing nanofertilizers with controlled nutrient release and targeted delivery systems.
- Combining biological waste valorization with nanotechnology allows for the development of innovative fertilization solutions that optimize nutrient use efficiency, reduce environmental impacts, and promote sustainable agricultural practices.

9.16.2 Nanomaterials for Biomass Conversion

Nanotechnology provides novel approaches for converting biomass residues, including crop residues, animal manure, and food waste, into value-added products. Nanocatalysts enhance the efficiency of biochemical and thermochemical processes such as pyrolysis, hydrothermal conversion, and anaerobic digestion, leading to higher yields of biofuels, biochar, and biogas. Furthermore, nanomaterial-based catalysts facilitate the conversion of lignocellulosic biomass into bio-based chemicals and platform molecules, contributing to the development of a bio-based economy and reducing dependence on fossil fuels.

9.16.3 Nanofertilizers and Soil Amendments

Nanotechnology enables the development of precision agriculture practices through the targeted delivery of nutrients and growth-promoting substances to crops. Nanofertilizers, formulated with nano-sized nutrient particles, enhance nutrient uptake efficiency, reduce nutrient losses through leaching and volatilization, and promote plant growth and productivity. Similarly, nanostructured soil amendments, such as Nano composites and nanogels, improve soil fertility, water retention, and nutrient cycling, enhancing crop yields while minimizing environmental impacts.

9.16.4 Bioactives from Agro Waste: Micro/Nano Formulation and Food Application

The primary distinction between microencapsulation and nano-encapsulation lies in particle size, typically ranging from 1 micrometer (μm) to 1 millimeter (mm), and the duration of time microencapsulation has been utilized [51]. Microencapsulation has found applications across various industrial sectors, spanning agriculture, environmental, household, and personal care products. The utilization of supercritical carbon dioxide (CO_2) has shown promising benefits in ensuring precise particle size design in both microencapsulated and nano-encapsulated pharmaceuticals, along with the ability to control drug-loading processes under different temperature, pressure, and flow conditions [38]. One of the most attractive features of nano-encapsulation is its ability to protect active pharmaceutical ingredients from degradation [26].

In agriculture, nano-fertilizers offer advantages in nutrient management by significantly improving nutrient utilization efficiency. Nano-dimensional adsorbents bond with nutrients either individually or in combination, leading to slower nutrient release compared to conventional fertilizers [44]. Nano-encapsulation presents a promising approach to safeguarding various food components from specific physiological conditions or deterioration while also masking undesirable aromas and flavors. Reducing the particle size of bioactive compounds (BACs) could enhance their stability, solubility, bioavailability, and delivery, thereby augmenting their functional efficacy [18].

Through nano-encapsulation, food scientists and engineers can develop protective barriers capable of withstanding harsh environments, concealing unpleasant flavors and odors, and improving nutrient absorption, release, and delivery, while also enhancing the solubility of lipophilic bioactive substances in water [24].

9.17 Biofungicides and Valorization

Biofungicides and nanofungicides represent two different approaches to controlling plant diseases, with biofungicides being derived from living organisms or natural products, and nanofungicides incorporating nanotechnology to enhance fungicidal properties. The valorization of these technologies involves maximizing their effectiveness, scalability, and commercial viability.

Biofungicides are derived from living organisms such as bacteria, fungi, or viruses, as well as natural compounds with fungicidal properties. They work through various mechanisms including competition, antibiosis, parasitism, and induced resistance. Valorization of biofungicides includes optimizing production processes, enhancing efficacy through strain selection or genetic modification, and integrating them into sustainable agriculture practices. Nanofungicides utilize nanotechnology to improve the delivery and effectiveness of fungicidal agents. Nanoparticles can encapsulate or deliver active ingredients, enhancing their stability, solubility, and bioavailability. Nanofungicides may have increased efficacy against fungal pathogens, reduced environmental impact, and prolonged activity compared to traditional fungicides. Valorization involves optimizing nanoparticle synthesis, characterizing their interactions with pathogens and plants, and ensuring their safety for use in agriculture.

Biofungicides are biological agents used to control plant diseases caused by fungi. They are derived from living organisms or natural products and are considered environmentally friendly alternatives to synthetic chemical fungicides. These biofungicides work through various mechanisms such as competing for resources with pathogenic fungi, producing antifungal compounds, inducing systemic resistance in plants, or directly parasitizing the pathogen.

Valorization, in the context of biofungicides, refers to the process of enhancing the value of these biological agents, both economically and environmentally. This can include improving their efficacy, increasing their production efficiency, finding new applications, or optimizing their integration into agricultural systems.

9.17.1 Bio-Fungicides and their Valorization

Eco-Friendly Alternative: Biofungicides offer a sustainable approach to disease management, reducing reliance on synthetic chemicals that can have adverse effects on the environment, non-target organisms, and human health.

Biological Agents: Biofungicides can consist of beneficial microorganisms such as bacteria, fungi, or viruses, as well as natural compounds derived from plants or other sources with fungicidal properties.

Mechanisms of Action: These agents work through various mechanisms, including competition, antibiosis (production of antimicrobial compounds), parasitism, induced resistance, and alteration of the environment to make it less favorable for pathogen growth.

Research and Development: Valorization involves ongoing research and development efforts to improve the efficacy, stability, and applicability of biofungicides. This includes screening for new strains or compounds with potent antifungal activity, optimizing production processes, and conducting field trials to evaluate performance under different conditions.

Integration into Integrated Pest Management (IPM) Programs: Biofungicides are often used as part of integrated pest management programs, alongside other control measures such as cultural practices, crop rotation, and the use of resistant varieties. Valorization involves identifying the most effective combinations of control strategies for specific crops and diseases.

Commercialization and Market Development: Valorization also involves efforts to commercialize biofungicides and expand their market reach. This includes educating growers about their benefits, obtaining regulatory approvals, developing formulations suitable for different application methods, and establishing distribution channels.

Sustainable Agriculture: By reducing reliance on synthetic chemicals and promoting environmentally friendly practices, the valorization of biofungicides contributes to the promotion of sustainable agriculture, enhancing soil health, biodiversity, and ecosystem resilience.

9.18 Nanofungicides

Nanofungicides utilize nanotechnology to improve the delivery and effectiveness of fungicidal agents. Nanoparticles can encapsulate or deliver active ingredients, enhancing their stability, solubility, and bioavailability. Nanofungicides may have increased efficacy against fungal pathogens, reduced environmental impact, and prolonged activity compared to traditional fungicides. Valorization involves optimizing nanoparticle synthesis, characterizing their interactions with pathogens and plants, and ensuring their safety for use in agriculture.

Biofungicides are biological agents used to control plant diseases caused by fungi. They are derived from living organisms or natural products and are considered environmentally friendly alternatives to synthetic chemical fungicides. These biofungicides work through various mechanisms such as competing for resources with pathogenic fungi, producing antifungal compounds, inducing systemic resistance in plants, or directly parasitizing the pathogen. Valorization, in the context of biofungicides, refers to the process of enhancing the value of these biological agents, both economically and environmentally. This can include improving their efficacy, increasing their production efficiency, finding new applications, or optimizing their integration into agricultural systems.

9.19 Nanopesticides and Plant Protection

In pest and disease management, nanotechnology offers sustainable alternatives to conventional pesticides by enabling the controlled release and targeted delivery of active ingredients. Nano formulations enhance the efficacy and stability of pesticides, reducing the need for frequent applications and minimizing environmental contamination. Additionally, nanomaterials-based sensors and delivery systems enable real-time monitoring of pest populations and disease outbreaks, enabling timely intervention and reducing crop losses.

9.20 Valorization of Biowaste into Biopolymers and Bioplastics

Biopolymers, which are significant products derived from food wastes and by-products, encompass a diverse range of applications across various industries such as medicine, cosmetics, pharmaceuticals, food processing, water treatment, biosensor development, industrial plastics, and textile manufacturing. This broad utilization stems from their desirable properties, including biodegradability, biofunctionality, biostability, and biocompatibility [92].

Food waste is also employed in the production of bioplastics, including polyhydroxyalkanoates (PHA) and polyhydroxybutyrate (PHB), which are organic polymers capable of complete degradation into carbon dioxide and water within a matter of months when buried. Therefore, the production of bioplastics from food waste contributes to the reduction of both plastic waste and food waste [90]. Bio-based molecules, natural biopolymers, and phytochemicals can be acquired directly through extraction rather than synthesizing them from petroleum-based chemicals [61]. Once these valuable bio-based products are extracted, they can serve as high-value components such as food additives, nutraceuticals, therapeutics, and cosmetics [75]. However, the economic viability of extracting these high-value components needs consideration. To ensure economic feasibility, the desired components should be extracted using suitable methods to fully exploit the potential of the waste material [8].

Extraction techniques can be categorized into conventional and non-conventional methods [100]. Conventional methods, including solvent extraction, Soxhlet, maceration, and hydrodistillation, typically involve factors such as temperature, agitation, and the use of organic solvents like methanol, ethanol, and acetone [75, 100].

To optimize the extraction process and enhance the resistance of bioactive components, parameters such as temperature, contact time, pH, particle size, solid-to-liquid ratio, and stirring rate should be carefully selected. However, conventional techniques have drawbacks, including extended extraction times, large solvent volumes leading to increased toxic waste generation, and the need for additional steps like isolation or clarification to remove solvent residues or impurities due to the use of toxic and costly organic solvents [75, 106].

Reducing solvent consumption and extraction time, improving extraction efficiency, and utilizing greener solvents contribute to the effectiveness, cleanliness, and environmental friendliness of modern or nonconventional extraction techniques, resulting in decreased energy usage and reduced reliance on organic solvents [100, 106]. Novel thermal and non-thermal extraction methods, including microwave-assisted, ultrasound-assisted, pressurized liquid, supercritical fluid, pulsed electric field-assisted, and enzyme-assisted extractions, have been extensively researched. Compared to conventional methods, these advanced extraction techniques demonstrate higher effectiveness due to their increased extraction efficiency, lower solvent consumption, reduced extraction time, and lower energy costs [37, 118]. Conventional solvents used in waste valorization have drawbacks such as high cost, toxicity, and melting points, prompting the search for greener and more cost-effective alternatives like deep eutectic solvents (DESs) and their bioanalogs, natural deep eutectic solvents (NADESs) [35, 48]. DESs and NADESs are environmentally friendly and innovative solvents with the capacity to dissolve biomasses effectively, enabling the valorization of food wastes. They have been utilized to extract valuable components from various sources including vegetable oils, dairy products, beverages, as well as natural raw materials like lignocellulosic biomass, bark, wood, and algae [35, 48].

Bioplastics have emerged as a sustainable alternative to conventional plastics derived from fossil fuels, offering potential solutions to environmental pollution and resource depletion. This abstract explores the valorization of biowaste as a feedstock for the production of bioplastics, highlighting the feasibility, benefits, challenges, and applications of this approach. Biowaste, including agricultural residues, food waste, algae, and other organic materials, represents a vast and underutilized resource that can be transformed into bioplastics through various conversion processes. These processes typically involve the extraction or fermentation of biopolymers from biowaste, followed by polymerization or processing into bioplastic products.

9.20.1 Key Aspects of the Valorization of Biowaste into Bioplastics

- **Biopolymer Extraction:** Biowaste contains biopolymers such as starch, cellulose, lignin, and proteins, which can serve as raw materials for bioplastic production. Extraction methods, including physical, chemical, and enzymatic processes, are employed to isolate these biopolymers from the complex matrix of biowaste.
- **Biopolymer Modification:** Biopolymers extracted from biowaste often require modification to improve their processing properties, mechanical strength, and biodegradability. Chemical or enzymatic treatments can be applied to modify biopolymer structures and enhance their performance as bioplastic materials.
- **Polymerization and Processing:** Extracted and modified biopolymers are polymerized or processed into bioplastic products using techniques such as extrusion, injection molding, or 3D printing. Additives may be incorporated to achieve desired properties such as flexibility, durability, and biodegradability.
- **Applications:** Bioplastics derived from bio-waste find applications in various industries, including packaging, agriculture, automotive, textiles, and biomedical devices. They can be used to manufacture packaging materials, mulch films, disposable utensils, biodegradable bags, and medical implants, among other products.
- **Environmental Benefits:** Valorization of biowaste into bioplastics offers several environmental benefits, including reduced dependence on fossil fuels, decreased greenhouse

gas emissions, and diversion of organic waste from landfills. Bioplastics derived from biowaste are biodegradable and compostable, contributing to the reduction of plastic pollution and promoting a circular economy. Challenges associated with the valorization of biowaste into bioplastics include technological limitations, cost competitiveness, scale-up challenges, and regulatory considerations. Addressing these challenges requires interdisciplinary research, technological innovation, and collaboration among stakeholders from academia, industry, government, and civil society.

9.21 Valorization of Biowaste into Biosurfactants

Biowaste, comprising organic residues from agricultural, industrial, and municipal sources, poses significant challenges for waste management and environmental sustainability. However, recent research has demonstrated the potential of biowaste valorization through the production of biosurfactants, which are surface-active compounds with diverse industrial applications. This abstract explores the feasibility, benefits, and applications of converting biowaste into biosurfactants.

Biosurfactants, produced by microbial fermentation or enzymatic processes using biowaste as a substrate, offer several advantages over conventional surfactants derived from petrochemical sources. These advantages include biodegradability, low toxicity, and compatibility with a wide range of environmental conditions. Additionally, biosurfactants exhibit diverse functionalities, including emulsification, foaming, wetting, and solubilization, making them valuable in various industries such as agriculture, food, pharmaceuticals, cosmetics, and environmental remediation.

9.21.1 Key Aspects of the Valorization of Biowaste into Biosurfactants

- **Biowaste Selection**: Various types of biowaste, including agricultural residues, food waste, wastewater sludge, and industrial by-products, can serve as substrates for biosurfactant production. The choice of biowaste depends on factors such as availability, composition, and suitability for microbial fermentation.
- **Microbial Production**: Microorganisms such as bacteria, yeast, and fungi can produce biosurfactants through

metabolic pathways utilizing biowaste components as carbon and energy sources. Optimization of fermentation conditions, including pH, temperature, and nutrient supply, is essential to maximize biosurfactant yield and quality.

- **Characterization and Application**: Biosurfactants derived from biowaste must undergo characterization to evaluate their physicochemical properties, surface activity, stability, and performance in specific applications. Common techniques include surface tension measurement, emulsification index determination, and stability testing under various environmental conditions.
- **Industrial Applications**: Biosurfactants produced from biowaste find applications in diverse industrial sectors, including agriculture (as biopesticides and soil amendments), petroleum (enhanced oil recovery), cosmetics (emulsifiers and foaming agents), and environmental remediation (oil spill cleanup and wastewater treatment).
- **Environmental Sustainability**: Valorization of biowaste into biosurfactants contributes to environmental sustainability by reducing reliance on fossil fuels, minimizing waste generation, and promoting circular economy principles. Biosurfactants derived from biowaste are biodegradable and eco-friendly, offering potential solutions to environmental pollution and resource depletion.

9.22 Valorization Strategies

Research and Development: Continuously improving the effectiveness and applicability of both biofungicides and nanofungicides through research and development efforts.

Formulation Development: Developing formulations that enhance stability, shelf life, and efficacy of fungicidal agents, ensuring easy application and compatibility with existing agricultural practices.

Integration into Agricultural Systems: Identifying the most effective strategies for incorporating biofungicides and nanofungicides into integrated pest management (IPM) programs, considering factors such as crop rotation, timing of application, and compatibility with other control methods.

Regulatory Approval: Obtaining regulatory approvals for biofungicides and nanofungicides, ensuring compliance with safety and environmental regulations.

Market Development: Educating growers about the benefits of biofungicides and nanofungicides, establishing distribution channels, and promoting their adoption through marketing and outreach efforts.

Sustainability: Ensuring that the production and use of biofungicides and nanofungicides contribute to sustainable agriculture practices, minimizing environmental impact and preserving ecosystem health.

9.23 Environmental Impacts of Biological Solid Waste

- Analysis of the environmental consequences of untreated biological solid waste, including soil degradation, water contamination, and greenhouse gas emissions.
- Exploration of the potential ecological benefits of properly managed waste valorization practices.

9.24 Economic, Policy Considerations, Challenges and Opportunities in Biowaste Valorization

- Technical challenges and barriers to adoption.
- Economic feasibility and market potential.
- Policy and regulatory considerations.
- Analysis of the economic viability of bio-waste valorization in agricultural sectors.
- Discussion of the policy frameworks and incentives needed to promote the adoption of waste valorization practices.

9.25 Sustainable Solid Waste Management

An integrated approach to solid waste management involves the comprehensive handling of waste through multiple coordinated strategies to reduce, recycle, and properly dispose of waste in an environmentally sustainable manner. This approach combines several techniques such as waste minimization, recycling, composting, energy recovery, and landfilling, ensuring that waste is managed at every stage of its lifecycle. It emphasizes the 3Rs—reduce, reuse, and recycle—to minimize waste generation and promote resource recovery. Waste is segregated at the source to facilitate recycling of materials like plastics, paper, and metals, while organic waste is composted or processed through anaerobic digestion to produce biogas and nutrient-rich compost. Non-recyclable, non-compostable waste may be incinerated for energy recovery, while only the residual waste is sent to landfills. The integrated approach also incorporates strong governance, community engagement, and technological innovation, along with strict environmental regulations to ensure waste is managed safely and efficiently. This method helps reduce the environmental impact of waste, conserve natural resources, lower greenhouse gas emissions, and promote a circular economy where resources are continuously cycled back into the system. Sustainable agricultural practices that utilize biological waste focus on minimizing environmental impact while enhancing resource efficiency. Biological waste, such as crop residues, animal manure, and food processing by-products, can be repurposed in various ways to improve soil health, reduce reliance on chemical inputs, and promote circularity in farming systems. Effective management of biological solid waste is crucial for maintaining environmental sustainability and public health. Biological solid waste, including organic residues from agriculture, food processing, and animal husbandry, can be managed through various sustainable practices that recycle nutrients, reduce environmental impact, and generate value-added products. Below are six key methods for managing biological solid waste (Figure 9.6).

Sustainable agriculture is vital for meeting global food demands while minimizing environmental impacts. Waste valorization, the process of converting agricultural residues and by-products into valuable resources, is a key strategy for enhancing sustainability in agriculture. Nanotechnology offers innovative solutions for waste valorization by enabling the efficient conversion of biomass residues into high-value products such as fertilizers, soil amendments, and nanomaterial-based agricultural inputs. This explores the applications, benefits, challenges, and future prospects of nanotechnology in waste valorization for sustainable agriculture.

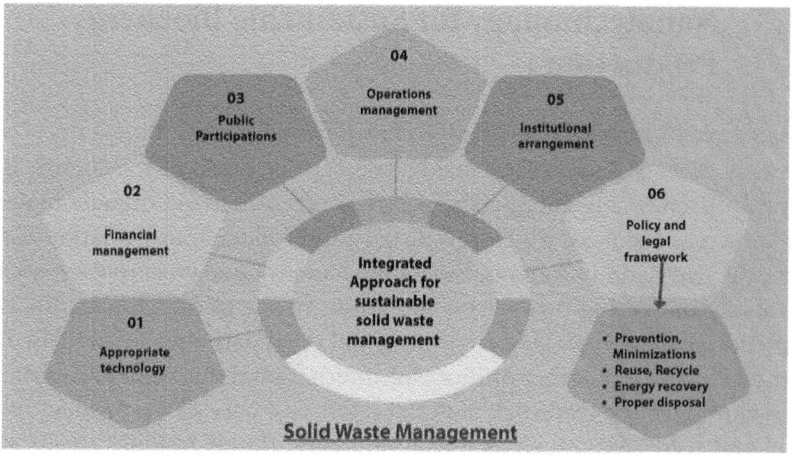

Figure 9.6 Integrated approach for solid waste management.

9.26 Benefits of Valorizing Biological Solid Waste for Agricultural Applications

I. **Resource Recovery**: Valorization allows for the recovery of valuable resources from waste streams, reducing the need for virgin materials and minimizing environmental pollution.

II. **Soil Health Improvement**: The resulting products, such as compost and biochar, enhance soil health by improving structure, water retention, and nutrient content, thereby promoting sustainable agricultural practices and enhancing crop productivity.

III. **Green Energy Production**: Anaerobic digestion of biological waste produces biogas, a renewable energy source that can be used for electricity generation, heating, or vehicle fuel, reducing reliance on fossil fuels and mitigating greenhouse gas emissions.

IV. **Circular Economy**: Valorization contributes to the principles of a circular economy by closing the loop on waste streams, creating value from materials that would otherwise be discarded and reducing the burden on landfills.

V. **Climate Change Mitigation**: By diverting organic waste from landfills and utilizing it for beneficial purposes, valorization helps mitigate greenhouse gas emissions associated with waste decomposition and fossil fuel use.

9.27 Nanotechnology for Sustainable Bioenergy Production

Bioenergy is a renewable energy produced from biomass. Bioenergy is one of many diverse resources available to meet our demand for energy. It is a form of renewable energy derived from recently living organic materials known as biomass, which can be used to produce transportation fuels, heat, electricity, and products. Bioenergy has attracted great interest owing to rapidly increasing fuel prices, alterations in the global climate, the fast depletion of fossil fuels, and environmental degradation by fossil fuels. The manufacturing of another product for international marketing in substantial volumes of the feedstock material refers as a base of the product from food and feed crops, such as wastes and agricultural residues (fruits and vegetable waste, forestry, fishing wheat straw, rice straw, municipal waste), non-food crops (weeds, litter, miscanthus and short rotation coppice) and algae, etc. Their key inputs serve as biomass feedstock, the production of bioenergy. Biomass feedstock is available from various sources such as agricultural or energy crops and waste fuel. This feedstock is sourced from organic matter.

9.28 Future Prospective and Limitations

In the future, biotechnological methods will be utilized to extract high-value bioproducts from agricultural waste. Because of their low energy requirements and low cost, integrated nano and biotechnological techniques are preferred for industrial waste valorization [15]. Therefore, it is necessary to design eco-friendly and cost-effective cascade conversion techniques. In order to preserve the physicochemical and microbiological stability of bioproducts during the valorization of these agricultural wastes, the unitary drying process is required to prevent microbial hazards [33]. Agro-industrial food waste can be a useful low-cost feedstock for sustainable production of industrial products [98]. Thus, an appropriate green technology is required for bioproduct development.

In summary, the valorization process involves several key steps, including collection, sorting, preprocessing, and conversion. Agricultural residues such as crop residues, straw, and husks can be processed into biochar, bioenergy, or organic fertilizers through techniques such as pyrolysis, anaerobic digestion, or composting. Food scraps and green waste can similarly be composted or converted into biogas through anaerobic digestion, providing renewable energy and nutrient-rich soil amendments.

9.29 Conclusion

This chapter provides a comprehensive overview of the valorization of biological solid waste for agricultural sectors, emphasizing its potential to promote sustainable agriculture and environmental stewardship. Biological solid waste generated from various sources, including agricultural residues, food scraps, and green waste, presents both a challenge and an opportunity for sustainable resource management. The biological waste valorization serves as a platform for the production of biofertilizers and nanofertilizers, which play important roles in sustainable agriculture by recycling nutrients from organic waste materials and enhancing nutrient management practices. These technologies contribute to resource conservation, environmental sustainability, and the development of more efficient and eco-friendly fertilization strategies. The valorization of biowaste into biofertilizer holds great promise for sustainable agriculture and waste management. Through continued research, technological innovation, and stakeholder engagement, this approach can contribute to building resilient agricultural systems, enhancing food security, and advancing environmental sustainability on a global scale. The valorization of bio-waste into bioplastics represents a promising approach for sustainable waste management and the development of eco-friendly materials. Further research, investment, and policy support are needed to unlock the full potential of this approach and accelerate the transition towards a circular economy based on renewable resources. The valorization of biowaste into biosurfactants represents a promising avenue for sustainable waste management and the development of environmentally friendly products. Further research, technological innovation, and collaboration between academia, industry, and government agencies are necessary to realize the full potential of this approach and address global challenges related to waste management and environmental sustainability. Managing biological solid waste through these sustainable methods not only minimizes environmental impact but also transforms waste into valuable resources. Practices like composting, anaerobic digestion, and biochar production promote nutrient recycling, reduce pollution, and contribute to renewable energy generation. These solutions are essential for achieving sustainability in waste management, agriculture, and energy production.

References

1. Afolalu, S.A., Oladipupo, S., Bose, M.E., Abioye, A.A., Adejuyigbe, S.B., Ajayi, O.O., Ongbali, S.O., Agro Waste a Sustainable Source for Steel Reinforcement - Review. *J. Phys. Conf. Ser.*, 1378, 3, 032032, 2019, December, IOP Publishing.
2. Alcazar-Ruiz, A., Ortiz, M.L., Sanchez-Silva, L., Dorado, F., Catalytic effect of alkali and alkaline earth metals on fast pyrolysis pre-treatment of agricultural waste. *Biofuels, Bioprod. Biorefin.*, 15, 1473–1484, 2021.
3. Anupama, and Ravindra, P., Value-added food: Single-cell protein. *Biotechnol. Adv.*, 18, 6, 459–479, 2000, https://doi.org/10.1016/S0734-9750(00)00041-1.
4. Aravani, V.P., Sun, H., Yang, Z., Liu, G., Wang, W., Anagnostopoulos, G., Syriopoulos, G., Charisiou, N.D., Goula, M.A., Kornaros, M., *et al.*, Agricultural and livestock sector's residues in Greece & China: Comparative qualitative and quantitative characterization for assessing their potential for biogas production. *Renew. Sustain. Energy Rev.*, 154, 111821, 2022.
5. Aristizábal-Marulanda, V., Solarte-Toro, J.C., Cardona Alzate, C.A., Study of biorefneries based on experimental data: production of bioethanol, biogas, syngas, and electricity using cofee-cut stems as raw material. *Environ. Sci. Pollut. Res.*, 28, 24590–24604, 2021.
6. Awasthi, P., Shrivastava, S., Kharkwal, A.C., Varma, A., Biofuel from agricultural waste: a review. *Int. J. Curr. Microbiol. Appl. Sci.*, 4, 470–7, 2015.
7. Awogbemi, O. and VonKallon, D.V., Pretreatment techniques for agricultural waste. *Case Stud. Chem. Environ. Eng.*, 6, 100229, 2022.
8. Baiano, A., Recovery of biomolecules from food wastes-a review. *Molecules*, 19, 9, 14821–14842, 2014.
9. Bala, S., Sharma, M., Dashora, K., Siddiqui, S., Diwan, D., Tripathi, M., Nanomaterials based sustainable bioenergy production systems: Current trends and future prospects. *Nanofabrication*, 7, 314–324, 2022.
10. Balan, A., Murthy, V.V., Kadeppagari, R.K., Immobilized enzymes for bioconversion of waste to wealth, in: *Biotechnology for Zero Waste: Emerging Waste Management Techniques*, C.M. Hussain and R.K. Kadeppagari (Eds.), pp. 33–46, Wiley, 2022.
11. Banwo, K., Olojede, A.O., Adesulu-Dahunsi, A.T., Verma, D.K., Thakur, M., Tripathy, S., Singh, S., Patel, A.R., Gupta, A.K., Aguilar, C.N., *et al.*, Functional importance of bioactive compounds of foods with potential health benefits: A review on recent trends. *Food Biosci.*, 43, 101320, 2021.
12. Benítez, V., Mollá, E., Martín-Cabrejas, M.A., Aguilera, Y., López-Andréu, F.J., Cools, K., Terry, L.A., Esteban, R.M., Characterization of industrial onion wastes (Allium cepa L.): dietary fibre and bioactive compounds. *Plant Foods Hum. Nutr.*, 66, 48, 2011.
13. Bjerre, A.B., Olesen, A.B., Fernqvist, T., Pretreatment of wheat straw using combined wet oxidation & alkaline hydrolysis resulting in convertible cellulose & hemicellulose. *Biotechnol. Bioeng.*, 49, 586e677, 1996.

14. Canete-Rodriguez, A.M., Santos-Duenas, I.M., Jimenez-Hornero, J.E., Ehrenreich, A., Liebl, W., Garcia-Garcia, I., Gluconic acid: properties, production methods and applications—an excellent opportunity for agro-industrial by-products and waste bio-valorization. *Process Biochem.*, 51, 12, 1891–1903, 2016.
15. Capanoglu, E., Nemli, E., Tomas-Barberan, F., Novel Approaches in the Valorization of Agricultural Wastes and Their Applications. *J. Agric. Food. Chem.*, 70, 23, 6787–6804, 2022, https://doi.org/10.1021/acs.jafc.1c07104.
16. Caruso, M.C., Braghieri, A., Capece, A., Napolitano, F., Romano, P., Galgano, F., Altieri, G., Genovese, F., Recent updates on the use of agro-food waste for biogas production. *Appl. Sci.*, 9, 6, 1217, 2019.
17. Castillo-Quimbo, M. and Mamaril, C.P., Organic fertilizer efficacy and financial viability in upland rice (Oryza sativa L.) production. *Philipp. Agric. Sci.*, 98, 174, 2015.
18. Chaudhari, A.K., Singh, V.K., Das, S., Dubey, N.K., Nanoencapsulation of Essential Oils and Their Bioactive Constituents: A Novel Strategy to Control Mycotoxin Contamination in Food System. *Food Chem. Toxicol.*, 149, 112019, 2021.
19. Cheng, F. and Brewer, C., Conversion of protein-rich lignocellulosic wastes to bio-energy: Review and recommendations for hydrolysis + fermentation and anaerobic digestion. *Renew. Sustain. Energy Rev.*, 146, 111167, 2021.
20. Cheng, H.-H. and Whang, L.-M., Resource recovery from lignocellulosic wastes *via* biological technologies: Advancements and prospects. *Bioresour. Technol.*, 343, 126097, 2022.
21. Chilakamarry, C.R., Sakinah, A.M., Zularisam, A., Sirohi, R., Khilji, I.A., Ahmad, N., Pandey, A., Advances in solid-state fermentation for bioconversion of agricultural wastes to value-added products: Opportunities and challenges. *Bioresour. Technol.*, 343, 126065, 2022.
22. Choi, I.S., Cho, E.J., Moon, J.H., Bae, H.J., Onion skin waste as a valorization resource for the by-product quercetin and biosugar. *Food Chem.*, 188, 537, 2015.
23. Chojnacka, K., Gorazda, K., Witek-Krowiak, A., Moustakas, K., Recovery of fertilizer nutrients from materials-Contradictions, mistakes and future trends. *Renew. Sustain. Energy Rev.*, 110, 485–498, 2019.
24. Comunian, T., Babazadeh, A., Rehman, A., Shaddel, R., Akbari-Alavijeh, S., Boostani, S., Jafari, S., Protection and controlled release of vitamin C by different micro/nanocarriers. *Crit. Rev. Food Sci. Nutr.*, 62, 17, 3301–3322, 2022.
25. Daliri, E.B.M. and Lee, B.H., Current trends and future perspectives on functional foods and nutraceuticals, in: *Beneficial Microorganisms in Food and Nutraceuticals*, pp. 221–244, Springer International Publishing, 2015.
26. Das, C.A., Kumar, V.G., Dhas, T.S., Karthick, V., Kumar, C.V., Nanomaterials in Anticancer Applications and Their Mechanism of Action - A Review. *Nanomed. Nanotechnol. Biol. Med.*, 47, 102613, 2022.

27. Das, D. and Deka, H., Vermicomposting of harvested waste biomass of potato crop employing Eisenia fetida: Changes in nutrient profile and assessment of the maturity of the end products. *Environ. Sci. Pollut. Res.*, 28, 35717–35727, 2021.
28. Das, L., Achinivu, E.C., Barcelos, C.A., Sundstrom, E., Amer, B., Baidoo, E.E., Simmons, B.A., Sun, N., Gladden, J.M., Deconstruction of woody biomass *via* protic and aprotic ionic liquid pretreatment for ethanol production. *ACS Sustain. Chem. Eng.*, 9, 4422–4432, 2021.
29. de Souza, L. and Shivakumar, S., Conversion of agro-industrial wastes for the manufacture of bio-based plastics, in: *Bioplastics for Sustainable Development*, pp. 177–204, Springer, 2021.
30. Diacono, M., Persiani, A., Testani, E., Montemurro, F., Ciaccia, C., Recycling agricultural wastes and by-products in organic farming: biofertilizer production, yield performance, and carbon footprint analysis. *Sustainability*, 11, 1, 3824, 2019.
31. Dittmar, H., Drach, M., Vosskamp, R., Trenkel, M.E., Gutser, R., Steffens, G., *Ullmann's Encyclopedia of Industrial Chemistry*, Wiley-VCH Verlag GmbH & Co. KGaA, 2009.
32. Donner, M., Gohier, R., de Vries, H., A new circular business model typology for creating value from agro-waste. *Sci. Total Environ.*, 716, 137065, 2020.
33. Federici, F., Fava, F., Kalogerakis, N., Mantzavinos, D., Valorisation of agro-industrial by-products, effluents and waste: Concept, opportunities and the case of olive mill wastewaters. *J. Chem. Technol. Biotechnol.*, 84, 7, 895–900, 2009.
34. Fermoso, F.G., Serrano, A., Alonso-Fariñas, B., Fernández-Bolaños, J., Borja, R., Rodríguez-Gutiérrez, G., Valuable Compound Extraction, Anaerobic Digestion, and Composting: A Leading Biorefinery Approach for Agricultural Wastes. *J. Agric. Food. Chem.*, 66, 8451–8468, 2018.
35. Florindo, C., Oliveira, M.M., Branco, L.C., Marrucho, I.M., Carbohydrates-based deep eutectic solvents: Thermophysical properties and rice straw dissolution. *J. Mol. Liq.*, 247, 441–447, 2017, https://doi.org/10.1016/j.molliq.2017.09.024.
36. Francioli, D., Schulz, E., Lentendu, G., Wubet, T., Buscot, F., Reitz, T., Mineral vs. organic amendments: microbial community structure, activity, and abundance of agriculturally relevant microbes are driven by long-term fertilization strategies. *Front. Microbiol.*, 7, 1–26, 2016.
37. Galanakis, C.M., Recovery of high added-value components from food wastes: conventional; emerging technologies and commercialized applications. *Trends Food Sci. Technol.*, 26, 2, 68–87, 2012, https://doi.org/10.1016/j.tifs.2012.02.005.
38. Ghasemi, K., Tasnim, S., Mahmud, S., PCM, Nano/Microencapsulation and Slurries: A Review of Fundamentals, Categories, Fabrication, Numerical Models and Applications. *Sustain. Energy Technol. Assess.*, 52, 102084, 2022.

39. Gómez-Muñoz, B., Magid, J., Jensen, L.S., Nitrogen turnover, crop use efficiency, and soil fertility in a long-term field experiment amended with different qualities of urban and agricultural waste. *Agric. Ecosyst. Environ.*, 240, 300, 2017.
40. Guía-García, J.L., Charles-Rodríguez, A.V., Reyes-Valdés, M.H., Ramírez-Godina, F., Robledo-Olivo, A., García-Osuna, H.T., Cerqueira, M.A., Flores-López, M.L., Micro and nanoencapsulation of bioactive compounds for agri-food applications: A review. *Ind. Crops Prod.*, 186, 115198, 2022.
41. Gupta, V.K., Nguyen, Q.D., Liu, S., Taherzadeh, M.J., Sirohi, R., Microbes in valorisation of biomass to value-added products (MVBVAP). *Bioresour. Technol.*, 347, 126738, 2022.
42. Hamelinck, C.N., Hooijdonk, G.V., Faaij, A.P.C., Ethanol from lignocellulosic biomass, techno-economic performance in short, middle, and long term. *Biomass Bioenergy*, 28, 384–410, 2005.
43. Hepbasli, A., Utlu, Z., Akdeniz, R.C., Energetic and exergetic aspects of cotton stalk production in establishing energy policies. *Energy Policy*, 35, 3015, 2007.
44. Hosseini, H. and Jafari, S.M., Introducing Nano/Microencapsulated Bioactive Ingredients for Extending the Shelf-Life of Food Products. *Adv. Colloid Interface Sci.*, 282, 102210, 2020.
45. Hsieh, Y.Y., Tsai, Y.C., He, J.R., Yang, P.F., Lin, H.P., Hsu, C.H., Loganathan, A., Rice husk agricultural waste-derived low ionic content carbon–silica nanocomposite for green reinforced epoxy resin electronic packaging material. *J. Taiwan Inst. Chem. Eng.*, 78, 493, 2017.
46. Indriani, D.W. and Susilo, B., The effect of microwave power on lignocellulose content, physical and chemical characteristics of biomass: A review. *IOP Conf. Ser.: Earth Environ. Sci.*, 924, 012069, 2021.
47. Isah, S. and Ozbay, G., Valorization of food loss and wastes: Feedstocks for biofuels and valuable chemicals. *Front. Sustain. Food Syst.*, 4, 82, 2020, https://doi.org/10.3389/fsufs.2020.00082.
48. Jablonský, M., Škulcová, A., Malvis, A., Šima, J., Extraction of value-added components from food industry based and agro-forest biowastes by deep eutectic solvents. *J. Biotechnol.*, 282, 46–66, 2018, https://doi.org/10.1016/j.jbiotec.2018.06.007.
49. Jahirul, M.I., Rasul, M.G., Chowdhury, A.A., Ashwath, N., Biofuels production through biomass pyrolysis—a technological review. *Energies*, 5, 12, 4952–5001, 2012.
50. Jayasuriya, H.P.W., Soni, P., Steele, P., AGS Divisional Project 214A1 "Small Farmer Livelihoods", 2003.
51. Jiang, Y., Lv, Y., Wu, R., Sui, Y., Chen, C., Xin, F., Zhou, J., Dong, W., Jiang, M., Current Status and Perspectives on Biobutanol Production Using Lignocellulosic Feedstocks. *Bioresour. Technol. Rep.*, 7, 100245, 2019.

52. Jin, E., Lim, S.O., Kim, S.O., Park, Y., Optimization of various extraction methods for quercetin from onion skin using response surface methodology. *Food Sci. Biotechnol.*, 20, 1727, 2011.
53. Joshi, S.M. and Gogate, P.R., Intensifying the biogas production from food waste using ultrasound: Understanding into effect of operating parameters. *Ultrason. Sonochem.*, 59, 104755, 2019.
54. Kantarelis, E. and Zabaniotou, A., Valorization of cotton stalks by fast pyrolysis and fixed bed air gasification for syngas production as precursor of second generation biofuels and sustainable agriculture. *Bioresour. Technol.*, 100, 942, 2009.
55. Karmee, S.K. and Lin, C.S.K., Valorisation of food waste to biofuel: current trends and technological challenges. *Sustain. Chem. Processes*, 2, 1, 1–4, 2014, https://doi.org/10.1186/s40508-014-0015-5.
56. Karthick, C. and Nanthagopal, K., A comprehensive review on ecological approaches of waste to wealth strategies for production of sustainable biobutanol and its suitability in automotive applications. *Energy Convers. Manage.*, 239, 114219, 2021.
57. Kaur, N., Kaur, A., Sridhar, K., Sharma, M., Singh, T.P., Kumar, S., Development and quality characteristics of functional Kulfi fortified with microencapsulated betalains. *Int. J. Food Sci. Technol.*, 56, 5362–5370, 2021.
58. Kavitha, S., Kannah, R.Y., Kumar, G., Gunasekaran, M., Banu, J.R., Introduction: sources and characterization of food waste and food industry wastes, in: *Food Waste to Valuable Resources*, J.R. Banu, G. Kumar, M. Gunasekaran, S. Kavitha (Eds.), pp. 1–13, Academic Press, 2020.
59. Khan, M.U., Usman, M., Ashraf, M.A., Dutta, N., Luo, G., Zhang, S., A review of recent advancements in pretreatment techniques of lignocellulosic materials for biogas production: Opportunities and limitations. *Chem. Eng. J. Adv.*, 10, 100263, 2022.
60. Kim, D.Y., Kadam, A., Shinde, S., Saratale, R.G., Patra, J.K., Ghodake, G., Recent developments in nanotechnology transforming the agricultural sector: A transition replete with opportunities. *J. Sci. Food Agric.*, 98, 849–864, 2018.
61. Kover, A., Kraljić, D., Marinaro, R., Rene, E.R., Processes for the valorization of food and agricultural wastes to value-added products: recent practices and perspectives. *Syst. Microbiol. Biomanuf.*, 2, 50, 2022.
62. Kumar, N., Gautam, R., Stallings, J.D., Coty, G.G., Lynam, J.G., Secondary agriculture residues pretreatment using deep eutectic solvents. *Waste Biomass Valorization*, 12, 2259–2269, 2021.
63. Lehmann, J. and Joseph, S., Biochar for environmental management: an introduction, in: *Biochar for Environmental Management: Science, Technology and Implementation*, J. Lehmann and S. Joseph (Eds.), pp. 1–14, Taylor and Francis Group, Routledge, 2015.

64. Leni, G., Caligiani, A., Sforza, S., Bioconversion of agri-food waste and by-products through insects: A new valorization opportunity, in: *Valorization of Agri-Food Wastes and By-Products*, pp. 809–828, Academic Press, 2021.
65. Lim, S.F. and Matu, S.U., Utilization of agro-wastes to produce biofertilizer. *Int. J. Energy Environ. Eng.*, 6, 31–35, 2015.
66. Lin, C.S.K., Koutinas, A.A., Stamatelatou, K., Mubofu, E.B., Matharu, A.S., Kopsahelis, N., Pfaltzgraff, L.A., Clark, J.H., Papanikolaou, S., Kwan, T.H., Luque, R., Current and future trends in food waste valorization for the production of chemicals; materials and fuels: a global perspective. *Biofuels, Bioprod. Biorefin.*, 8, 5, 686–715, 2014.
67. Magee, K., Halstead, J., Small, R., Young, I., Valorisation of Organic Waste By-Products Using Black Soldier Fly (Hermetia illucens) as a Bio-Convertor. *Sustainability*, 13, 15, 8345, 2021, https://doi.org/10.3390/su13158345.
68. Maiti, S., Sarma, S.J., Brar, S.K., Le Bihan, Y., Drogui, P., Buelna, G., Verma, M., Agro-industrial wastes as feedstock for sustainable bio-production of butanol by Clostridium beijerinckii. *Food Bioprod. Process.*, 98, 217–226, 2016.
69. Malolan, R., Gopinath, K.P., Vo, D.V.N., Jayaraman, R.S., Adithya, S., Ajay, P.S., Arun, J., Green ionic liquids and deep eutectic solvents for desulphurization, denitrification, biomass, biodiesel, bioethanol and hydrogen fuels: A review. *Environ. Chem. Lett.*, 19, 1001–1023, 2021.
70. Montemurro, F., Ciaccia, C., Leogrande, R., Ceglie, F., Diacono, M., Suitability of different organic amendments from agro-industrial wastes in organic lettuce crops. *Nutr. Cycling Agroecosyst.*, 102, 243, 2015.
71. Motamedi, A., Jafarpour, M., Oshaghi, M., Improving the vermicompost quality by using horticultural and agronomic residues. *J. Plant Nutr.*, 45, 727–738, 2022.
72. Mudzakir, A., Rizky, K.M., Munawaroh, H.S.H., Puspitasari, D., Oil palm empty fruit bunch waste pretreatment with benzotriazolium-based ionic liquids for cellulose conversion to glucose: Experiments with computational bibliometric analysis. *Indones. J. Sci. Technol.*, 7, 291–310, 2022.
73. Naik, G.P., Poonia, A.K., Chaudhari, P.K., Pretreatment of lignocellulosic agricultural waste for delignification, rapid hydrolysis, and enhanced biogas production: A review. *J. Indian Chem. Soc.*, 98, 100147, 2021.
74. Nasseri, A.T., Rasoul-Ami, S., Morowvat, M.H., Ghasemi, Y., Single-cell protein: production and process. *Am. J. Food Technol.*, 6, 2, 103–116, 2011, https://doi.org/10.3923/ajft.2011.103.116.
75. Nayak, A. and Bhushan, B., An overview of the recent trends on the waste valorization techniques for food wastes. *J. Environ. Manage.*, 233, 352–370, 2019, https://doi.org/10.1016/j.jenvman.2018.12.051.
76. Ndayisenga, F., Yu, Z., Zheng, J., Wang, B., Liang, H., Phulpoto, I.A., Habiyakare, T., Zhou, D., Microbial electrohydrogenesis cell and dark fermentation integrated system enhances biohydrogen production from

lignocellulosic agricultural wastes: Substrate pretreatment towards optimization. *Renew. Sustain. Energy Rev.*, 145, 111078, 2021.

77. Neves, M.A., Kimura, T., Shimizu, N., Nakajima, M., State of the Art and Future Trends of Bioethanol Production, in: *Dynamic Biochemistry, Process Biotechnology & Molecular Biology*, pp. 1–13, Global Science Books, 2007.
78. O'Connor, J., Hoang, S.A., Bradney, L., Dutta, S., Xiong, X., Tsang, D.C., Ramadass, K., Vinu, A., Kirkham, M.B., Bolan, N.S., A review on the valorisation of food waste as a nutrient source and soil amendment. *Environ. Pollut.*, 272, 115985, 2021.
79. Oreopoulou, V. and Russ, W., *Utilization of by-products and treatment of waste in the food industry*, Springer US, New York and Philadelphia, 2007.
80. Passoth, V. and Sandgren, M., Biofuel production from straw hydrolysates: Current achievements and perspectives. *Appl. Microbiol. Biotechnol.*, 103, 5105–5116, 2019.
81. Pateiro, M., Gómez, B., Munekata, P., Barba, F., Putnik, P., Kovačević, D., Lorenzo, J., Nanoencapsulation of promising bioactive compounds to improve their absorption, stability, functionality and the appearance of the final food products. *Molecules*, 26, 1547, 2021.
82. Pathak, N., Singh, S., Singh, P., Singh, P.K., Singh, R., Bala, S., Thirumalesh, B.V., Gaur, R., Tripathi, M., Valorization of jackfruit waste into value-added products and their potential applications. *Front. Nutr.*, 9, 1061098, 2022.
83. Pattanaik, L., Pattnaik, F., Saxena, D.K., Naik, S.N., Biofuels from agricultural wastes, in: *Second and Third Generation of Feedstocks*, pp. 103–142, Elsevier, 2019.
84. Periyasamy, S., Karthik, V., Kumar, P.S., Isabel, J.B., Temesgen, T., Hunegnaw, B.M., Melese, B.B., Mohamed, B.A., Vo, D.V., Chemical, physical and biological methods to convert lignocellulosic waste into value-added products. A review. *Environ. Chem. Lett.*, 20, 1129–1152, 2022.
85. Petruccioli, M., Raviv, M., Di Silvestro, R., Dinelli, G., Agriculture and agro-industrial wastes, byproducts, and wastewaters: origin, characteristics, and potential in bio-based compounds production, 2011.
86. Prasad, B.R., Padhi, R.K., Ghosh, G., A review on key pretreatment approaches for lignocellulosic biomass to produce biofuel and value-added products. *Int. J. Environ. Sci. Technol.*, 1–16, 2022.
87. Putri, D.N., Sahlan, M., Montastruc, L., Meyer, M., Negny, S., Hermansyah, H., Progress of fermentation methods for bio-succinic acid production using agro-industrial waste by Actinobacillus succinogenes. *Energy Rep.*, 6, 234–239, 2020.
88. Rahman, S.M.E., Islam, M.A., Rahman, M.M., Oh, D.H., Effect of cattle slurry on growth, biomass yield and chemical composition of maize fodder. *Asian-Australas. J. Anim. Sci.*, 21, 11, 1592–1598, 2008.
89. Rahmani, A.M., Gahlot, P., Moustakas, K., Kazmi, A.A., Ojha, C.S.P., Tyagi, V.K., Pretreatment methods to enhance solubilization and anaerobic

biodegradability of lignocellulosic biomass (wheat straw): Progress and challenges. *Fuel*, 319, 123726, 2022.

90. Ramadhan, M.O. and Handayani, M.N., The potential of food waste as bioplastic material to promote environmental sustainability: A review. *IOP Conf. Ser.: Mater. Sci. Eng.*, 980, 012082, 2020, https://doi.org/10.1088/1757-899X/980/1/012034.
91. Ramnarain, Y.I., Ansari, A.A., Ori, L., Vermicomposting of different organic materials using the epigeic earthworm Eisenia foetida. *Int. J. Recyl. Org. Waste Agric.*, 8, 23–36, 2022.
92. Ranganathan, S., Dutta, S., Moses, J.A., Anandharamakrishnan, C., Utilization of food waste streams for the production of biopolymers. *Heliyon*, 6, 9, e04891, 2020, https://doi.org/10.1016/j.heliyon.2020.e04891.
93. Rasid, N.S.A., Shamjuddin, A., Amin, N.A.S., Chemical and structural changes of ozonated empty fruit bunch (EFB) in a ribbon-mixer reactor. *Bull. Chem. React. Eng. Catal.*, 16, 383–395, 2021.
94. Ruanglek, V., Maneewatthana, D., Tripetchkul, S., Evaluation of Thai agro-industrial wastes for bio-ethanol production by Zymomonas mobilis. *Process Biochem.*, 41, 6, 1432–1437, 2006.
95. Rogers, J.N., Stokes, B., Dunn, J., Cai, H., Wu, M., Haq, Z., Baumes, H., An assessment of the potential products and economic and environmental impacts resulting from a billion-ton bioeconomy. *Biofuels, Bioprod. Biorefin.*, 11, 110e128, 2016, https://doi.org/10.1002/bbb.1728.
96. Sadh, P.K., Duhan, S., Duhan, J.S., Agro-industrial wastes and their utilization using solid state fermentation: a review. *Bioresour. Bioprocess.*, 5, 1, 1, 2018.
97. Saini, J.K., Saini, R., Tewari, L., Lignocellulosic agriculture wastes as biomass feedstocks for second-generation bioethanol production: concepts and recent developments. *3 Biotech*, 5, 4, 337–353, 2015.
98. Sharma, V., Tsai, M.L., Nargotra, P., Chen, C.W., Kuo, C.H., Sun, P.P., Dong, C.D., Agro-Industrial Food Waste as a Low-Cost Substrate for Sustainable Production of Industrial Enzymes: A Critical Review. *Catalysts*, 12, 11, 1373, 2022.
99. Sigurnjak, I., Brienza, C., Snauwaert, E., De Dobbelaere, A., De Mey, J., Vaneeckhaute, C., Michels, E., Schoumans, O., Adani, F., Meers, E., Production and performance of bio-based mineral fertilizers from agricultural waste using ammonia (stripping-) scrubbing technology. *Waste Manage. (Oxford)*, 89, 265–274, 2019.
100. Soquetta, M.B., Terra, L.D.M., Bastos, C.P., Green technologies for the extraction of bioactive compounds in fruits and vegetables. *CyTA J. Food*, 16, 1, 400–412, 2018.
101. Sorathiya, L.M., Fulsoundar, A.B., Tyagi, K.K., Patel, M.D., Singh, R., Eco-friendly and modern methods of livestock waste recycling for enhancing farm profitability. *Int. J. Recyl. Org. Waste Agric.*, 3, 1–7, 2014.

102. Teixeira, S.R., de Souza, A.E., de Almeida Santos, G.T., Vilche Peña, A.F., Miguel, A.-G., Sugarcane bagasse ash as a potential quartz replacement in red ceramic. *J. Am. Ceram. Soc.*, 91, 1883, 2008.
103. Teixeira, S.R., Magalhães, R.S., Arenales, A., Souza, A.E., Romero, M., Rincón, J.M., Valorization of sugarcane bagasse ash: Producing glass-ceramic materials. *J. Environ. Manage.*, 134, 15, 2014.
104. Teixeira, S.R., Pena, A.F.V., Miguel, A.G., Briquetting of charcoal from sugarcane bagasse fly ash (scbfa) as an alternative fuel. *Waste Manage. (Oxford)*, 30, 804, 2010.
105. Torres-Climent, A., Martin-Mata, J., Marhuenda-Egea, F., Moral, R., Barber, X., Perez-Murcia, M.D., Composting of the solid phase of digestate from biogas production: optimization of the moisture, C/N ratio, and pH conditions. *Commun. Soil Sci. Plant Anal.*, 46, 197, 2015.
106. Torres-Valenzuela, L.S., Ballesteros-Gómez, A., Rubio, S., Green solvents for the extraction of high added-value compounds from agri-food waste. *Food Eng. Rev.*, 12, 1, 83–100, 2020.
107. Ugheoke, I.B. and Mamat, O., A critical assessment and new research directions of rice husk silica processing methods and properties. *Maejo Int. J. Sci. Technol.*, 6, 430, 2012.
108. Usmani, Z., Sharma, M., Diwan, D., Tripathi, M., Whale, E., Jayakody, L.N., Moreau, B., Thakur, V.K., Tuohy, M., Gupta, V.K., Valorization of sugar beet pulp to value-added products: A review. *Bioresour. Technol.*, 346, 126580, 2022.
109. Usmani, Z., Sharma, M., Tripathi, M., Nizami, A.S., Gong, L., Nguyen, Q.D., Reddy, M.S., Thakur, V.K., Gupta, V.K., Converting biowaste streams into energy – leveraging microwave-assisted valorization technologies for enhanced conversion. *J. Energy Inst.*, 107, 101161, 2023.
110. Verma, A., Kumar, S., Jain, P.K., Key pretreatment technologies on cellulosic ethanol production. *J. Sci. Res.*, 55, 57–63, 2011.
111. Vilas-Boas, A.A., Pintado, M., Oliveira, A.L., Natural bioactive compounds from food waste: Toxicity and safety concerns. *Foods*, 10, 1564, 2021.
112. Wang, R., Wang, Q., Dong, L., Zhang, J., Cleaner agricultural production in drinking-water source areas for the control of non-point source pollution in China. *J. Environ. Manage.*, 285, 112096, 2021.
113. Xiong, X., Yu, I.K.M., Tsang, D.C., Bolan, N.S., Ok, Y.S., Igalavithana, A.D., Kirkham, M.B., Kim, K., Vikrant, K., Value-added chemicals from food supply chain wastes: State-of-the-art review and future prospects. *Chem. Eng. J.*, 375, 121983, 2019.
114. Yin, X., Wei, L., Pan, X., Liu, C., Jiang, J., Wang, K., The pretreatment of lignocelluloses with green solvents as biorefinery pre-process: A minor review. *Front. Plant Sci.*, 12, 670061, 2021.
115. Zhang, D., Jiang, B., Luo, Y., Fu, X., Kong, H., Shan, Y., Ding, S., Effects of ultrasonic and ozone pretreatment on the structural and functional

properties of soluble dietary fiber from lemon peel. *J. Food Process Eng.*, 45, e13916, 2022.
116. Zheng, Y., Zhang, Q., Zhang, Z., Jing, Y., Hu, J., He, C., Lu, C., A review on biological recycling in agricultural waste-based biohydrogen production: Recent developments. *Bioresour. Technol.*, 347, 126595, 2021.
117. Zhou, Z., Ouyang, D., Liu, D., Zhao, X., Oxidative pretreatment of lignocellulosic biomass for enzymatic hydrolysis: Progress and challenges. *Bioresour. Technol.*, 367, 128208, 2022.
118. Zia, S., Khan, M.R., Shabbir, M.A., Aslam Maan, A., Khan, M.K.I., Nadeem, M., Khalil, A.A., Din, A., Aadil, R.M., An inclusive overview of advanced thermal and nonthermal extraction techniques for bioactive compounds in food and food-related matrices. *Food Rev. Int.*, 1–31, 2020, https://doi.org/10.1080/87559129.2020.1853265.
119. Taneja, T., Chopra, M., Sharma, I., Bioenergy: biomass sources, production, and applications, in: *Green Approach to Alternative Fuel for a Sustainable Future*, pp. 443–451, Elsevier, 2023 Jan 1.

10

Nanotechnology for Waste Valorization of Sustainable Agriculture

Vandana Jhalora, Sonika Kashyap, Riddhi Garg and Renu Bist*

Department of Zoology, Centre of Advanced Studies, University of Rajasthan, Jaipur, India

Abstract

Nanotechnology has emerged as a promising avenue for advancing sustainable agriculture by addressing waste valorization challenges. The integration of nanotechnological approaches for the efficient utilization of agricultural waste is explained in the current chapter, emphasizing its potential in sustainable farming practices. The increasing global demand for food necessitates innovative solutions to manage agricultural residues effectively. Nanomaterials offer unique properties, enabling targeted and controlled interactions with waste streams. Through nanotechnology, waste valorization in sustainable agriculture encompasses various aspects such as nutrient recovery, soil enhancement, and pest control. Nanoparticles (NPs) enable the controlled release of nutrients from agricultural residues, fostering improved soil fertility and crop yield. Additionally, nanomaterials facilitate the remediation of contaminated soils, contributing to the overall health of agricultural ecosystems. Nanopesticides, with reduced environmental impact, provide an efficient means of pest management. Furthermore, nanosensors aid in the real-time monitoring of soil conditions and plant health, enabling precision agriculture practices. While nanotechnology holds immense promise for sustainable waste valorization in agriculture, careful consideration must be given to potential environmental and human health implications. The transformative potential of nanotechnology in converting agricultural waste into valuable resources, promoting ecological balance, and advancing the sustainability of modern farming practices is elucidated.

***Keywords*:** Nanotechnology, nanosensors, nano pesticides, sustainable agriculture, precision agriculture

**Corresponding author*: renu_bisht22@yahoo.co.in; ORCID: 0000-0003-2993-7983

Priyanka Singh, Pooja Agarwal and V. Vivekanand (eds.) Solid Waste Management: Challenges, Sustainability and Advancements, (335–360) © 2026 Scrivener Publishing LLC

10.1 Introduction

Agricultural food waste and other by-products from farming hugely affect the world by negatively impacting its economy and environment. To deal with this challenge, there is a need to minimize the waste generated from food. There have been recent developments in nanotechnology and biotechnology which have offered a glimmer of hope for turning agricultural food waste into valuable substances that are more stable and useful as inputs for sustainable agriculture as shown in Figure 10.1 [17]. With the rapid development in science, efforts have focused on nanoagriculture aimed at improving crops and animal yields by using nano-sized particles [13]. These nanomaterials possess unique features that are suitable for creating new tools to be used in support of sustainable agriculture practices [34].

The nanotechnology in agriculture represents several applications among which the development of nanofertilizers and nanopesticides has been highly effective in reducing water and soil contamination by enabling controlled delivery of nutrients and pest control [84]. It has been indicated that these nano-formulations have more advantages in terms of stability, distribution, and solubility; however, they can also reduce the volume of active ingredients and depress the release of non-target toxicity substances, hence mitigating resistance development [51]. Furthermore, sensor-based approaches for plant health and soil quality monitoring are one of the other facets of sustainable agricultural practices through nanotechnology [84]. With the growing world population,

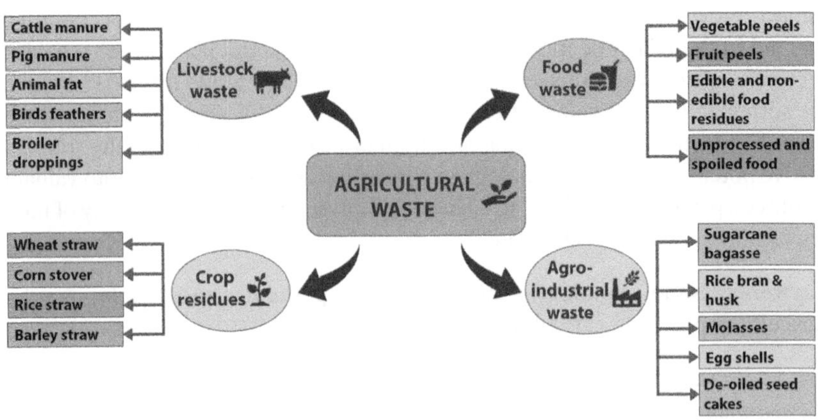

Figure 10.1 Major types of agricultural waste.

the need for large-scale farming output and smart agricultural methods is also rising. Therefore, the application of nanomaterials in agriculture is increasing [37]. This integration promises to completely transform agriculture by combining smart nanotools with high-tech agriculture practices. The goal is to improve both the quality and quantity of yields and at the same time lessen the negative impact on our natural resources [98]. Nanotechnology, having the capability to protect plants, diagnose diseases, and escalate food productivity and quality, will play a significant role in supporting sustainable intensification efforts [14]. The impact of nanotechnology on sustainable agriculture development is substantial, aiming to escalate crop production while reducing inputs like pesticides, herbicides, and fertilizers *via* environmental vigilance and pin-point interventions [34].

The use of nanosensors in agriculture has increased, for the purposes of quick and robust monitoring of soil and water contamination. Several detection technologies based on nano-materials are important to detect ultra-trace levels of heavy metals. These sensors are outstanding at detecting minute quantities of contaminants, pests, and nutrients, as well as stress factors such as nutrient deficiency, drought, or pathogens. Moreover, nanocarriers guarantee the accurate delivery of drugs or nutrients to crops and animals thus improving agricultural productivity [49].

10.2 Sustainable Agriculture through Advanced Waste Management

Agriculture faces the challenge of increasing supply during the 21st century to meet the growing demand for food crops. It must do this without losing the quality of soil or even improve it and in maintaining the balance of agroecosystems. Given the so-far discussed fact, large-scale commercialization of biochar appears as a very favorable strategy given sustainable agriculture and little adverse impact on ecosystems in general [93]. Biochar is essentially an organic material that is produced from the waste biomass through thermo-chemical conversion at high temperature, where either the oxygen is not available or in very scanty supply [110]. This transformation along with long-time carbon storage offer had some significant benefits like improved soil fertility and reduced greenhouse emissions; in addition, bioenergy would get produced [44, 126]. In particular, nano-biochar (nano-BC) stands out as a more advanced form over conventional biochar because it has higher adsorption capacity, mobility property, and specific surface area on soil which then helps improve soil attributes, crop productivity, and ecological restoration efforts [64, 106]. This method has the potential to address the need of enhancing agricultural output and ensuring environmental sustainability.

In addition, soil and underground water pollution with hazardous chemicals constitutes serious threats to human health and our ecosystem [109]. Nanoscience appears as a potential green way of coping with these global concerns, especially those that have to do with soil protection and conservation of water resources. There is hope that nanotechnology may provide solutions for these problems through low-cost, innovative methods targeting soil pollutants like dyes, organic contaminants, metalloids, and heavy metals [41].

Furthermore, new types of fertilizers combine chitosan-encapsulated hydrogel supported by biomass-immobilized micronutrients to enable precision agriculture which minimizes nutrient losses. For slow-release advantages and increased nutrient bioavailability, these formulations incorporate waste biomass along with Cu2+-enriched cores within the alginate-carboxymethyl cellulose matrix [105].

In addition, before agricultural irrigation is done for agronomic purposes NPs exhibit potential in effective treatment of wastewater [8]. For example, Titanium dioxide NPs (TiO_2-NPs) improve the vigour of many plant species in poor conditions, thereby boosting their growth rate [1]. Nanotechnology offers clever solutions for instance, nano-sorbents, nano-filters and nanosensors that are some of the conventional remedies applied in wastewater treatment [38].

Food loss and waste remain a major problem worldwide, but it additionally offers an excellent opportunity to make high-value materials. These materials can range from food packaging to advanced components in drug delivery systems, capacitors, and sensors which can be derived from food waste leading to more structured and systemic food production and supply chain needed for supporting the ever-growing human population [78]. Bread waste is an example of such waste which has been used as raw material to prepare high-quality single-layer graphene exemplifying how food waste could be transformed into low-cost nanomaterials for sensing applications like their available commercial equivalents [79]. Additionally, fruit and vegetables wastes have been utilized for the fabrication of different types of sensors. The grape stalk was used as both a reducing agent and stabilizer through which metallic NPs were synthesized for screen-printed electrodes applied in the voltammetric determination of Pb(II) and Cd(II) [12]. Further, vegetable parts and biomass residues find use in the creation of organic electronics, such as flexible organic light-emitting diodes (FOLEDs) made from onion pulp (*Allium cepa L.*) as a green/ecological substrate [31].

10.3 Emerging Trends in Nanotechnology for Greener Farming Practices in Sustainable Agriculture

In recent years, the population has expanded tremendously throughout the world, which has led to a growing urgency for upgraded production in agriculture with highly efficient farming practices. Subsequently, the trend has shifted towards the integration of agriculture with nanotechnology to improve farming techniques [7]. The use of nanotechnology has enabled significant advancements in farming methods, resulting in higher productivity, better use of resources, and sustainability [85]. By utilizing the distinctive characteristics of nanomaterials and nanodevices, farmers and researchers are transforming agricultural methods worldwide [83]. The field of nanotechnology in agriculture includes various uses, from precision farming and smart delivery systems to soil remediation and post-harvest preservation [48, 74, 86]. The use of nanotechnology in agriculture aims to improve resource efficiency, reduce environmental harm, and guarantee food security for an expanding world population.

Sustainable farming is the use of farming systems and practices which maintain or enhance the economic viability of agricultural production, the natural resource base, and other ecosystem which are influenced by agricultural practices [114]. Nanotechnology provides several benefits in agriculture such as enhancing the yield of crops, delivering nutrients to the soil, developing stress-resistant crops, and delivering pesticides with accuracy [94]. This can be accomplished through the utilization of different methods like nano-fertilizers, nano-membranes, nano-catalysts, nano-biosensors, nano-composites, and nano-pesticides (Figure 10.2) [15, 21, 65, 76, 108, 111].

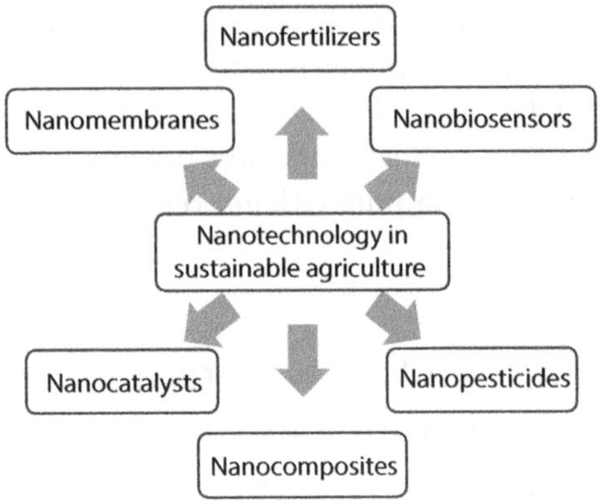

Figure 10.2 Schematic representation of various applications of nanotechnology in sustainable agriculture.

10.3.1 Nano-Fertilizers

The combination of fertilizers and nanomaterials to create nano-fertilizers enhances the transfer and effectiveness of nutrients in plants [76]. They provide many benefits over traditional fertilizers in reducing environmental pollution, increasing crop yield, decreasing nutrient leaching, and enhancing soil fertility. Furthermore, they specifically deliver nutrients to plants by controlled release and encapsulation, thereby facilitating sustainable agriculture (Figure 10.3) [99]. The use of nano-fertilizers is increasing tremendously because they efficiently increase crop yield while reducing environmental pollution simultaneously. Nano-fertilizers are broadly divided into three types based on their composition:

a) Nanoscale fertilizers: These types of fertilizers are wholly made up of NPs comprising nutrients. Examples: Nano urea NPs, nanoscale zeolites, hydroxyapatite nano-fertilizers [29, 101, 113].
b) Nanoscale additive fertilizers: These are the traditional fertilizers that are complemented with NPs. Examples: Nano-titanium dioxide, nano-silica, multiwall carbon nanotubes (MWCNT) [47, 68, 103].
c) Nanoscale coating fertilizers: These are also traditional fertilizers that are coated with a layer of NPs. Example: Urea-loaded hydroxyapatite [19].

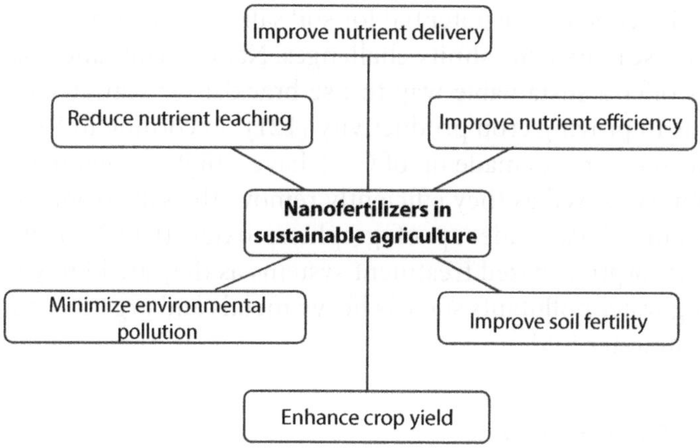

Figure 10.3 Advancing sustainable agriculture with nano-fertilizers by boosting yields with minimal environmental impacts.

Silica nanoparticles (SiNPs) are frequently used to supply nutrients to plants by efficiently trapping and slowly releasing nutrients, which helps maintain a consistent nutrient supply [39]. The CNTs are considered a novel fertilizer as they gradually release nutrients and stimulate plant growth simultaneously [116].

They have the capability to enter the plant cell wall due to which they are efficient in improving the delivery of nutrients to the crops and boosting their uptake [4]. Furthermore, the nano-fertilizers made up of a calcium phosphate compound, known as hydroxyapatite, enhance the uptake of phosphorus in plants and reduce the phosphorus runoff [29]. According to a study, nano-fertilizers made up of biodegradable polymers such as chitosan and poly-(lactic-co-glycolic acid) (PLGA) act as carriers for the controlled release of the fertilizers [82].

10.3.2 Nano-Membranes

These are the ultra-thin membranes made up of graphene and carbon nanotubes that play an important role in the protection of crops, and the purification of water. They also play an important role in the filtration system in the field of water management by purifying the irrigation water and removing the water pollutants, salts, and pathogens to provide clean water to crops [111]. This enhances both crop health and yield while reducing the potential for soil salinization, especially in areas with water scarcity and salinity challenges. Nano-membrane-based technologies offer a sustainable way to use brackish or seawater for irrigation, thus improving crop productivity [121]. According to Sarkar *et al.*, the nano-membranes made up of CNT have a higher potential for water permeability as well as they efficiently remove the salts from water [95]. Furthermore, NPs made up of graphene oxide (GO-NP) are widely used in adsorption-based treatment systems as they are highly potent in removing water pollutants such as heavy metals and organic compounds from wastewater [25].

10.3.3 Nano-Catalysts

These catalysts are widely famous due to their high surface area to volume ratio which imparts them strong catalytic activity [90]. They play an important role in eliminating the pollutants such as pesticides, and heavy metals from the soil and improving the fertility and quality of soil. According to Wang *et al.*, CNTs are the most potent nano-catalysts as they can form a stable CNT-contaminant complex *via* the formation of chemical bonds and physical adsorption, thereby reducing their presence and motility in the soil [120]. Ferric oxide (Fe_2O_3) nano-catalysts can soak up and trap heavy metals such as cadmium and arsenic [75]. Titanium oxide (TiO_2) nano-catalysts exhibit photocatalytic activity, enabling them to degrade organic pollutants under ultraviolet (UV) light exposure [75]. Moreover, nano-catalysts degrade various types of dyes from the soil and water such as silver nano-catalysts can degrade malachite green, congo red, and methylene blue dyes [5].

10.3.4 Nano-Biosensors

Nano-biosensors are sophisticated instruments created to detect and analyze biological substances at the nanoscale with outstanding sensitivity and precision. They use NPs combined with biological recognition components for accurate and quick identification of insecticides, pesticides, and herbicides on crops [73]. According to a study, incredible microcantilever biosensors have been developed which are made up of tiny cantilever beams. When these beams come in contact with specific molecules such as pesticides, or insecticides, they tend to bend producing measurable signals [42]. As per the study done by Erdem *et al.*, nano-biosensors can also be fabricated using 3D printing technology using eco-friendly organic materials that can be used to detect specific insecticides [30]. Additionally, other sensitive biosensors such as nylon nano-compounds and CNTs are also enormously used as they can easily detect the presence of minute insecticides by a change in optical or electrical properties [33, 91].

10.3.5 Nano-Composites

A nano-composite is a solid material made up of several phases, with one phase being less than 100 nanometres (nm) in size [16]. Nano-composites are made up of a base material, typically a polymer or ceramic, that is strengthened by nanoparticles or nanofillers [40]. Moreover, nano-composite water filtration membranes and irrigation systems enhance water quality and efficiency in agricultural processes [92]. In addition, nanocomposite packaging materials provide improved barrier characteristics and antimicrobial effects, guaranteeing the security and freshness of agricultural products during both storage and transport [45]. MXenes are a recent type of 2D nanocomposite sensors able to detect different agricultural contaminants and pathogens [87]. Furthermore, a nanocomposite called Cobalt phthalocyanine multiwalled carbon nanotubes (CoPc-MWCNTs) can identify essential nutrients such as nitrogen, phosphorus, and potassium in soil or plant tissues [10]. Their distinct construction enables them to capture particular light frequencies, which could potentially improve photosynthesis in plants [10].

10.3.6 Nano-Pesticides

Nano-pesticides in comparison to such conventional pesticides have many advantages including higher efficacy, increased target specificity, lower environmental hazard, and lower application rates [21]. They are preferred over traditional pesticides as they target the specific pathogen or pest without harming other organisms that successfully aid in pest control. Recently, a new formulation of 4-chloro-2-methylphenoxyacetic acid-(MCPA) herbicide was introduced using clay and gelatine carriers [6]. This herbicide reduces the leaching of MCPA into the environment, provides targeted delivery to plants over time, and is environmentally friendly due to the natural components used. Furthermore, mesoporous silica NPs (MSNs) possess a porous structure that allows them to encapsulate pesticides such as mildamine within their cavities showing enhanced uptake by cucumber plant [124]. The GO-chlorpyrifos and GO-cypermethrin nano pesticides also showed improved pesticidal efficiency [119]. Furthermore, Zheng *et al.*, have reported the insecticidal properties of chitosan-based rotenone nano pesticides against *Solenopsis invicta* [127]. Additionally, Lakshmeesha *et al.* utilized the iron nanorods as an effective nano-fungicide to inhibit the growth of various phytopathogenic fungi and zinc oxide NPs to control the fungal pathogens of soybean [60]. The comparative studies of chitosan and chitosan-metal as nano-fungicides using triazolyl dithiocarbamate were done against the *Fusarium* wilt of *Cicer arietinum* L. [54]. Silica nano pesticides were used to treat insect-infected *Vicia faba* [107].

10.4 Successful Applications of Nanotechnology in Agricultural By-Products Valorization

In recent years, waste products from the agricultural and food industries have been recognized and utilized increasingly as valuable sources of renewable energy. The residual materials formed after harvesting of primary products and processing of crops and livestock which include crop residues, husks, shells, stems and other waste materials are defined as agricultural by-products [24, 72]. These by-products have the potential to address both environmental and economic concerns; therefore their valorization is seeking attention rapidly. Rather than discarding them as waste, these can be transformed to form valorized goods which can help in maintaining a sustainable environment. Transformation of waste resulted in the creation of biofuels, bioplastics, animal feed, fertilizers rich in nutrients, and even functional drinks and bioactive chemicals by employing sustainable procedures and technology (Figure 10.4) [17, 56]. This encourages a circular economy strategy by reducing waste, adding value to the agricultural chain, and creating new opportunities for sustainable economic development and waste management in rural areas [11, 43]. To boost overall sector productivity, implications of materials and techniques at the nanoscale to improve the characteristics and value, NPs and nanotechnology may be used for this transformation. It can boost multidisciplinary research to have partnerships with innovation and technology in the fields of biotechnology, engineering, material science, and agriculture, hence vitalizing the establishment of resilient, efficient, and sustainable

Figure 10.4 Successful applications of agricultural by-products towards sustainability and circular economy.

agricultural systems. These sustainable systems then be able to satisfy present and future demands of the environment while protecting the natural resources making it a necessary tool for valorization [100].

10.4.1 How Does Nanotechnology Contribute to Circular Economy Models in Agriculture

For several reasons like innovation, modernization, and technological revolution, nanotechnology holds significant importance on a global scale in every domain. It makes far-reaching changes in agriculture by maximizing and enhancing resource efficiency and productivity, promoting sustainability, generating economic value, and aiding environmental remediation. Resource efficiency using nanotechnology has been enhanced by using nanoscale materials to extract and transform biomass residues, which improves their surface area and catalytic capabilities hence increasing value as a valorized product. This encourages a shift towards a circular economy model and decreases environmental impacts. To improve environmental sustainability, nanotechnology should be combined with garbage disposal, which decreases the waste volume and pollutant emission also. Nanoscale innovations enhance resource recovery and recycling to make a transition towards regenerative upcoming economic systems. It upgrades the agronomic inputs such as fertilizers and crop protecting agents, upgrading soil health, increasing pest resistance, and improving nutrient bioavailability, significantly enhancing plant growth and meeting global food requirements. These low-value resources can be turned into valuable products using nanotechnology, opening up new agricultural revenue streams. This economic stimulant improves financial stability, which promotes long-term growth and a dynamic edge in local and international markets. In the agriculture sector, nanotechnology contributes in ecological renovation by reducing nutrient leaching, soil contamination, and harmful emissions of greenhouse gases. Heavy metals and organic contaminants as pollutants have been targeted by nanomaterials, forwarding toward ecosystem health and biodiversity. Along with providing a complete answer to major agricultural problems, nanotechnology also provides a key for environmentally responsible sustainable development and growth [63, 85, 90].

By exploiting the use of nanoscale materials, the agricultural sector can enhance resource utilization, and minimize environmental impacts to meet the demands of a rapidly evolving global food system and population.

10.4.2 Greener Solutions Promoting Environmental Health

10.4.2.1 Biosurfactants as Nature's Versatile Emulsifiers

Amphiphilic molecules, possessing both hydrophobic and hydrophilic moieties, are produced by microorganisms and are known as biosurfactants. These amphipathic moieties decrease surface tensions occurred between immiscible phases like water and oil, which helps in the emulsification and dispersion of hydrophobic compounds [67]. Biosurfactants offer several other advantages like biodegradability and low toxicity along with diversification in structures including glycolipids and lipopeptides. Due to their biocompatibility and other unique properties, they found several applications in bioremediation, pharmaceuticals, enhanced oil recovery, food, cosmetics, and agriculture [36, 96, 117].

10.4.2.2 Biofuels as Renewable Energy Sources

Renewable fuels referred to as biofuels generally recovered from organic materials like remnants of plant biomass and animal waste, using various conversion processes and methods. As these fuels are produced using sustainable methods, they can serve as an efficient choice to conventional fossil fuels [53]. Common types of biofuels include bioethanol, biodiesel, and biogas [52]. Bioethanol, usually obtained *via* sugar or starch fermentation using crops like corn or sugarcane [27, 62, 88]. Biodiesel is obtained from animal fats or vegetable oils *via* a method called transesterification [115]. Biogas obtained through anaerobic digestion of organic matter, such as agricultural waste, sewage, and municipal solid waste [20, 50, 52, 57]. Due to their potential to ease the emission of greenhouse gases and decrease dependence on fossil fuels, biofuels have gained significant attention, contributing to energy security and environmental sustainability.

10.4.2.3 Bioplastics as Sustainable Alternatives to Conventional Plastics

Bioplastics are transformed renewable plant biomass like sugarcane, vegetable fats and oils, maize, or starch, which unlike traditional plastics are not harmful and non-sustainable that are obtained from non-renewable fuels like petroleum [3, 9]. Depending on their composition these can be classified as biodegradable or non-biodegradable bioplastics. Since they may decompose under certain circumstances into natural components, these biodegradable bioplastics have a lower environmental effect, when compared to conventional plastics. Bioplastics, which have the potential to minimize dependence on fossil fuels and reduce plastic pollution, have significantly become popular due to their ability to degrade naturally and require processing in industrial composting facilities [18, 118].

10.4.2.4 Synthesis of NPs from Agro-Industrial Waste

In the race towards achieving environmental conservation and sustainable development, searching for alternative resources for NPs synthesis has gained remarkable attention. Agro-industrial valorized waste that comes as a by-product of agricultural and industrial processes presents a favorable pathway for the production of NPs (Table 10.1). This innovative approach not only addresses the challenge of waste management but also harnesses the untapped potential of organic compounds within these residues to synthesize valuable nanomaterials.

Table 10.1 Synthesis of NPs from agro-industrial waste and their diverse applications.

Agro-waste	Nanomaterial	Application	References
Coconut shell	Silver NPs	Antibacterial activity and cell wall degradation against human pathogens	[104]
Fruit peel extracts of *Annona squamosal*, Leaves extract of *Annona reticulata*	Silver NPs	Acaricidal, insecticidal, larvicidal efficacy and antimicrobial activity with biocompatibility to RBC	[58, 80]
Mango peel extracts	Gold NPs	Non-cytotoxic effects on African green monkey kidney cells (CV-1) and human fetal fibroblast cells (WI-38)	[125]
Waste black tea leaves	Silver NPs	New amperometric glucose biosensor, biofuel production, soil remediation	[26, 81]
Spent mushroom substrate, microalgae *Chlorella vulgaris*	Silver-chloride NPs	Antibacterial agent, antimicrobial activity	[23, 70]
Wasted olive leaf extract	NiO NPs	Antiparasitic and antimicrobial activity	[2]

(*Continued*)

Table 10.1 Synthesis of NPs from agro-industrial waste and their diverse applications. (*Continued*)

Agro-waste	Nanomaterial	Application	References
Rice husk as silica reservoir	Silica/polypyrrole nanocomposite, crystalline nano-silica	Corrosion-resistant coating agents, carriers for cancer drug delivery	[22, 59, 66, 102]
Eggshell	Hydroxyapatite NPs and bio nanocomposites	Biomedical applications, food packaging, wastewater treatment, soil remediation	[69, 112, 122]
Harvested algae *Spirogyra*	ZnO NPs	Bioethanol production	[46]
Macadamia nutshells, potato peels, and orange peels	Mesoporous carbon-TiO_2 nanocomposite, Nitrogen carbon-ZnO_2 nanosheet and carbon hollow-Al_2O_3 nanosheet	Heavy metals removal in wastewater and Blood fingerprint detection	[77]
Rice straw	Nitrogen-doped carbon dots	Ant counterfeiting stamps, self-healing inks	[71]
Agricultural waste straw	Carbon dots	Fluorescent sensors for the detection of Fe^{3+} ions and biological cell imaging	[28]

10.5 Conclusion

The integration of nanotechnology into agriculture presents a transformative approach toward achieving sustainable farming futures. By harnessing the unique properties of nanomaterials, farmers can optimize resource utilization, enhance crop health and productivity, and mitigate environmental impact. Nanotechnology enables precise delivery of fertilizers, pesticides, and water to crops, minimizing waste and maximizing efficiency. Smart nanomaterials facilitate targeted release of nutrients and growth regulators, leading to healthier plants and increased yields. Additionally, nanoscale sensors and remediation techniques offer innovative solutions for monitoring soil health, detecting contaminants, and mitigating environmental pollution. Precision farming technologies empowered by nanotechnology, such as nanosensors, enable real-time monitoring and management of crops, promoting sustainable land management practices. Furthermore, nanotechnology applications ensure food safety and quality by detecting contaminants and pathogens while extending the shelf life of perishable foods. However, to realize the full potential of nanotechnology in agriculture, collaborative efforts are necessary to address safety, ethical, and regulatory considerations, ensuring responsible and sustainable deployment of these technologies for the benefit of future generations.

Bibliography

1. Abdel Latef, A.A.H., Srivastava, A.K., El-sadek, M.S.A., Kordrostami, M., Tran, L.S.P., Titanium dioxide nanoparticles improve growth and enhance tolerance of broad bean plants under saline soil conditions. *Land Degrad. Dev.*, *29*, 4, 1065–1073, 2018.
2. Alghamdi, S.Q., Alotaibi, N., Al-Ghamdi, S.N., Alqarni, L.S., Amna, T., Moustafa, S.M., Alsohaimi, I.H., Alruwaili, I., Nassar, A., High Antiparasitic and Antimicrobial Performance of Biosynthesized NiO Nanoparticles *via* Wasted Olive Leaf Extract. *Int. J. Nanomed.*, 19, 1469–1485, 2024, https://doi.org/10.2147/IJN.S443965.
3. Ali, S.S., Elsamahy, T., Abdelkarim, E.A., Abdelfattah, A., Ramadan, H., Mostafa, S., Metwally, S.M., Sun, J., 12 - Engineered yeast for the production of bioplastics, in: *Advances in Yeast Biotechnology for Biofuels and Sustainability*, A. Daverey, K. Dutta, S. Joshi, T. Gea, (Eds.), pp. 277–296, Elsevier, 2023, https://doi.org/10.1016/B978-0-323-95449-5.00017-5.
4. Alimohammadi, M., *et al.*, Physiological responses induced in tomato plants by a two-component nanostructural system composed of carbon nanotubes conjugated with quantum dots and its *in vivo* multimodal detection. *Nanotechnology*, 22, 29, 295101, 2011, https://doi.org/10.1088/0957-4484/22/29/295101.
5. Allam, N.G., *et al.*, Biosynthesis of silver nanoparticles by cell-free extracts from some bacteria species for dye removal from wastewater. *Biotechnol. Lett.*, 41, 3, 379–389, 2019, https://doi.org/10.1007/s10529-019-02652-y.
6. Alromeed, A.A., *et al.*, Slow-release formulations of the herbicide MCPA by using clay–protein composites. *Pest. Manage. Sci.*, 71, 9, 1303–1310, 2015, https://doi.org/10.1002/ps.3929.
7. An, C., *et al.*, Nanomaterials and nanotechnology for the delivery of agrochemicals: strategies towards sustainable agriculture. *J. Nanobiotechnol.*, 20, 1, 1–19, 2022, https://doi.org/10.1186/s12951-021-01214-7.
8. Ashraf, S.A., Siddiqui, A.J., Abd Elmoneim, O.E., Khan, M.I., Patel, M., Alreshidi, M., Moin, A., Singh, R., Snoussi, M., Adnan, M., Innovations in

nanoscience for the sustainable development of food and agriculture with implications on health and environment. *Sci. Total Environ., 768*, 144990, 2021, https://doi.org/10.1016/j.scitotenv.2021.144990.

9. Ashter, S.A., 10 - New Developments, in: *Introduction to Bioplastics Engineering (Plastics Design Library)*, S.A. Ashter, (Ed.), pp. 251–274, William Andrew Publishing, Oxford, 2016, https://doi.org/10.1016/B978-0-323-39396-6.00010-5.
10. Balogun, S.A. and Fayemi, O.E., Recent Advances in the Use of CoPc-MWCNTs Nanocomposites as Electrochemical Sensing Materials. *Biosensors*, 12, 10, 850, 2022, https://doi.org/10.3390/bios12100850.
11. Barros, M.V., Salvador, R., de Francisco, A.C., Piekarski, C.M., Mapping of research lines on circular economy practices in agriculture: From waste to energy. *Renew. Sustain. Energy Rev.*, 131, 109958, 2020, https://doi.org/10.1016/j.rser.2020.109958.
12. Bastos-Arrieta, J., Florido, A., Pérez-Ràfols, C., Serrano, N., Fiol, N., Poch, J., Villaescusa, I., Green synthesis of Ag nanoparticles using grape stalk waste extract for the modification of screen-printed electrodes. *Nanomaterials*, 8, 11, 946, 2018, https://doi.org/10.3390/nano8110946.
13. Batsmanova, L.M., Gonchar, L.M., Taran, N.Y., Okanenko, A.A., Using a colloidal solution of metal nanoparticles as micronutrient fertiliser for cereals, in: *Proceedings of the International Conference Nanomaterials: Applications and Properties*, Crimea, Ukraine, September 16–21, 2013, Available from: http://nap.sumdu.edu.ua/index.php/nap/nap2013/paper/view/1097/504. Accessed April 18, 2014.
14. Biswal, S.K., Nayak, A.K., Parida, U.K., Nayak, P.L., Applications of nanotechnology in agriculture and food sciences. 2012.
15. Bruckmann, F.S., *et al.*, Adsorption and Photocatalytic Degradation of Pesticides into Nanocomposites: A Review. *Molecules (Basel, Switzerland)*, 27, 19, 6261, 2022, https://doi.org/10.3390/molecules27196261.
16. Camargo, P.H.C., Satyanarayana, K.G., Wypych, F., Nanocomposites: synthesis, structure, properties and new application opportunities. *Mater. Res.*, 12, 1–39, 2009, https://doi.org/10.1590/S1516-14392009000100002.
17. Capanoglu, E., Nemli, E., Tomas-Barberan, F., Novel Approaches in the Valorization of Agricultural Wastes and Their Applications. *J. Agric. Food Chem.*, 70, 6787–6804, 2022, https://doi.org/10.1021/acs.jafc.1c07104.
18. Censi, V., Saiano, F., Bongiorno, D., Indelicato, S., Napoli, A., Piazzese, D., Bioplastics: A new analytical challenge. *Front. Chem.*, 10, 2022, https://doi.org/10.3389%2Ffchem.2022.971792.
19. Channab, B.-E., *et al.*, Urea-Loaded Hydroxyapatite–Carboxylated Cellulose Composites as Slow-Release N Fertilizer Pellets for Efficient Delivery of Nitrogen. *Ind. Eng. Chem. Res.*, 62, 37, 14853–14865, 2023, https://doi.org/10.1021/acs.iecr.3c00737.
20. Charcosset, C., 2 - Combining membrane processes with renewable energy technologies: perspectives on membrane desalination, biofuels and biogas production, and microbial fuel cells, in: *Membranes for Clean and Renewable*

Power Applications, A. Gugliuzza, and A. Basile (Eds.), pp. 44–62, Woodhead Publishing, 2014, https://doi.org/10.1533/9780857098658.1.44.

21. Chaud, M., *et al.*, Nanopesticides in Agriculture: Benefits and Challenge in Agricultural Productivity, Toxicological Risks to Human Health and Environment. *Toxics*, 9, 6, 131, 2021, https://doi.org/10.3390/toxics9060131.
22. Chen, S.-Y., Jian, J.-Y., Lin, H.-M., Functionalization of rice husk-derived mesoporous silica nanoparticles for targeted and imaging in cancer drug delivery. *J. Sci. Food Agric.*, 104, 2120–2129, 2024, https://doi.org/10.1002/jsfa.13165.
23. da Silva Ferreira, V., ConzFerreira, M.E., Lima, L.M.T.R., Frasés, S., de Souza, W., Sant'Anna, C., Green production of microalgae-based silver chloride nanoparticles with antimicrobial activity against pathogenic bacteria. *Enzyme Microb. Technol.*, 97, 114–121, 2017, https://doi.org/10.1016/j.enzmictec.2016.10.018.
24. Damian, C.S., Devarajan, Y., Jayabal, R., A comprehensive review of the resource efficiency and sustainability in biofuel production from industrial and agricultural waste. *J. Mater. Cycles Waste Manage.*, 2024, https://doi.org/10.1007/s10163-024-01918-6.
25. Dayana Priyadharshini, S., *et al.*, Graphene oxide-based nanomaterials for the treatment of pollutants in the aquatic environment: Recent trends and perspectives – A review. *Environ. Pollut.*, 306, 119377, 2022, https://doi.org/10.1016/j.envpol.2022.119377.
26. Debnath, B. and Purkait, M.K., Sustainable Utilization of Tea Waste, in: *Agricultural Waste: Environmental Impact, Useful Metabolites and Energy Production, Sustainable Development and Biodiversity*, K.G. Ramawat, J.-M. Mérillon, J. Arora, (Eds.), pp. 245–275, Springer Nature, Singapore, 2023, https://doi.org/10.1007/978-981-19-8774-8_11.
27. Dodić, S., Popov, S., Dodić, J., Ranković, J., Zavargo, Z., Jevtić Mučibabić, R., Bioethanol production from thick juice as intermediate of sugar beet processing. *Biomass Bioenergy*, 33, 822–827, 2009, https://doi.org/10.1016/j.biombioe.2009.01.002.
28. Du, Y., Li, Y., Liu, Y., Liu, N., Cheng, Y., Shi, Q., Liu, X., Tao, Z., Guo, Y., Zhang, J., Askaria, N., Li, H., Stalk-derived carbon dots as nanosensors for $Fe3+$ ions detection and biological cell imaging. *Front. Bioeng. Biotechnol.*, 11, 2023, https://doi.org/10.3389/fbioe.2023.1187632.
29. Elsayed, A.A.A., *et al.*, Hydroxyapatite nanoparticles as novel nano-fertilizer for production of rosemary plants. *Sci. Hortic.*, 295, 110851, 2022, https://doi.org/10.1016/j.scienta.2021.110851.
30. Erdem, A., *et al.*, Implementation of 3D printing technologies to electrochemical and optical biosensors developed for biomedical and pharmaceutical analysis. *J. Pharm. Biomed. Anal.*, 230, 115385, 2023, https://doi.org/10.1016/j.jpba.2023.115385.
31. Faraco, T.A., de OX Silva, H., da S. Barud, H., Maciel, I.O., Da Silva, R.R., Quirino, W.G., Fragneaud, B., Ribeiro, C.A., dos S. Dias, D., G. Pandoli, O.,

Cremona, M., Ecological biosubstrates obtained from onion pulp (*Allium cepa* L.) for flexible organic light-emitting diodes. *ACS Appl. Mater. Interfaces*, 11, 45, 42420–42428, 2019, https://doi.org/10.1021/acsami.9b14029.
32. Farrell, M. and Jones, D.L., Critical evaluation of municipal solid waste composting and potential compost markets. *Bioresour. Technol.*, 100, 19, 4301–4310, 2009, https://doi.org/10.1016/j.biortech.2009.04.029.
33. Ferrier, D.C. and Honeychurch, K.C., Carbon Nanotube (CNT)-Based Biosensors. *Biosensors*, 11, 12, 2021, https://doi.org/10.3390/bios11120486.
34. Fraceto, L.F., Grillo, R., de Medeiros, G.A., Scognamiglio, V., Rea, G., Bartolucci, C., Nanotechnology in agriculture: which innovation potential does it have? *Front. Environ. Sci.*, 4, 20, 2016, https://doi.org/10.3389/fenvs.2016.00020.
35. Fuchs, W., Wager, F., Kirchmayr, R., Braun, R., Dorsg, B., Digestate treatment: comparison and assessment of existing technologies, in: *Third International Symposium on Energy from Biomass and Waste*, CISA Venice, Italy, 2010.
36. Gaur, V.K., Sharma, P., Sirohi, R., Varjani, S., Taherzadeh, M.J., Chang, J.-S., Yong Ng, H., Wong, J.W.C., Kim, S.-H., Production of biosurfactants from agro-industrial waste and waste cooking oil in a circular bioeconomy: An overview. *Bioresour. Technol.*, 343, 126059, 2022, https://doi.org/10.1016/j.biortech.2021.126059.
37. Gogos, A., Knauer, K., Bucheli, T.D., Nanomaterials in plant protection and fertilization: current state, foreseen applications, and research priorities. *J. Agric. Food. Chem.*, 60, 39, 9781–9792, 2012, https://doi.org/10.1021/jf302154y.
38. Gómez-Pastora, J., Dominguez, S., Bringas, E., Rivero, M.J., Ortiz, I., Dionysiou, D.D., Review and perspectives on the use of magnetic nanophotocatalysts (MNPCs) in water treatment. *Chem. Eng. J.*, *310*, 407–427, 2017, https://doi.org/10.1016/j. cej.2016.04.140.
39. Goswami, P., Mathur, J., Srivastava, N., Silica nanoparticles as novel sustainable approach for plant growth and crop protection. *Heliyon*, 8, 7, e09908, 2022, https://doi.org/10.1016/j.heliyon.2022.e09908.
40. Gupta, H., Role of Nanocomposites in Agriculture. *Nano Hybrids Compos.*, 20, 81–89, 2018, https://doi.org/10.4028/www.scientific.net/NHC.20.81.
41. Gupta, V.K., Kumar, R., Nayak, A., Saleh, T.A., Barakat, M.A., Adsorptive removal of dyes from aqueous solution onto carbon nanotubes: a review. *Adv. Colloid Interface Sci.*, 193, 24–34, 2013, https://doi.org/10.1016/j.cis.2013.03.003.
42. Hansen, K.M. and Thundat, T., Microcantilever biosensors. *Methods*, 37, 1, 57–64, 2005, https://doi.org/10.1016/j.ymeth.2005.05.011.
43. Haque, F., Fan, C., Lee, Y.-Y., From waste to value: Addressing the relevance of waste recovery to agricultural sector in line with circular economy. *J. Clean. Prod.*, 415, 137873, 2023, https://doi.org/10.1016/j.jclepro.2023.137873.
44. He, M., Xu, Z., Sun, Y., Chan, P.S., Lui, I., Tsang, D.C., Critical impacts of pyrolysis conditions and activation methods on application-oriented

production of wood waste-derived biochar. *Bioresour. Technol.*, *341*, 125811, 2021, https://doi.org/10.1016/j.biortech.2021.125811.

45. Honarvar, Z., Hadian, Z., Mashayekh, M., Nanocomposites in food packaging applications and their risk assessment for health. *Electron. Physician*, 8, 6, 2531–2538, 2016, https://doi.org/10.19082/2531.
46. Huang, X., Kainat, I., Hasan, M., Zafar, A., Tariq, T., Ahmad, K., Hassan, S.G., Javed, H.U., Shu, X., Ghorbanpour, M., Investigation of pretreatment parameters for bioethanol production from Spirogyra using ZnO nanoparticles. *Biomass Convers. Biorefin.*, 2023, https://doi.org/10.1007/s13399-023-05024-9.
47. Janmohammadi, M., *et al.*, Impact of foliar application of nano micronutrient fertilizers and titanium dioxide nanoparticles on the growth and yield components of barley under supplemental irrigation. *Acta Agric. Slov.*, 107, 2, 265–276, 2016, https://doi.org/10.14720/aas.2016.107.2.01.
48. Jasrotia, P., *et al.*, Nanomaterials for Postharvest Management of Insect Pests: Current State and Future Perspectives. *Front. Nanotechnol.*, 3, 2022, https://www.frontiersin.org/articles/10.3389/fnano.2021.811056.
49. Johnston, C.T., Probing the nanoscale architecture of clay minerals. *Clay Miner.*, 45, 3, 245–279, 2010, https://doi.org/10.1180/claymin.2010.045.3.245.
50. Joshi, S.M. and Gogate, P.R., Intensifying the biogas production from food waste using ultrasound: Understanding into effect of operating parameters. *Ultrason. Sonochem.*, 59, 104755, 2019, https://doi.org/10.1016/j.ultsonch.2019.104755.
51. Kah, M. and Hofmann, T., Nanopesticide research: current trends and future priorities. *Environ. Int.*, 63, 224–235, 2014, https://doi.org/10.1016/j.envint.2013.11.015.
52. Kaparaju, P., Serrano, M., Thomsen, A.B., Kongjan, P., Angelidaki, I., Bioethanol, biohydrogen and biogas production from wheat straw in a biorefinery concept. *Bioresour. Technol.*, 100, 2562–2568, 2009, https://doi.org/10.1016/j.biortech.2008.11.011.
53. Karmee, S.K. and Lin, C.S.K., Valorisation of food waste to biofuel: current trends and technological challenges. *Sustain. Chem. Processes*, 2, 22, 2014, https://doi.org/10.1186/s40508-014-0022-1.
54. Kaur, P., Duhan, J.S., Thakur, R., Comparative pot studies of chitosan and chitosan-metal nanocomposites as nano-agrochemicals against fusarium wilt of chickpea (*Cicer arietinum* L.). *Biocatal. Agric. Biotechnol.*, 14, 466–471, 2018, https://doi.org/10.1016/j.bcab.2018.04.014.
55. Kavitha, S., Kannah, R.Y., Kumar, G., Gunasekaran, M., Banu, J.R., Introduction: sources and characterization of food waste and food industry wastes, in: *Food Waste to Valuable Resources: Applications and Management*, pp. 1–13, Elsevier, 2020, https://doi.org/10.1016/B978-0-12-818353-3.00001-8.
56. Koul, B., Yakoob, M., Shah, M.P., Agricultural waste management strategies for environmental sustainability. *Environ. Res.*, 206, 112285, 2022, https://doi.org/10.1016/j.envres.2021.112285.

57. Kristoferson, L.A. and Bokalders, V., 9 - Biogas, in: *Renewable Energy Technologies*, L.A. Kristoferson, and V. Bokalders, (Eds.), pp. 100–113, Pergamon, 1986, https://doi.org/10.1016/B978-0-08-034061-6.50015-9.
58. Kumar, R., Roopan, S.M., Prabhakarn, A., Khanna, V.G., Chakroborty, S., Agricultural waste *Annona squamosa* peel extract: Biosynthesis of silver nanoparticles. *Spectrochim. Acta, Part A*, 90, 173–176, 2012, https://doi.org/10.1016/j.saa.2012.01.029.
59. Kumari, A., Singh, R.K., Kumar, N., Kumari, R., Monalisa, Sharma, S., Green synthesis and physical properties of crystalline silica engineering nanomaterial from rice husk (agriculture waste) at different annealing temperatures for its varied applications. *J. Indian Chem. Soc.*, 100, 100982, 2023, https://doi.org/10.1016/j.jics.2023.100982.
60. Lakshmeesha, T.R., *et al.*, Biofabrication of zinc oxide nanoparticles from *Melia azedarach* and its potential in controlling soybean seed-borne phytopathogenic fungi. *Saudi J. Biol. Sci.*, 27, 8, 1923–1930, 2020, https://doi.org/10.1016/j.sjbs.2020.06.013.
61. Li, Z., Lu, H., Ren, L., He, L., Experimental and modeling approaches for food waste composting: A review. *Chemosphere*, 93, 7, 1247–1257, 2013, https://doi.org/10.1016/j.chemosphere.2013.06.064.
62. Lin, Y. and Tanaka, S., Ethanol fermentation from biomass resources: current state and prospects. *Appl. Microbiol. Biotechnol.*, 69, 627–642, 2006, https://doi.org/10.1007/s00253-005-0229-x.
63. Liu, R. and Lal, R., Potentials of engineered nanoparticles as fertilizers for increasing agronomic productions. *Sci. Total Environ.*, 514, 131–139, 2015, https://doi.org/10.1016/j.scitotenv.2015.01.104.
64. Liu, Y.Y., Zhao, Y.H., Zhou, Y., Guo, X.L., Chen, Z.T., Zhang, W.J., Zhang, Y., Chen, J., Wang, Z.M., Sun, L.T., Zhang, T., High-efficient catalytic reduction of 4-nitrophenol based on reusable Ag nanoparticles/graphene-loading loofah sponge hybrid. *Nanotechnology, 29*, 31, 315702, 2018, https://doi.org/10.1088/1361-6528/aac3e8.
65. Lou, J., *et al.*, Agricultural waste valorization towards (nano)catalysts for the production of chemicals and materials. *Fuel*, 351, 128935, 2023, https://doi.org/10.1016/j.fuel.2023.128935.
66. Mahalingam, V., Sivaraju, M., Kumar, G.S., Lalithambigai, K., Priyan, S.R., Alam, M.M., Synergistic effect of silica/polypyrrole nanocomposites synthesized by hydrothermal method using rice husk as a silica source for corrosion protection. *Mater. Chem. Phys.*, 313, 128752, 2024, https://doi.org/10.1016/j.matchemphys.2023.128752.
67. Makkar, R.S., Cameotra, S.S., Banat, I.M., Advances in utilization of renewable substrates for biosurfactant production. *AMB Express*, 1, 5, 2011, https://doi.org/10.1186/2191-0855-1-5.
68. Mathur, P. and Roy, S., Nanosilica facilitates silica uptake, growth and stress tolerance in plants. *Plant Physiol. Biochem.*, 157, 114–127, 2020, https://doi.org/10.1016/j.plaphy.2020.10.011.

69. Milla, L.E., Hartami, E., Priyanto, R., Adrian, A.G.W., Synthesis and Characterization of Eggshell Based Hydroxyapatite Nanoparticles: X-ray Diffraction and Spectroscopy Analysis. *Malays. J. Med. Health Sci.*, 19, 2023.
70. Mo, W., Yao, C., Chen, H., Nassor, A.K., Gui, F., Hong, C., Huang, T., Guan, X., Xu, L., Pan, X., Biosynthesis of high antibacterial silver chloride nanoparticles against Ralstonia solanacearum using spent mushroom substrate extract. *Nano Express*, 2024, https://doi.org/10.1088/2632-959X/ad2b81.
71. Mogharbel, A.T., Hameed, A., Sayqal, A.A., Katouah, H.A., Al-Qahtani, S.D., Saad, F.A., El-Metwaly, N.M., Preparation of carbon dots-embedded fluorescent carboxymethyl cellulose hydrogel for anticounterfeiting applications. *Int. J. Biol. Macromol.*, 238, 124028, 2023, https://doi.org/10.1016/j.ijbiomac.2023.124028.
72. Mohite, A.S., Jagtap, A.R., Avhad, M.S., More, A.P., Recycling of major agriculture crop residues and its application in polymer industry: A review in the context of waste to energy nexus. *Energy Nexus*, 7, 100134, 2022, https://doi.org/10.1016/j.nexus.2022.100134.
73. Mondal, R., *et al.*, Potential of nanobiosensor in sustainable agriculture: the state-of-art. *Heliyon*, 8, 12, e12207, 2022, https://doi.org/10.1016/j.heliyon.2022.e12207.
74. Nandini, B. and Geetha, N., Smart delivery mechanisms of nanofertilizers and nanocides in crop biotechology, in: *Advances in Nano-Fertilizers and Nano-Pesticides in Agriculture*, S. Jogaiah, *et al.* (Eds.), pp. 385–414, Woodhead Publishing, Woodhead Publishing Series in Food Science, Technology and Nutrition, 2021, https://doi.org/10.1016/B978-0-12-820092-6.00015-X.
75. Ningthoujam, R., *et al.*, Nanocatalyst in remediating environmental pollutants. *Chem. Phys. Impact*, 4, 100064, 2022, https://doi.org/10.1016/j.chphi.2022.100064.
76. Nongbet, A., *et al.*, Nanofertilizers: A Smart and Sustainable Attribute to Modern Agriculture. *Plants*, 11, 19, 2587, 2022, https://doi.org/10.3390/plants11192587.
77. Nthwane, Y.B., Synthesis, Characterization of Agricultural Waste Materials for Removal of Heavy Metals in Wastewater and Reuse in Blood Fingerprint Development, 2022, https://hdl.handle.net/10210/504002.
78. Otoni, C.G., Azeredo, H.M., Mattos, B.D., Beaumont, M., Correa, D.S., Rojas, O.J., The food–materials nexus: Next generation bioplastics and advanced materials from agri-food residues. *Adv. Mater.*, 33, 43, 2102520, 2021, https://doi.org/10.1002/adma.202102520.
79. Panahi-Kalamuei, M., Amiri, O., Salavati-Niasari, M., Green hydrothermal synthesis of high quality single and few layers graphene sheets by bread waste as precursor. *J. Mater. Res. Technol.*, 9, 3, 2679–2690, 2020, https://doi.org/10.1016/j.jmrt.2020.01.001.
80. Parthiban, E., Manivannan, N., Ramanibai, R., Mathivanan, N., Green synthesis of silver-nanoparticles from *Annona reticulata* leaves aqueous extract and its mosquito larvicidal and anti-microbial activity on human pathogens.

Biotechnol. Rep., 21, e00297, 2019, https://doi.org/10.1016/j.btre.2018.e00297.

81. Pektaş, S.Ü., Keskin, M., Bodur, O.C., Arslan, F., Green Synthesis of Silver Nanoparticles and Designing A New Amperometric Biosentsor to Determine Glucose Levels. *J. Food Compos. Anal.*, 106133, 2024, https://doi.org/10.1016/j.jfca.2024.106133.
82. Pereira, A.E.S., Oliveira, H.C., Fraceto, L.F., Polymeric nanoparticles as an alternative for application of gibberellic acid in sustainable agriculture: a field study. *Sci. Rep.*, 9, 1, 7135, 2019, https://doi.org/10.1038/s41598-019-43494-y.
83. Pérez-de-Luque, A., Interaction of Nanomaterials with Plants: What Do We Need for Real Applications in Agriculture? *Front. Environ. Sci.*, 5, 2017, https://www.frontiersin.org/articles/10.3389/fenvs.2017.00012.
84. Prasad, R., Bhattacharyya, A., Nguyen, Q.D., Nanotechnology in sustainable agriculture: recent developments, challenges, and perspectives. *Front. Microbiol.*, 8, 1014, 2017, https://doi.org/10.3389/fmicb.2017.01014.
85. Pulizzi, F., Nano in the future of crops. *Nat. Nanotechnol.*, 14, 6, 507, 2019, https://doi.org/10.1038/s41565-019-0475-1.
86. Raliya, R., *et al.*, Nanofertilizer for Precision and Sustainable Agriculture: Current State and Future Perspectives. *J. Agric. Food. Chem.*, 66, 26, 6487–6503, 2018, https://doi.org/10.1021/acs.jafc.7b02178.
87. Riazi, H., Taghizadeh, G., Soroush, M., MXene-Based Nanocomposite Sensors. *ACS Omega*, 6, 17, 11103–11112, 2021, https://doi.org/10.1021/acsomega.0c05828.
88. Rončević, Z., Bajić, B., Dodić, S., Grahovac, J., Pajović-Šćepanović, R., Dodić, J., Optimization of bioethanol production from soybean molasses using different strains of Saccharomyces cerevisiae. *Hem. Ind.*, 73, 1–12, 2019, http://dx.doi.org/10.2298/HEMIND180713004R.
89. Rosso, D., Fan, J., Montoneri, E., Negre, M., Clark, J., Mainero, D., Conventional and microwave assisted hydrolysis of urban biowastes to added value lignin-like products. *Green Chem.*, *17*, 6, 3424–3435, 2015, https://doi.org/10.1039/c5gc00357a.
90. Roy, A., Elzaki, A., Tirth, V., Kajoak, S., Osman, H., Algahtani, A., Islam, S., Faizo, N.L., Khandaker, M.U., Islam, M.N., Emran, T.B., Bilal, M., Biological Synthesis of Nanocatalysts and Their Applications. *Catalysts*, 11, 1494, 2021, https://doi.org/10.3390/catal11121494.
91. Samal, S., *et al.*, Implications of biosensors and nanobiosensors for the eco-friendly detection of public health and agro-based insecticides: A comprehensive review. *Heliyon*, 9, 5, 2023, https://doi.org/10.1016/j.heliyon.2023.e15848.
92. Sahu, A., *et al.*, Advanced Polymeric Nanocomposite Membranes for Water and Wastewater Treatment: A Comprehensive Review. *Polymers*, 15, 3, 540, 2023, https://doi.org/10.3390/polym15030540.
93. Sani, M.N.H., Amin, M., Siddique, A.B., Nasif, S.O., Ghaley, B.B., Ge, L., Wang, F., Yong, J.W.H., Waste-derived nanobiochar: A new avenue towards

sustainable agriculture, environment, and circular bioeconomy. *Sci. Total Environ.*, 166881, 2023, https://doi.org/10.1016/j.scitotenv.2023.166881.

94. Saritha, G.N.G., Anju, T., Kumar, A., Nanotechnology - Big impact: How nanotechnology is changing the future of agriculture? *J. Agric. Food Res.*, 10, 100457, 2022, https://doi.org/10.1016/j.jafr.2022.100457.
95. Sarkar, B., *et al.*, Designer carbon nanotubes for contaminant removal in water and wastewater: A critical review. *Sci. Total Environ.*, 612, 561–581, 2018, https://doi.org/10.1016/j.scitotenv.2017.08.132.
96. Satpute, S.K., Płaza, G.A., Banpurkar, A.G., Biosurfactants' Production from Renewable Natural Resources: Example of Innovative and Smart Technology in Circular Bioeconomy. *Manage. Syst. Prod. Eng.*, 25, 46–54, 2017, http://dx.doi.org/10.1515/mspe-2017-0007.
97. Scott, N. and Chen, H., Nanoscale Science and Engineering for Agriculture and Food Systems. *Ind. Biotechnol.*, 9, 17–18, 2013, https://doi.org/10.1089/ind.2013.1555.
98. Sekhon, B., Nanotechnology in agri-food production: an overview. *Nanotechn. Sci. Appl.*, 7, 31–53, 2014. https://doi.org/10.2147/NSA.S39406
99. Sharma, B., *et al.*, Nano-biofertilizers as bio-emerging strategies for sustainable agriculture development: Potentiality and their limitations. *Sci. Total Environ.*, 860, 160476, 2023, https://doi.org/10.1016/j.scitotenv.2022.160476.
100. Sharma, R., Lata, S., Garg, R., Valorisation of agricultural waste and their role in green synthesis of value-added nanoparticles. *Environ. Technol. Rev.*, 13, 40–59, 2024, https://doi.org/10.1080/21622515.2023.2283412.
101. Sharma, V., *et al.*, Zeolites as Carriers of Nano-Fertilizers: From Structures and Principles to Prospects and Challenges. *Appl. Nano*, 3, 3, 163–186, 2022, https://doi.org/10.3390/applnano3030013.
102. Shrestha, D., Nayaju, T., Kandel, M.R., Pradhananga, R.R., Park, C.H., Kim, C.S., Rice husk-derived mesoporous biogenic silica nanoparticles for gravity chromatography. *Heliyon*, 9, e15142, 2023, https://doi.org/10.1016/j.heliyon.2023.e15142.
103. Singh, G., *et al.*, Role of multi-walled carbon nanotubes as a growth regulator for Basella alba (Malabar spinach) plant and its soil microbiota. *Chem. Biol. Technol. Agric.*, 9, 1, 71, 2022, https://doi.org/10.1186/s40538-022-00337-9.
104. Sinsinwar, S., Sarkar, M.K., Suriya, K.R., Nithyanand, P., Vadivel, V., Use of agricultural waste (coconut shell) for the synthesis of silver nanoparticles and evaluation of their antibacterial activity against selected human pathogens. *Microb. Pathogen.*, 124, 30–37, 2018, https://doi.org/10.1016/j.micpath.2018.08.025.
105. Skrzypczak, D., Mikula, K., Izydorczyk, G., Dawiec-Liśniewska, A., Moustakas, K., Chojnacka, K., Witek-Krowiak, A., New directions for agricultural wastes valorization as hydrogel biocomposite fertilizers. *J. Environ. Manage.*, *299*, 113480, 2021, https://doi.org/10.1016/j.jenvman.2021.113480.

106. Song, B., Cao, X., Gao, W., Aziz, S., Gao, S., Lam, C.H., Lin, R., Preparation of nano-biochar from conventional biorefineries for high-value applications. *Renew. Sustain. Energy Rev.*, *157*, 112057, 2022.
107. Thabet, A.F., *et al.*, Silica nanoparticles as pesticide against insects of different feeding types and their non-target attraction of predators. *Sci. Rep.*, 11, 1, 14484, 2021, https://doi.org/10.1038/s41598-021-93518-9.
108. Thakur, M., Wang, B., Verma, M.L., Development and applications of nanobiosensors for sustainable agricultural and food industries: Recent developments, challenges and perspectives. *Environ. Technol. Innov.*, 26, 102371, 2022, https://doi.org/10.1016/j.eti.2022.102371.
109. Thomé, A., Reddy, K.R., Reginatto, C., Cecchin, I., Review of nanotechnology for soil and groundwater remediation: Brazilian perspectives. *Water Air Soil Pollut.*, *226*, 1–20, 2015.
110. Tomczyk, A., Sokołowska, Z., Boguta, P., Biochar physicochemical properties: pyrolysis temperature and feedstock kind effects. *Rev. Environ. Sci. Bio/Technol.*, 19, 191–215, 2020, https://doi.org/10.1007/s11157-020-09523-3.
111. Tripathy, D.B. and Gupta, A., Nanomembranes-Affiliated Water Remediation: Chronology, Properties, Classification, Challenges and Future Prospects. *Membranes*, 13, 8, 713, 2023, https://doi.org/10.3390/membranes13080713.
112. Upadhyay, P., Synthesis of Eggshell-derived Nanoparticles and Bionanocomposites Thereof for Food Packaging, ERA, 2023, https://doi.org/10.7939/r3-sepg-x464.
113. Upadhyay, P.K., *et al.*, Conjoint application of nano-urea with conventional fertilizers: An energy efficient and environmentally robust approach for sustainable crop production. *PLoS One*, 18, 7, e0284009, 2023, https://doi.org/10.1371/journal.pone.0284009.
114. Velten, S., *et al.*, What Is Sustainable Agriculture? A Systematic Review. *Sustainability*, 7, 6, 7833–7865, 2015, https://doi.org/10.3390/su7067833.
115. Viswanathan, B., Chapter 2 - Petroleum, in: *Energy Sources*, B. Viswanathan, (Ed.), pp. 29–57, Elsevier, Amsterdam, 2017, https://doi.org/10.1016/B978-0-444-56353-8.00002-2.
116. Vithanage, M., *et al.*, Contrasting effects of engineered carbon nanotubes on plants: a review. *Environ. Geochem. Health*, 39, 6, 1421–1439, 2017, https://doi.org/10.1007/s10653-017-9957-y.
117. Vučurović, D., Bajić, B., Trivunović, Z., Dodić, J., Zeljko, M., Jevtić-Mučibabić, R., Dodić, S., Biotechnological Utilization of Agro-Industrial Residues and By-Products—Sustainable Production of Biosurfactants. *Foods*, 13, 711, 2024, https://doi.org/10.3390/foods13050711.
118. Meereboer, K.W., Misra, M., Mohanty, A.K., Review of recent advances in the biodegradability of polyhydroxyalkanoate (PHA) bioplastics and their composites. *Green Chem.*, 22, 5519–5558, 2020, https://doi.org/10.1039/D0GC01647K.

119. Wang, X., *et al.*, Graphene oxide as a pesticide delivery vector for enhancing acaricidal activity against spider mites. *Colloids Surf., B*, 173, 632–638, 2019, https://doi.org/10.1016/j.colsurfb.2018.10.010.
120. Wang, Y., *et al.*, Environmental Remediation Applications of Carbon Nanotubes and Graphene Oxide: Adsorption and Catalysis. *Nanomaterials*, 9, 3, 439, 2019, https://doi.org/10.3390/nano9030439.
121. Wei, H., *et al.*, The future of freshwater access: functional material-based nano-membranes for desalination. *Mater. Today Energy*, 22, 100856, 2021, https://doi.org/10.1016/j.mtener.2021.100856.
122. Wu, B., Zhang, G., Shuang, S., Choi, M.M.F., Biosensors for determination of glucose with glucose oxidase immobilized on an eggshell membrane. *Talanta*, 64, 546–553, 2004, https://doi.org/10.1016/j.talanta.2004.03.050.
123. Xiong, X., Iris, K.M., Tsang, D.C., Bolan, N.S., Ok, Y.S., Igalavithana, A.D., Kirkham, M.B., Kim, K.H., Vikrant, K., Value-added chemicals from food supply chain wastes: State-of-the-art review and future prospects. *Chem. Eng. J.*, 375, 121983, 2019, https://doi.org/10.1016/j.cej.2019.121983.
124. Xu, Y., *et al.*, Size Effect of Mesoporous Silica Nanoparticles on Pesticide Loading, Release, and Delivery in Cucumber Plants. *Appl. Sci.*, 11, 2, 575, 2021, https://doi.org/10.3390/app11020575.
125. Yang, N., WeiHong, L., Hao, L., Biosynthesis of Au nanoparticles using agricultural waste mango peel extract and its *in vitro* cytotoxic effect on two normal cells. *Mater. Lett.*, 134, 67–70, 2014, https://doi.org/10.1016/j.matlet.2014.07.025.
126. Zhang, M., Liang, W., Tu, Z., Li, R., Zhang, Z., Ali, A., Xiao, R., Succession of bacterial community during composting: dissimilarity between compost mixture and biochar additive. *Biochar*, 3, 229–237, 2021, https://doi.org/10.1007/s42773-020-00078-8.
127. Zheng, Q., *et al.*, Novel application of biodegradable chitosan in agriculture: Using green nanopesticides to control *Solenopsis invicta*. *Int. J. Biol. Macromol.*, 220, 193–203, 2022, https://doi.org/10.1016/j.ijbiomac.2022.08.066.

11

Eco-Friendly Nanomaterial: A Sustainable Approach for Heavy Metals Removal from Industrial Wastewater

Mahak Kushwaha[1], Astha Bisht[1], Yanshi Agrawal[1], Swati Agarwal[2] and Suphiya Khan[1,2,3]*

[1]*Department of Bioscience and Biotechnology, Banasthali Vidyapith, Banasthali Tonk, Rajasthan, India*
[2]*Drumlins Water Technologies Pvt. Ltd., Jaipur, Rajasthan, India*
[3]*Shriram Institute for Industrial Research, Gurugram, Haryana, India*

Abstract

Nowadays, heavy metal pollution is the main environmental threat. Heavy metals are persistent and recalcitrant in the environment, and handling them is very important. Many techniques for the removal of heavy metals from wastewater have been thoroughly researched in the past few years. Among the methods that are being used to evaluate and treat wastewater that contains heavy metals are ion exchange, adsorption, membrane filtration, coagulation, flocculation, flotation, and chemical precipitation. Nanotechnology has been important in the cleanup of contaminated water through the development of nano-adsorbents. The water sector has seen a sharp increase in demand for nano-adsorbents in recent years. Comparing nanotechnology to other techniques yields several benefits. Cost and volumetric efficiency are higher for nanomaterials, and several processes operate more effectively and quickly at the nanoscale. Even though nanoparticles have previously shown their worth in the field of water technology, some problems with their stability, toxicity, and recovery have given rise to solutions. Adsorbents are regenerated using a variety of methods, including thermal desorption, chemical desorption, microwave-assisted adsorbent regeneration, and bio regeneration. The following chapter briefly describes conventional methods as well as the use of nanomaterials with their regeneration methods for heavy metal removal from wastewater.

**Corresponding author*: suphiyakhan@gmail.com

Priyanka Singh, Pooja Agarwal and V. Vivekanand (eds.) Solid Waste Management: Challenges, Sustainability and Advancements, (361–396) © 2026 Scrivener Publishing LLC

Keywords: Heavy metals, wastewater, nanomaterials, adsorbents, regeneration

11.1 Introduction

Water is the most abundant substance on Earth. It is essential for the survival of species in all kingdoms. Water is a critical resource for humans, and water issues have a direct influence on society and the quality of life. Water contamination may be caused by both natural and man-made factors, such as industrial and agricultural operations, poor waste management, and urbanization. Water pollution may have a variety of negative consequences for both individuals and the environment.

Heavy metals are those having metallic elements with high atomic weights, such as lead, mercury, cadmium, zinc, arsenic, and chromium. Excessive levels of some metals in groundwater, on the other hand, can be harmful to human health. In Table 11.1 permissible limits of heavy metals

Table 11.1 Permissible limit of heavy metals in wastewater [38].

S. no.	Heavy metals in wastewater (WHO)	Permissible limit (ppm)
1.	Mercury (Hg)	0.001 ppm
2.	Lead (Pb)	0.01 ppm
3.	Chromium (Cr)	0.05 ppm
4.	Cadmium (Cd)	0.003 ppm
5.	Nickel (Ni)	0.007 ppm
6.	Cobalt (Co)	0.005 ppm
7.	Copper (Cu)	2.00 ppm
8.	Arsenic (As)	0.05 ppm
9.	Zinc (Zn)	5.00 ppm
10.	Iron (Fe)	5.00 ppm
11.	Manganese (Mn)	5.00 ppm
12.	Antimony (Sb)	1.00 ppm
13.	Selenium (Se)	1.00 ppm

in wastewater including lead, mercury, cadmium, zinc, and arsenic are listed. To prevent and reduce metal contamination in water, it is important to implement regulations and best practices to limit the release of heavy metals into water bodies. This entails keeping an eye on and assessing the quality of the water, cutting back on or doing away with the use of heavy metals in industrial operations, and appropriately disposing of hazardous waste. Efforts can also be made to remediate areas that have been affected by heavy metal pollution, such as cleaning up contaminated sediment and restoring degraded ecosystems. Water bodies become host to several hazardous metals and compounds that harm the ecosystem, including lead, mercury, arsenic, cobalt, nickel, and others.

Drinking water contamination can lead to diarrhea, cholera, and typhoid fever. Polluted water also harms aquatic plants and animals, causing population and biodiversity losses. Water pollutants might be chemical (such as pesticides, heavy metals, and therapeutic purposes), biological (such as bacteria and viruses), or physical [3].

For sustainable development, nanotechnology plays a significant role by providing cost-effective, efficient, and eco-friendly solutions to global challenges such as water scarcity, food security, energy resources, healthcare, and environmental protection. Traditional methods of synthesizing nanomaterials using harsh chemicals pose risks to human health and the environment. However, the use of green nanotechnology, which incorporates biological methods and green chemistry principles, offers a sustainable and environmentally friendly approach to nanomaterial synthesis. Nanotechnology contributes to sustainable development by introducing zero-waste tools and technologies in various industries, including automobiles, water, and agriculture. Overall, nanotechnology has a favorable impact on sustainable development, addressing global challenges and promoting a clean and green environment.

Conventional techniques are expensive, and it was observed that these treatments are inefficient for water treatment. Usually, industrial wastewater contains toxic heavy metals [11]. Several studies have focused on the fabrication of nanomaterials for wastewater treatment. In this chapter, we will go through the different types of nanomaterials for the treatment of heavy metals that are being disposed of by different kinds of industries and how the restoration and renewal of nanomaterials is done.

11.2 Industrial Wastewater

Heavy metals are present in industrial effluent streams released by various industries, and metal surface treatment processes create huge amounts of wastewater containing heavy metals (like cadmium, zinc, lead, chromium, nickel, copper, vanadium, platinum, silver, and titanium) from an assortment of utilizations. These incorporate electroplating, electroless affidavits, transformation covering, anodizing, cleaning, processing, and carving. Wood handling industries' effluent contains arsenic metals waste which are produced by chromated copper-arsenate wood treatment; inorganic shade fabricating, delivering colors containing chromium mixtures and cadmium sulfide. These produce nickel, vanadium, and chromium-tainted transformation impetuses; additionally, they produce visual chores that supply large quantities of silver and ferrocyanide in film. Large volumes of effluent, buildup, and slime are produced by these generators, which might be classified as hazardous wastes needing extensive waste treatment [81].

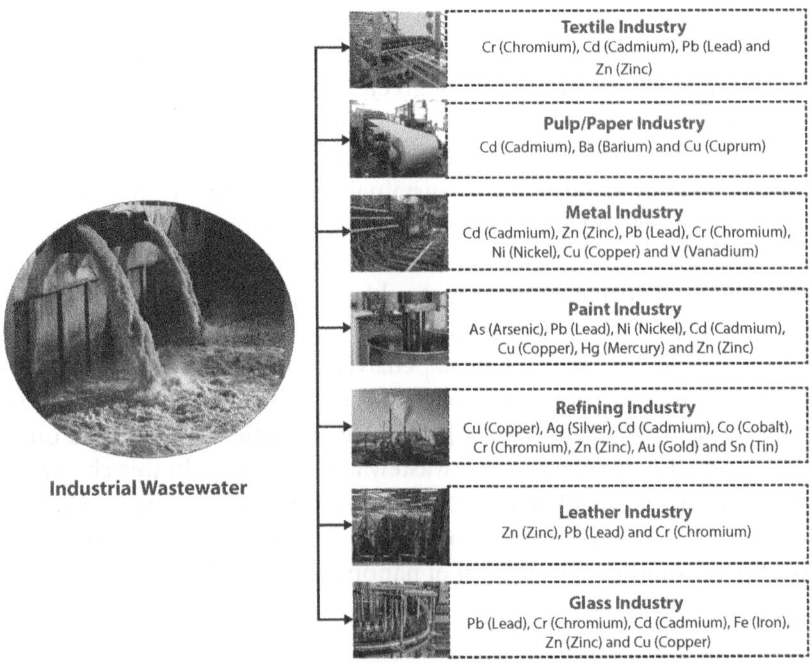

Figure 11.1 Industrial wastewater classification with different heavy metal presence in it.

Several industries discharge heavy metals in different concentrations (Figure 11.1). Amongst all the heavy metals, chromium is the most widely utilized and released to the climate from various sources. In any case, the metal is generally risky to living organic entities. Considerably more harmful are cadmium, lead, and mercury. These exhibit extreme preferences for sulfur and affect the ability of proteins to form bonds with sulfur groups in compounds. Heavy metals also limit amino (-NH_2) and carboxylic corrosive (-CO_2H) groups in proteins.

11.3 Techniques Used to Remove Heavy Metals

Different reported methods were devoted to the removal of heavy metal ions from various wastewater sources. These methods could be classified into adsorption, membrane filtration, chemical, electrocoagulation, electrolysis, ion exchange, and photocatalytic-based treatments (Figure 11.2). Detailed discussion on these methods is as follows.

11.3.1 Adsorption

Although extensive usage of adsorption has a relatively short history in comparison to other water treatment technologies, it has been used for over 4,000 years, according to Sanskrit writings [25]. A particular substance

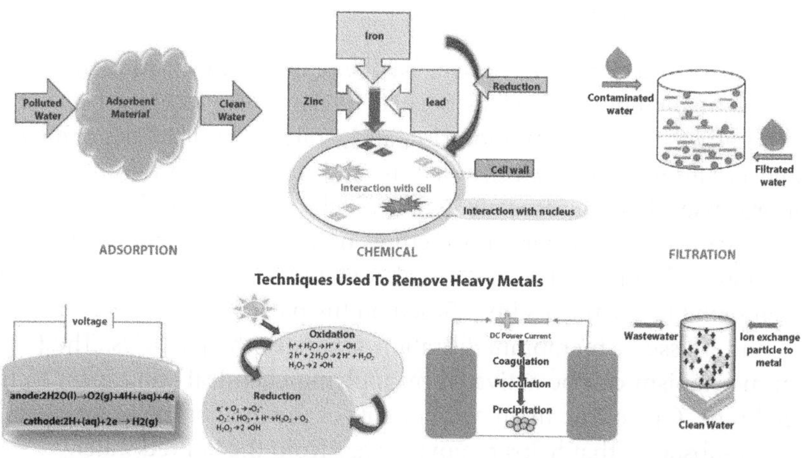

Figure 11.2 Different techniques used to remove heavy metals from wastewater.

clinging to the surface of a solid object by physical or chemical forces is known as adsorption, which is a surface phenomenon. An adsorbent is a solid surface, whereas an adsorbate is a material (pollutant) that binds to a solid surface. Three crucial steps are involved in the sorption of a pollutant onto an adsorbent: (i) the pollutant's entry into the adsorbent structure, (ii) its adsorption on the adsorbent surface, and (iii) its penetration from the bulk solution to the adsorbent surface [12].

For the removal of heavy metal ions, adsorption is typically used because of its great efficiency, amazing simplicity, and low cost. Activated carbon, microbial biomasses and clay minerals [27]. Adsorbents like metal oxide and polymeric polymers are frequently used to remove heavy metal ions from aqueous solutions [5]. Adsorption is affected by several variables, including temperature, the kind of adsorbent and adsorbate. Sometimes pre-filtration is required because the presence of oils, greases, and suspended particles reduces the efficiency of the operation. Adsorption is widely recognized as a workable and reasonably priced heavy metal wastewater treatment method.

The adsorption technique is the most efficient way to remove heavy metals from wastewater when compared to other methods [39]. It offers high-quality treated effluent. This procedure involves mass conversion. Due to the reversible nature of the desorption process and the many uses of the regenerated adsorbent, adsorbents can be rebuilt in the adsorption process [92].

11.3.2 Filtration

The field of heavy metal removal has benefited greatly from the development of membranes and associated filtering technologies because of their high efficiency, small footprint, and straightforward design. Reverse osmosis, ultrafiltration, electrodialysis, and nanofiltration are the different processes that are associated with membrane filtration [7]. One modern pressure-driven separation method used in wastewater treatment is membrane filtration. This method also includes disinfection in addition to removing heavy metals [49]. Based on the particle's size, concentration, pH, and pressure, membrane filtration separates the particles. The filtration mechanism can be activated by applying chemical substances to the membrane. One of the main components of the membrane is a particular porous substance that helps remove metals from the tainted water.

There are two categories of membrane materials: ceramic and polymer. Ceramics, as opposed to polymers, are often utilized for their chemical resilience properties in the treatment of industrial wastewater. Ceramics are also capable of becoming hydrophobic. The primary disadvantage of ceramic materials is their weakness and high construction costs [65]. The three main phases in using these techniques to remove heavy metals are as follows. The initial stage is preparatory to therapy; to reduce potential fouling feed water must be treated before entering the system [40]. The mechanical, chemical, and thermal characteristics of positively charged polyethylene crosslinked polyimide nanofiltration membranes are good.

11.3.3 Chemical

To remove metals, inorganic chemicals, oils, fats, and organic components from wastewater, chemical techniques are widely utilized in water treatment. Pollutants that are suspended or dissolved in solution go through a process called precipitation that turns them into a solid that may be later removed from the liquid. A significant precipitate makes it simple to extract the ions and particles. The little suspended particles are gathered into bigger precipitates by coagulants, such as polymers. The polymer molecules can be neutral or charged (cationic, anionic, or non-ionic). The polymers operate as either bond between particles or particle neutralizers in solution due to the interaction between ions and charged particles in a liquid.

Depending on the pollutants that need to be eliminated, several precipitation techniques can be used [16]. Insoluble metal precipitation is formed by reacting dissolved metals in the solution with a precipitant. To remove small particles in precipitants some techniques like chemical precipitants, coagulants, and flocculation are used which enlarges the size of the precipitant in the sludge for easy disposal [29].

11.3.4 Electrocoagulation

A straightforward and effective method utilized in the wastewater treatment industry is electrocoagulation (EC). Because of its poor systematic reactor design and issues with electrode dependability, it was never considered a trustworthy approach. EC has gained recognition recently as a small-scale wastewater treatment technique with enhanced technological approaches [17]. By using electrolysis, bothersome contaminants are eliminated. Several wastewater types have been treated using the straightforward and effective technique of electrocoagulation (EC), including electroplating wastewater, laundry wastewater, restaurant wastewater, and poultry abattoir wastewater. Pollutants have been effectively removed from a variety of industrial wastewater using EC [1].

The use of electrochemical wastewater systems has been limited due to their high initial cost and requirement for a costly electrical source [88]. However, during the past 20 years, electrochemical technologies such as electrocoagulation, electroflotation, and electrodeposition have become more significant globally due to strict environmental restrictions governing wastewater disposal. In electrocoagulation (EC), iron or aluminum ions are electrically dissolved from iron or aluminum electrodes to produce coagulants *in situ* [15]. The flocculated particles may be able to float out of the water with the aid of hydrogen gas [33].

11.3.5 Electrolysis

Electrolytic recovery is one method for eliminating metals from wastewater streams. Through the use of electricity, a current is sent through an aqueous metal-bearing solution that has an insoluble anode and a cathode plate. To create electricity, electrons can go from one element to another. Electrochemical treatment of wastewater containing heavy metals includes letting the metals precipitate as hydroxides in a neutralized or mildly acidic catholyte. The electrochemical procedures that are used to treat wastewater are electro-deposition, electrocoagulation, electroflotation, and electro-oxidation [79]. Coagulation and precipitation by hydroxide creation to acceptable levels is the electro-destabilization of colloids.

The simplicity of usage, lack of need for chemicals, and less sludge output are the characteristics of this procedure. To lower metal levels to a level suitable for release, chemical precipitation needs a lot of chemicals. Significant sludge generation, poor settling, sluggish metal precipitation, and agglomeration of metal are other disadvantages [9].

11.3.6 Ion Exchange

Ion exchange is an old method that dates back more than a century. Since then, this method of softening water has been used in an unimaginably vast range of situations and is now a crucial component of innovative industrial and technological procedures [84]. The ion exchange technique is an economical approach that often uses inexpensive materials and easy-to-use procedures [73].

Ion exchange is the reversible exchange of ions between the solid and liquid phases caused by an insoluble material (resin) that, without altering the resin's structural integrity, takes ions out of an electrolytic solution and releases them in a chemically comparable quantity [18]. Valuable heavy metals may also be extracted from inorganic wastewater *via* ion exchange. The metal is recovered in a more concentrated form by elution using the appropriate reagents following the separation of the loaded resin [73]. Synthetic or natural solid resin that is used for ion exchange has the unique capacity to exchange cations with heavy metals [58].

11.3.7 Photocatalytic

Photocatalysis is a highly inventive approach to the swift and effective elimination of environmental contaminants. This technique requires light energy above the semiconductor bandgap to illuminate a semiconductor-electrolyte interface, leading to the formation of electron-hole (e^-/h^+) pairs in the semiconductors' valence and conduction bands [34]. Among the semiconductors that are frequently investigated are TiO_2, ZnO, CeO_2, SnO_2, CdS, and ZnS [62]. A trap is thereafter required for the electron-hole pairs that are created during the photocatalysis process over semiconductor particles to prevent their recombination [37]. Instead of using chemical substances, this technology makes use of non-toxic semiconductors that capture light at the right wavelength [100].

There are five steps involved in introducing a pollutant-containing solution into a photocatalytic device [13]. Subsequently, the semiconductor surface absorbs them. Photocatalytic processes that take place in the absorbed phase constitute the third stage. Ultimately, the products break down and are taken out of the interface area [19]. The bandgap (Eg), which forms in semiconductors [60].

11.4 Types of Nanomaterials

Nowadays nanomaterials are popularly employed in water purification and sustainable environment missions. Industries employ nanomaterials to enhance the adsorption of wastewater effluents. Table 11.2 summarises different nanomaterials used to remove heavy meats from wastewater. Nanomaterials come in a variety of forms, which we shall address below.

Table 11.2 Different nanomaterials used for the removal of heavy metals from wastewater.

Heavy metals	Nanomaterials	Experimental conditions	Experiment outcomes	References
Lead (Pb)	Modified magnetic nanocomposite with SiO_2 and Zincon	pH: 5.4 Conc.: 4 mg l^{-1}	Maximum capacity uptake for lead was 513.5 mg g^{-1}	[71]
	Titanium dioxide (TiO_2 NPs)	Temp.: 38°C Time: 12 hr Conc.: 0.6 g l^{-1}	Lead removal 82.53%	[76]
	Magnetic nanoparticles (Fe_3O_4@ TiO_2-CN)	pH: 5.5 Time: 160 min Conc.: 100 mg l^{-1}	High adsorption capacity 92.59 mg g^{-1}	[56]
	Magnetite Oxide	pH: 4 and 6 Temp.: 30°C Conc.: 2 g l^{-1}	Lead removal 100%	[82]
	Heat-inactivated hybrid biosorbent (HI HB)	pH: 2 to 4.5 Temp.: 30-70°C Conc.: 100 mg l^{-1}	High biosorption capacity 365.9 mg g^{-1}	[31]

(Continued)

Table 11.2 Different nanomaterials used for the removal of heavy metals from wastewater. (*Continued*)

Heavy metals	Nanomaterials	Experimental conditions	Experiment outcomes	References
	Polyvinyl chloride and cellulose acetate nanocomposite modified with Fe_3O_4 nanoparticles	Conc.: 0.1%	Membrane with 0.1% Fe_3O_4 showed better flux and rejection of lead	[30]
	Cellulose nanocrystals decorated with silver or zinc oxide nanoparticle	pH: 2.1 to 8.1 Temp.: 25-50°C Time: 150 min Conc.: 2.5 mg ml^{-1}	Lead removal more than 94%	[10]
Zinc (Zn)	Nano Bentonite	pH: 3 to 7 Temp.: 25-30°C Time: 5 to 120 min Conc.: 0.05 to 1.5 g l^{-1}	Maximum adsorption of 98.9% was recorded	[2]
	Modified mesoporous mixed matrix PES NF membrane	0.3 wt% mesoporous Cur-SBA-15 incorporated mixed matrix membrane	Rejection over 97% against Zn^{2+} was recorded	[55]

(*Continued*)

Table 11.2 Different nanomaterials used for the removal of heavy metals from wastewater. (*Continued*)

Heavy metals	Nanomaterials	Experimental conditions	Experiment outcomes	References
	Magnetic 5 $MnFe_2O_4$ and $CoFe_2O_4$ spinel ferrite nanoparticles	pH: 6 Time: 120 min Temp.: 25-60°C	High adsorption capacities of 454.5 and 384.6 mg g^{-1} for $MnFe_2O_4$ and $CoFe_2O_4$, respectively	[8]
Chromium (Cr)	Zn-Ni Fe4O3@ pinecone biochar nanocomposite	pH: 2 to 11 Temp.: 25-55°C Conc.: 10 to 150 mg ml^{-1}	Maximum chromium adsorption of 95% was recorded	[54]
	Reduced graphene oxide and ZnO composite (rGO/ZnO)	pH: 2 Conc.: 2 g l^{-1}	Maximum chromium adsorption capacity 13.5 mg g^{-1}	[66]
Nickel (Ni)	Zinc oxide nanoparticles (ZnO NPs)	pH: 6 Time: 120 min Temp.: 60°C Conc.: 0.2 g l^{-1}	High percentage removal 92.5%	[61]
	Fe_2O_3 Nanoparticles	pH: 7 Time: 60 min Temp.: 25°C Conc.: 0.2 g l^{-1}	Maximum Nickel adsorption of 97.93% was recorded	[106]

(*Continued*)

Table 11.2 Different nanomaterials used for the removal of heavy metals from wastewater. (*Continued*)

Heavy metals	Nanomaterials	Experimental conditions	Experiment outcomes	References
	Acrylamide/hydrolyzed Luffa Cylindrica-$CoFe_2O_4$ composite	pH: 6 Temp.: room temperature Conc.: 1 g l^{-1}	High percentage removal 93%	[6]
Cobalt (Co)	Multi-walled carbon nanotubes (MWCNTs)	pH: 10 Time: 38.6 min Conc.: 1.57 mg l^{-1}	High percentage removal 93%	[20]
	Chitosan-magnetite nanocomposite (CMNC)	pH: 6 Time: 120 min Conc.: 2 g l^{-1}	Maximum cobalt adsorption capacity 53.19 mg g^{-1}	[77]
	Magnetite-alginate nanoparticles (M-AlgNPs)	pH: 6.5 Conc.: 2 mg l^{-1}	High percentage removal 95.01%	[23]
Mercury (Hg)	Magnetic adsorbent (BTA-MIL-125(Ti)@Fe_3O_4)	pH: 6 Conc.: 100 mg l^{-1}	High percentage removal 95.8%	[45]
	Sulfur-coated magnetic multi-walled carbon nanotube (S-M-MWCNT)	pH: 4.5 to 8 Time: 3 hr	The maximum adsorption capacity of S-M-MWCNT adsorbent was acquired as 62.11 mg g^{-1}	[26]

(*Continued*)

Table 11.2 Different nanomaterials used for the removal of heavy metals from wastewater. (*Continued*)

Heavy metals	Nanomaterials	Experimental conditions	Experiment outcomes	References
	1D/2D Polypyrrole nanotubes@pg-C_3N_4 binary nanocomposite	pH: 5 Time: 60 min Conc.: 0.5 g l^{-1}	The maximum mercury absorption capacity reached was 197 mg g^{-1}	[21]
	Molybdenum disulfide (MoS_2) nanosheets	pH: 2 to 9 Temp.: 25°C	Absorption capacity was found 253.6 mg g^{-1}	[46]
Copper (Cu)	Zwitterionic carbon quantum dots incorporated ultrafiltration membrane	pH: 10	Ultrafiltration achieved a 99.8% removal rate of copper ion	[95]
	Nano-hydroxyapatite (HAp@ wood composite)	-	The maximum adsorption capacity for Cu(II) ions were found 98.95 mg g^{-1}	[48]
	Nano magnetite (Fe_3O_4) modified biochar	pH: 5	90.27% removal efficiency for copper ion	[41]
	ZnO NPs	pH: 4 Conc.: 100-120 mg l^{-1} Time: 4 hr Temp.: 25°C	Copper removal of more than 95%	[68]

(*Continued*)

Table 11.2 Different nanomaterials used for the removal of heavy metals from wastewater. (*Continued*)

Heavy metals	Nanomaterials	Experimental conditions	Experiment outcomes	References
	Graphene Oxide–ZnO Nanocomposites	pH: 4 Conc.: 0.5 mg ml^{-1} Time: 20 hr Temp.: 22°C	92.9% removal efficiency for copper ion	[74]
Manganese (Mn)	MK/CNF/GO/ZnO composite	pH: 4.5 Temp.: 25°C	The maximum adsorption capacity for Mn(II) ions was found 4.84 mg g^{-1}	[32]
	Graphene Oxide–ZnO Nanocomposites	pH: 4 Time: 20 hr Temp.: 22-24°C	61.2% removal efficiency for manganese ion	[44]

11.4.1 Carbon-Based Nanomaterials

Carbon-based nanomaterials like carbon nanotubes (CNTs), carbon nanofibers (CNFs), graphene oxide (GO), graphene, and have emerged as a great material. CNTs, graphene-based materials, like graphene oxide membranes, usually have large surface areas and exceptional adsorption capacities, making it possible to remove heavy metal ions from aqueous solutions efficiently. Graphene oxide membranes have a significant observation for their ability to selectively permeate water molecules while rejecting heavy metal ions, giving a sustainable and energy-efficient approach to water purification. The recyclability of CNTs and graphene-based materials minimizes waste generation and reduces the environmental impact of heavy metal removal processes.

Metal nanoparticles (MNPs) functionalized nanocomposites, including graphene oxide (GOx), carbon nanotubes (CNTs), zeolites (ZLs), metal-organic frameworks (MOFs), and, graphene quantum dots (GQDs) are promising to remove heavy metals from wastewater. These nanocomposites offer enhanced adsorption capacity and selectivity due to the synergistic effects of MNPs and the host matrix. These sustainable materials offer a cost-effective solution for water purification while utilizing renewable resources [78].

11.4.2 Cellulose-Based Nanomaterials

Cellulose-based nanomaterials, such as nitrocellulose, cellulose nanocrystals (CNCs), cellulose nanofibers (CNFs), and cellulose aerogels, have been employed to eliminate heavy metals [63]. CNCs and CNFs, which have different morphology and particle sizes, can be synthesized chemically and mechanically [83].

Nanocellulose-based adsorbents, modified with functional groups like amino, carboxyl, aldehyde, and thiol, have shown the ability to remove single and multiple heavy metal ions from water. These cellulose-based nanoparticles provide an effective and environmentally friendly method of removing heavy metals from water treatment applications.

11.4.3 Polymer-Based Nanomaterials

Heavy metals from water have long been removed with the use of polymer-based nanoparticles. Because of their natural abundance, affordability, toughness, non-toxicity, stability, and strong sorption activity against both cationic and anionic pollutants, these materials have a lot of potential.

Adsorption-based remediation of wastewater has been achieved through the use of a variety of polymer-based nanocomposites and sorbents [72].

Some Polymer based nanomaterials include conducting polymers (CPs) such as polyaniline (PANI)-based nanocomposites [59], zeolites (ZLs), carbon nanotubes (CNTs), and polyethersulfone (PES) membranes modified with iron nanoparticles (M-Fe_3O_4). These polymer-based nanomaterials have shown promising removal efficiency for different heavy metals including mercury (Hg), zinc (Zn), cadmium (Cd), chromium (Cr), copper (Cu), and lead (Pb) nickel (Ni).

11.4.4 Chitosan-Based Nanomaterials

Chitosan-based nanomaterials exhibit high adsorption capacities and can effectively remove heavy metals under various conditions. Various kinds of chitosan-based nanomaterials have been created and investigated for their adsorption efficiency, hydrophilicity, and recyclability. These nanomaterials include chitosan-based magnetic nano-sorbents modified with sodium tripolyphosphate.

Magnetite nanoparticles coated with mesoporous silica layers and chitosan (Fe-NPs), and chitosan composites and nanocomposites. These nanomaterials have shown promising results in the removal of heavy metals such as Cd(II), Co(II), Cu(II), Pb(II), and Cr^{6+} from aqueous solutions.

11.4.5 Magnetic-Based Nanomaterials

Magnetic-based nanomaterials can be used for the removal and degradation of various pollutants, including heavy metals, pharmaceuticals, synthetic dyes, and oil. Nanoparticles such as magnetite have proven to be efficient adsorbents because of their high surface-to-volume ratio and tunable band edges. The presence of magnetic nanoparticles allows for easy separation of pollutants using external magnetic fields [14]. Furthermore, there are benefits to using magnetic nanoparticles in wastewater treatment, including their low cost, recyclable nature, and simplicity of removal when there's an outside magnetic field. All things considered, magnetic nanoparticles offer a highly efficient and reasonably priced option for the filtration of domestic and commercial wastewater.

11.4.6 Metal-Based Nanomaterials

Metal-based nanomaterials have shown great potential in wastewater treatment. MNPs-based nanocomposites have enhanced adsorption sites, and superior stability, higher active surface areas, making them good adsorbents for HMs removal [59]. Nanomaterials, including different carbon allotropes and metal oxide nanoparticles, have great properties, such as large surface area, high catalytic activity, small size high porosity, and tailorable physical and chemical properties, making them attractive for wastewater treatment [67]. Nanomaterials synthesized using biomolecules from plants and enzymes found in microbes have shown effectiveness in removing heavy metal ions from wastewater without adverse effects on living organisms [94]. These nanomaterials, such as graphene oxide (GOx), metal-organic frameworks (MOFs), zeolites (ZLs), carbon nanotubes (CNTs), metal nanoparticles (MNPs) functionalized nanocomposites, and graphene-quantum dots (GQDs) have been studied for their ability to remove heavy metals (HMs) from wastewater.

11.4.7 Silica-Based Nanomaterials

Silica-based nanomaterials have given good results in industrial wastewater treatment. Silica nanoparticles obtained from rice husk (RHSil) have been used as low-cost adsorbents for dye removal [51]. Silica nanomaterials have experimentally shown efficiency and selectivity in removing Pb_2+ ions from wastewater. Nanomaterials, including carbon-based nanomaterials, polymeric nanomaterials, and silica nanoparticles, are efficient in removing organo-phosphorus pesticides from contaminated water. Nanomaterials, such as different carbon allotropes and metal oxide nanoparticles, have been extensively studied for their potential in wastewater treatment. Functionalized silica derivatives are effective adsorbents for the removal of toxic dyes from wastewater [28].

11.5 Types of Heavy Metals

11.5.1 Zinc (Zn)

Zinc, an unavoidable heavy metal, is generally tracked down in industrial wastewater, starting with processes like galvanization, metal getting done, and metal plating. Its broad use in different modern applications contributes fundamentally to its presence in effluents released into water bodies. While zinc fills in as a fundamental micronutrient for organic capabilities in both plants and individuals, raised fixations in industrial effluents can have adverse ecological effects.

Moreover, human activities like mining and purifying add to the arrival of zinc into the climate, further worsening its presence in wastewater [47]. Exertions to address zinc defilement in wastewater include the execution of compelling treatment advancements to lessen zinc focus before release. Different treatment strategies, including synthetic precipitation and adsorption, particle trade, and film filtration, have been utilized to eliminate zinc from wastewater streams [89]. Manganese and cobalt spinel ferrites ($MnFe_2O_4$ and $CoFe_2O_4$) were combined using the co-precipitation approach in a prior study, and the resultant nanoparticles ranged in size from 20 to 80 nm. Divalent zinc particles were removed by the produced nanoparticles [8].

11.5.2 Cadmium (Cd)

Cadmium is a harmful heavy metal and presents natural and well-being risks when present in industrial wastewater. Normally, in some industries, for example, battery production, electroplating, and certain enterprises, cadmium tainting in water bodies can have serious environmental results. Cadmium is determined by the climate and can bioaccumulate in amphibian creatures, prompting poisonousness and biomagnification along the well-established pecking order.

Remediation advancements like substance precipitation, adsorption, and film filtration are utilized to eliminate cadmium from industrial effluents before release into the climate [99]. Prepared a 3D nanomaterial with a size of 20 nm by combining synthetic GO hydrogel with MnO_2 nanotubes. The reduced GO of the permeable 3D nanostructures and the combined action of MnO_2 nanotubes produced a high adsorption limit. In an additional investigation, volcanic stone was applied to α-Fe_2O_3 nanoparticles to facilitate the removal of divalent cadmium ions from an aqueous solution [104].

11.5.3 Copper (Cu)

Since copper is thought to be the most important and frequently used metal in a variety of applications, including metal polishing, electroplating, plastics, and carving, it is typically found in wastewater at high concentrations [105]. Additionally, because copper is a highly harmful element even in small quantities, effluent contaminated with copper should be treated before being released into the environment. The United States Environmental Protection Agency (USEPA) determined that the acceptable limit for copper particles in wastewater effluents was 1.3 mg/L [53]. In one of the tests, copper particles were successfully adsorbed on CNFs at an acidic pH, demonstrating the applicability of CNFs for copper particle elimination in multi-part circumstances [22]. In another study, Yu *et al.* [97] examined the adsorption of copper particles utilizing GO-based nanomaterials.

11.5.4 Nickel (Ni)

Nickel is the typical heavy metal particle in contaminated wastewater, which is highly toxic and carcinogenic. Nickel is normally present in wastewater, originating from different industries like chemical, electroplating, printing, coloring, drug, and metallurgical. The removal of nickel particles from water has been broadly talked about by researchers all over the world. At present, different treatment methods, including compound precipitation, molecule exchange, and film filtration, have been used to reduce the concentration of nickel ions in wastewater streams [90].

Furthermore, the improvement in existing remediation techniques, for example, adsorption onto nanomaterials and electrochemical treatment, shows guarantees in moderating nickel contamination in industrial effluents, which is truly significant [24]. However, progressing research is fundamental to enhancing these advancements and guaranteeing their expense adequacy and natural supportability in modern applications [70].

11.5.5 Cobalt (Co)

Cobalt is utilized in various industries, for example, battery fabricating, metal plating, and hardware preparation. The release of cobalt-containing effluents into water bodies can prompt extreme natural outcomes because of their toxicity and perseverance in the climate, which is startling stuff. The presence of cobalt in water causes severe health issues, such as respiratory problems, skin aggravation, cardiovascular issues, etc. [64]. Remediation methods like synthetic precipitation, adsorption, and layer filtration are utilized to remove cobalt from effluents [43].

Recent techniques used several adsorbent materials such as carbonaceous materials, minimal-cost adsorbents, Fe oxide/oxyhydroxide, Al-oxides, Mn-oxides, titanium oxides, mud minerals, zeolites, biopolymers, and random adsorbents. In a study, a group of researchers synthesized GO nanosheets for cobalt ion adsorption from an aqueous solution [102]. In another study, γ-alumina was prepared to adsorb cobalt particles from wastewater. The synthesized nano adsorbent can remove cobalt particles from wastewater in a limited time and with a lower concentration than other previous findings [20].

11.5.6 Lead (Pb)

Lead, a poisonous heavy metal typically tracked down in industrial wastewater, is released from different industrial sources, for example, battery fabrication, mining, metal refining, and plumbing materials. Its far-reaching use in different applications adds to its critical presence in wastewater effluents, presenting ecological and general wellness challenges [93].

Remediation innovations, for example, substance precipitation, particle trade, and film filtration, are utilized to eliminate lead from wastewater effluents. Furthermore, the ongoing research on other remediation techniques, including phytoremediation and electrochemical treatment, is fully intent on growing more effective for lead contamination in wastewater.

11.5.7 Arsenic (As)

Arsenic ranks fourteenth in seawater and twelfth in the human body, respectively, and is the twentieth most abundant element in the crust of Earth. Even though it's practical for a wide range of situations, for example, gadgets, the semiconductor industry, compound assembling, and horticulture, it is poisonous for human beings [52]. As indicated by the World Health Organization (WHO), skin sores and skin malignant growth are among the most widely recognized infections because of long-term arsenic harm.

A wide range of arsenic removal techniques has been developed, like precipitation, oxidation, ion exchange, adsorption, reverse, coagulation, nanofiltration, and osmosis. These techniques utilize a large number of materials, for example, activated carbon, activated alum, polymer membranes, ion exchange resins, and different kinds of nanomaterials [69]. Nowadays, several engineered nanomaterials are synthesized, for example, nanocomposites, nanowires, nanosheets, carbon-nano dots [4].

11.5.8 Uranium (U)

Uranium, a radioactive heavy metal, is present significantly in industrial wastewater. It is released from different sources, including atomic fuel creation, mining, and thermal energy stations. Uranium pollution can cause kidney harm, regenerative issues, and malignant growth [42]. Uranium enters the living body mostly by means of the breath or by some airborne residue particles through the ingestion of some food and drinks. Remediation advances, for example, synthetic precipitation, particle trade, and film filtration, are utilized to eliminate uranium from effluents [80].

Most commonly the adsorption method is used to eliminate uranium from wastewater. For adsorption purposes, several nanomaterials were synthesized with specific qualities such as high surface area, modest dimension fast dissolution, elevated reactivity, and solid adsorption. For example, carbon dots (CDs) [91].

11.6 Regeneration of Nanomaterials

Regeneration is a unique strategy that may restore the adsorption capacity of depleted adsorbents by the desorption of the pre-adsorbed contaminants. It is generally thought to be a better and more economical option than switching adsorbents (Table 11.3). The experimental temperature, pH, processing cycles needed, and time all affect the regeneration process. When desorption is caused by acid-base solutions like HCl, HNO_3, H_2SO_4, NaOH, etc., the proton exchange process is crucial. Various regeneration strategies, including chemical regeneration, biological regeneration, and photo-catalytic oxidation, have been effectively applied, either for the recovery of a specific metal or any other pollutants or on a specific adsorbent [36].

Similarly, there are some very good evaluations, although the majority of them concentrate on chemical regeneration (using acids and alkali) or biological regeneration (using microorganisms of various adsorbents). Adsorbents' chemical and physical characteristics can be altered by pH, occasionally leading to a change in their adsorbing ability [85]. The temperature is one of the most crucial factors in regeneration, and it must be managed carefully for optimal regeneration effectiveness.

Table 11.3 Regeneration of nanomaterials used for heavy metal adsorption from wastewater (adapted from Jain *et al.*, [36]).

Nanomaterials	Heavy metals	Regeneration method	Regeneration cycle	Observations	References
Acrylamide/ hydrolyzed Luffa Cylindrica-$CoFe_2O_4$ composite	Nickel (Ni^{2+})	Chemical regeneration	8	There had been no noticeable decline in the adsorption effectiveness.	[6]
Nano Zerovalent Iron (nZVI)/AC	Arsenic (As)	Chemical regeneration (NaOH)	8	Almost 100% desorption of metal was found with no loss in adsorption efficiency.	[103]
Magnetic 5 $MnFe_2O_4$ and $CoFe_2O_4$ spinel ferrite nanoparticles	Zinc (Zn)	Chemical regeneration (HCl)	3	More than 60% of adsorbed zinc by $CoFe_2O_4$ and 70% by $MnFe_2O_4$ were desorbed	[8]
Magnetite particles	Arsenic (As)	Chemical regeneration	-	Batch experiments were conducted with arsenic-free electrolytes to the aliquot, followed by 24 hr adsorption.	[96]

(Continued)

Table 11.3 Regeneration of nanomaterials used for heavy metal adsorption from wastewater (adapted from Jain *et al.*, [36]). (*Continued*)

Nanomaterials	Heavy metals	Regeneration method	Regeneration cycle	Observations	References
Cellulose nanocrystals decorated with silver or zinc oxide nanoparticle	Lead (Pb)	Chemical regeneration (HNO_3)	3	The regeneration efficiency is 97%, 93%, and 93% for CNCs, ZnO/CNC, and Ag/CNC, respectively.	[10]
Ferric Fe(III) Oxide/ hydroxide nanoparticles	Chromium (Cr)	Chemical regeneration (NaOH and $BaCl_2$)	-	100% recovery of adsorbent and desorbing agents as well.	[98]
Magnetic nanoparticles (Fe_3O_4@TiO_2-CN)	Nickel (Ni) and Lead (Pb)	Chemical regeneration	7	More than 90% recovery of adsorbent.	[56]
Nano Titanium oxide (TiO_2)	Selenium (Se)	Chemical regeneration (0.1M NaOH)	-	Various concentrations of NaOH were used. Amongst all, 2.0 ml, 0.1 mol L^{-1} readily desorbed selenium ions.	[101]

(*Continued*)

Table 11.3 Regeneration of nanomaterials used for heavy metal adsorption from wastewater (adapted from Jain *et al.*, [36]). (*Continued*)

Nanomaterials	Heavy metals	Regeneration method	Regeneration cycle	Observations	References
Titanium dioxide nanoparticles	Lead (Pb), Copper (Cu), and Zinc (Zn)	Chemical regeneration (0.01M $NaNO_3$, NaCl, $NaNO_3$, $NaHPO_4$, $NaHCO_3$ metal-free solution and EDTA)	8	Metal-free solutions proved ineffective desorbing agents with negligible desorption. Adsorption decreased for Pb, Zn, and Cu for 8 cycles.	[35]
Zinc oxide coated reduced Graphene oxide (RGO@ZnO)	-	Photocatalytic regeneration (sunlight radiation)	-	With sunlight radiation, regeneration was 77% of the original value, while without sunlight it was just 48%.	[87]
Kalolin/ZnO nanoparticles	Iron (Fe) and Chromium (Cr)	Chemical regeneration (NaOH)	-	The slight decrease in the regeneration efficiency.	[57]

(*Continued*)

Table 11.3 Regeneration of nanomaterials used for heavy metal adsorption from wastewater (adapted from Jain *et al.*, [36]). (*Continued*)

Nanomaterials	Heavy metals	Regeneration method	Regeneration cycle	Observations	References
Oxidized MWCNTs (o-MWCNT)	Cadmium (Cd)	Physical regeneration	-	About 1.5% of Cd^{2+} was desorbed from o-MWCNT at pH 5.5; this increased sharply at pH< 5.5 and reached a value of about 93% at pH 1.5.	[86]
Novel graphene oxide@ iron aluminum oxide composite	Arsenic (As)	Chemical regeneration (2.0M NaOH)	-	Only 50-60% regeneration of adsorbent with NaOH.	[50]

Adsorbent regeneration is influenced by several different parameters, which need to be maintained for regeneration processes to be successful. The amount of time needed for wasted adsorbents to regenerate has a significant impact on the adsorbent's ability for adsorption. Time management skills are crucial to ensure that the adsorbent [75]. Temperature and pH levels have a direct impact on how long it takes for desorption to occur.

Various types of adsorbents were investigated for regeneration such as aerogel, hydrogel, metal-based, carbon-based, and bio-adsorbent materials. On the other hand, there are certain environmental issues with the process that need to be overcome [36].

11.7 Conclusion

Water is the most essential resource for life. Ions of heavy metals are some of the most harmful contaminants. These days, there are increased risks to the environment and general well-being due to the heavy metal intake from wastewater. This chapter discusses different methods chemical, ion exchange, filtration, adsorption, electrolysis, electrocoagulation, and photocatalysis. Nanomaterials have strong adsorption capabilities, broad specific surface area and a number of sites that are active are characteristics of nanomaterials intended for use in water treatment applications. Wastewater treatment and purification have a lot of potential and prospects based on the application of nanotechnology. Additionally, adsorption is a cost-effective technique that has several benefits over traditional techniques, featuring high efficiency, cheap cost, and adsorbent renewal, in addition to the capacity to recover metal and minimize biological and chemical sludge.

Altogether, from the standpoint of the natural environment and human health, wastewater treatment for heavy metals is extremely important. The development of nanomaterials has given us a viable heavy metal removal option to more conventional adsorbents. Nonetheless, the majority of research on these nanoparticles' removal capacities uses stimulated water with quite straightforward components. Furthermore, as we produce nanomaterials, we cannot ignore their hazards and effects. The practical application of nanomaterials in heavy metal water treatment is still a long way off.

Acknowledgment

This study was funded by DBT BIG-BIRAC Ref No. BIRAC/CCAMP0723/BIG-13/18, DST major project titled "Low Cost-Renewable Energy Driven (LC-RED) Water Treatment Solution Centre." Ref. No. DST/TM/WTI/WIC/2K17/124 and B.B. and I.D. acknowledge the support of the ISF-NSFC joint research program (grant No. 3106/19).

References

1. Adhoum, N., Monser, L., Bellakhal, N., Belgaied, J.E., Treatment of electroplating wastewater containing Cu^{2+}, Zn^{2+} and Cr (VI) by electrocoagulation. *J. Hazard. Mater.*, *112*, 3, 207–213, 2004.
2. Ahmed, A.M., Ayad, M.I., Eledkawy, M.A., Darweesh, M.A., Elmelegy, E.M., Removal of iron, zinc, and nickel-ions using nano bentonite and its applications on power station wastewater. *Heliyon*, *7*, 2, e06315, 2021.
3. Akshhayya, C., Swedha, M., Elgorban, A.M., Bahkali, A.H., Varma, R.S., Younus, M., Khan, S.S., Intimate coupling of 3D $MnFe_2O_4$ cubes on 1D ZnO nanorods for sustainable photocatalysis under visible light: Computational analysis of reactive sites and degradation pathway. *J. Taiwan Inst. Chem. Eng.*, *141*, 104558, 2022.
4. Ali, I., New generation adsorbents for water treatment. *Chem. Rev.*, *112*, 10, 5073–5091, 2012.
5. Ali, I. and Gupta, V.K., Advances in water treatment by adsorption technology. *Nat. Protoc.*, *1*, 6, 2661–2667, 2006.
6. Alizadeh, M., Peighambardoust, S.J., Foroutan, R., Efficacious adsorption of divalent nickel ions over sodium alginate-g-poly (acrylamide)/hydrolyzed Luffa cylindrica-$CoFe_2O_4$ bionanocomposite hydrogel. *Int. J. Biol. Macromol.*, *254*, 127750, 2024.
7. Amuda, O.S., Giwa, A., Bello, I.A., Removal of heavy metal from industrial wastewater using modified activated coconut shell carbon. *Biochem. Eng. J.*, *36*, 2, 174–181, 2007.
8. Asadi, R., Abdollahi, H., Gharabaghi, M., Boroumand, Z., Effective removal of Zn (II) ions from aqueous solution by the magnetic $MnFe_2O_4$ and $CoFe_2O_4$ spinel ferrite nanoparticles with focuses on synthesis, characterization, adsorption, and desorption. *Adv. Powder Technol.*, *31*, 4, 1480–1489, 2020.
9. Aziz, H.A., Adlan, M.N., Ariffin, K.S., Heavy metals (Cd, Pb, Zn, Ni, Cu and Cr (III)) removal from water in Malaysia: post treatment by high quality limestone. *Bioresour. Technol.*, *99*, 6, 1578–1583, 2008.
10. Badawy, A.A., Ghanem, A.F., Yassin, M.A., Youssef, A.M., Rehim, M.H.A., Utilization and characterization of cellulose nanocrystals decorated with

silver and zinc oxide nanoparticles for removal of lead ion from wastewater. *Environ. Nanotechnol. Monit. Manage.*, *16*, 100501, 2021.

11. Banjare, M.K., Behera, K., Banjare, R.K., Carbon Allotropes in Other Metals (Cu, Zn, Fe etc.) Removal, in: *Carbon Allotropes and Composites: Materials for Environment Protection and Remediation*, pp. 113–154, 2023.
12. Barakat, M.A., New trends in removing heavy metals from industrial wastewater. *Arabian J. Chem.*, *4*, 4, 361–377, 2011.
13. Basile, A., Di Paola, L., Hai, F., Piemonte, V. (Eds.), *Membrane Reactors for Energy Applications and Basic Chemical Production*, Elsevier, 2015.
14. Chakrabarti, A., Banerjee, J., Chakravarty, S., Samanta, S., Nath, M., Chattopadhyay, S., Bandyopadhyay, A., Exploration of structural and magnetic aspects of biocompatible cobalt ferrite nanoparticles with canted spin configuration and assessment of their selective anti-leukemic efficacy. *J. Magn. Magn. Mater.*, *563*, 169957, 2022.
15. Chen, G., Electrochemical technologies in wastewater treatment. *Sep. Purif. Technol.*, *38*, 1, 11–41, 2004.
16. Chen, Q., Luo, Z., Hills, C., Xue, G., Tyrer, M., Precipitation of heavy metals from wastewater using simulated flue gas: sequent additions of fly ash, lime and carbon dioxide. *Water Res.*, *43*, 10, 2605–2614, 2009.
17. Chen, X., Chen, G., Yue, P.L., Separation of pollutants from restaurant wastewater by electrocoagulation. *Sep. Purif. Technol.*, *19*, 1–2, 65–76, 2000.
18. Dąbrowski, A., Hubicki, Z., Podkościelny, P., Robens, E., Selective removal of the heavy metal ions from waters and industrial wastewaters by ion-exchange method. *Chemosphere*, *56*, 2, 91–106, 2004.
19. Daghrir, R., Drogui, P., Robert, D., Photoelectrocatalytic technologies for environmental applications. *J. Photochem. Photobiol., A*, *238*, 41–52, 2012.
20. Dehghani, M.H., Yetilmezsoy, K., Salari, M., Heidarinejad, Z., Yousefi, M., Sillanpää, M., Adsorptive removal of cobalt (II) from aqueous solutions using multi-walled carbon nanotubes and γ-alumina as novel adsorbents: Modelling and optimization based on response surface methodology and artificial neural network. *J. Mol. Liq.*, *299*, 112154, 2020.
21. Diab, M.A., El-Sabban, H.A., Attia, S.Y., Moustafa, Y., Mohamed, S.G., Design of multifunctional 1D/2D polypyrrole nanotubes@ pg-C_3N_4 binary nanocomposite for removal of mercury (Hg^{2+}) from wastewater and supercapacitor applications. *J. Ind. Eng. Chem.*, *130*, 494–509, 2024.
22. Ding, C., Cheng, W., Wang, X., Wu, Z.Y., Sun, Y., Chen, C., Yu, S.H., Competitive sorption of Pb (II), Cu (II) and Ni (II) on carbonaceous nanofibers: a spectroscopic and modeling approach. *J. Hazard. Mater.*, *313*, 253–261, 2016.
23. El-Shamy, O.A., El-Azabawy, R.E., El-Azabawy, O., Synthesis and characterization of magnetite-alginate nanoparticles for enhancement of nickel and cobalt ion adsorption from wastewater. *J. Nanomater.*, *2019*, 2019.

24. Esmaeili, A. and Khoshnevisan, N., Optimization of process parameters for removal of heavy metals by biomass of Cu and Co-doped alginate-coated chitosan nanoparticles. *Bioresour. Technol.*, *218*, 650–658, 2016.
25. Esmaeili, H. and Foroutan, R., Investigation into ion exchange and adsorption methods for removing heavy metals from aqueous solutions. *Int. J. Biol. Pharm. Allied Sci.*, *4*, 620–629, 2015.
26. Fayazi, M., Removal of mercury (II) from wastewater using a new and effective composite: sulfur-coated magnetic carbon nanotubes. *Environ. Sci. Pollut. Res.*, *27*, 11, 12270–12279, 2020.
27. Fomina, M. and Gadd, G.M., Biosorption: current perspectives on concept, definition and application. *Bioresour. Technol.*, *160*, 3–14, 2014.
28. Fratilescu, I. and Fagadar-Cosma, E., Recovery of waste industrial waters containing red congo by multifunctionalized mesoporous silica nanomaterials. *Chem. Proc.*, *7*, 1, 19, 2022.
29. Fu, F. and Wang, Q., Removal of heavy metal ions from wastewaters: a review. *J. Environ. Manage.*, *92*, 3, 407–418, 2011.
30. Gholami, A., Moghadassi, A.R., Hosseini, S.M., Shabani, S., Gholami, F., Preparation and characterization of polyvinyl chloride based nanocomposite nanofiltration-membrane modified by iron oxide nanoparticles for lead removal from water. *J. Ind. Eng. Chem.*, *20*, 4, 1517–1522, 2014.
31. Hanif, A., Ali, S., Hanif, M.A., Rashid, U., Bhatti, H.N., Asghar, M., Giannakoudakis, D.A., A novel combined treatment process of hybrid biosorbent–nanofiltration for effective Pb (II) removal from wastewater. *Water*, *13*, 23, 3316, 2021.
32. Hegazy, S., Manninen, M., Tuomikoski, S., Lassi, U., Srivastava, V., Nano-impregnation on metakaolin backbone for enhanced removal of Cu (II) and Mn (II) ions in a binary system using fixed bed column. *Sep. Purif. Technol.*, *329*, 125163, 2024.
33. Heidmann, I. and Calmano, W., Removal of Zn (II), Cu (II), Ni (II), Ag (I) and Cr (VI) present in aqueous solutions by aluminium electrocoagulation. *J. Hazard. Mater.*, *152*, 3, 934–941, 2008.
34. Herrmann, J.M., Heterogeneous photocatalysis: fundamentals and applications to the removal of various types of aqueous pollutants. *Catal. Today*, *53*, 1, 115–129, 1999.
35. Hu, J. and Shipley, H.J., Regeneration of spent TiO_2 nanoparticles for Pb (II), Cu (II), and Zn (II) removal. *Environ. Sci. Pollut. Res.*, *20*, 5125–5137, 2013.
36. Jain, A., Kumari, S., Agarwal, S., Khan, S., Water purification *via* novel nano-adsorbents and their regeneration strategies. *Process Saf. Environ. Prot.*, *152*, 441–454, 2021.
37. Ji, S.M., Tiwari, A.P., Kim, H.Y., Graphene oxide coated zinc oxide core–shell nanofibers for enhanced photocatalytic performance and durability. *Coatings*, *10*, 12, 1183, 2020.

38. Kinuthia, G.K., Ngure, V., Beti, D., Lugalia, R., Wangila, A., Kamau, L., Levels of heavy metals in wastewater and soil samples from open drainage channels in Nairobi, Kenya: community health implication. *Sci. Rep.*, *10*, 1, 8434, 2020.
39. Kobya, M., Demirbas, E., Senturk, E., Ince, M., Adsorption of heavy metal ions from aqueous solutions by activated carbon prepared from apricot stone. *Bioresour. Technol.*, *96*, 13, 1518–1521, 2005.
40. Koyuncu, I., Sengur, R., Turken, T., Guclu, S., Pasaoglu, M.E., Advances in water treatment by microfiltration, ultrafiltration, and nanofiltration, in: *Advances in Membrane Technologies for Water Treatment*, pp. 83–128, Woodhead Publishing, 2015.
41. Kumar, P., Patel, A.K., Singhania, R.R., Chen, C.W., Saratale, R.G., Dong, C.D., Enhanced copper (II) bioremediation from wastewater using nano magnetite (Fe_3O_4) modified biochar of Ascophyllum nodosum. *Bioresour. Technol.*, *388*, 129654, 2023.
42. Lee, J., *et al.*, Uranium Contamination in Industrial Wastewater: Sources, Impacts, and Remediation. *J. Hazard. Mater.*, 25, 4, 890–907, 2020.
43. Lee, J., *et al.*, Sustainable Remediation Technologies for Cobalt in Industrial Wastewater. *J. Environ. Chem. Eng.*, 18, 6, 1123–1135, 2021.
44. Leiva, E., Tapia, C., Rodríguez, C., Removal of Mn (II) from acidic wastewaters using graphene Oxide–ZnO nanocomposites. *Molecules*, *26*, 9, 2713, 2021.
45. Li, J., Lin, G., Zeng, B., Wang, Z., Wang, S., Fu, L., Zhang, L., Synthetic of functionalized magnetic titanium-based metal–organic frameworks to efficiently remove Hg (II) from wastewater. *J. Colloid Interface Sci.*, *653*, 528–539, 2024.
46. Li, S., Yang, L., Zhang, K., Han, D., Yao, L., Wu, J., Jiang, G., Preparation of Macroporous Double-Network MoS_2-Based Beads for Efficient and Selective Removal of Hg (II) from Water. *ACS Environ. Sci. Technol. Eng.*, 2024.
47. Li, X., *et al.*, Zinc in Industrial Wastewater: Sources, Impacts, and Treatment Technologies. *J. Environ. Eng.*, 35, 9, 1234–1256, 2018.
48. Liang, Y., Jin, X., Xu, X., Wu, Y., Ghfar, A.A., Lam, S.S., Xia, C., A novel porous lignocellulosic standing hierarchical hydroxyapatite for enhanced aqueous copper (II) removal. *Sci. Total Environ.*, *912*, 168873, 2024.
49. Lyu, S., Chen, W., Zhang, W., Fan, Y., Jiao, W., Wastewater reclamation and reuse in China: Opportunities and challenges. *J. Environ. Sci.*, *39*, 86–96, 2016.
50. Maji, S., Ghosh, A., Gupta, K., Ghosh, A., Ghorai, U., Santra, A., Ghosh, U.C., Efficiency evaluation of arsenic (III) adsorption of novel graphene oxide@iron-aluminium oxide composite for the contaminated water purification. *Sep. Purif. Technol.*, *197*, 388–400, 2018.
51. Malafatti, J.O.D., Tavares, F.A., Neves, T.R., Mascarenhas, B.C., Quaranta, S., Paris, E.C., Modified Silica Nanoparticles from Rice Husk Supported on Polylactic Acid as Adsorptive Membranes for Dye Removal. *Materials*, *16*, 6, 2429, 2023.

52. Mandal, B.K. and Suzuki, K.T., Arsenic round the world: a review. *Talanta, 58*, 1, 201–235, 2002.
53. Manne, R., Kumaradoss, M.M.R.M., Iska, R.S.R., Devarajan, A., Mekala, N., Water quality and risk assessment of copper content in drinking water stored in copper container. *Appl. Water Sci., 12*, 3, 27, 2022.
54. Masuku, M., Nure, J.F., Atagana, H.I., Hlongwa, N., Nkambule, T.T., Advancing the development of nanocomposite adsorbent through zinc-doped nickel ferrite-pinecone biochar for removal of chromium (VI) from wastewater. *Sci. Total Environ., 908*, 168136, 2024.
55. Moradi, G., Zinadini, S., Rahimi, M., Shiri, F., Efficient Zn2+, Pb2+, and Ni2+ removal using antifouling mixed matrix nanofiltration membrane with curcumin modified mesoporous Santa Barbara Amorphous-15 (Cur-SBA-15) filler. *J. Environ. Chem. Eng., 10*, 3, 107302, 2022.
56. Mousavi, S.V., Bozorgian, A., Mokhtari, N., Gabris, M.A., Nodeh, H.R., Ibrahim, W.A.W., A novel cyanopropylsilane-functionalized titanium oxide magnetic nanoparticle for the adsorption of nickel and lead ions from industrial wastewater: Equilibrium, kinetic and thermodynamic studies. *Microchem. J., 145*, 914–920, 2019.
57. Mustapha, S., Tijani, J.O., Ndamitso, M.M., Abdulkareem, S.A., Shuaib, D.T., Mohammed, A.K., Sumaila, A.J.S.R., The role of kaolin and kaolin/ZnO nanoadsorbents in adsorption studies for tannery wastewater treatment. *Sci. Rep., 10*, 1, 13068, 2020.
58. Nabi, S.A., Naushad, M., Khan, A.M., Sorption studies of metal ions on naphthol blue–black modified Amberlite IR-400 anion exchange resin: Separation and determination of metal ion contents of pharmaceutical preparation. *Colloids Surf., A, 280*, 1–3, 66–70, 2006.
59. Natarajan, B., Kannan, P., Rather, J.A., Sheikh, R.A., Recent developments in metal nanoparticles functionalized nanocomposite adsorbents for heavy metals removal from wastewaters. *J. Taiwan Inst. Chem. Eng., 147*, 104942, 2023.
60. Oriol, R., Sirés, I., Brillas, E., De Andrade, A.R., A hybrid photoelectrocatalytic/photoelectro-Fenton treatment of Indigo Carmine in acidic aqueous solution using TiO_2 nanotube arrays as photoanode. *J. Electroanal. Chem., 847*, 113088, 2019.
61. Panneerselvam, A., Rajadurai, V., Anguraj, B.L., Removal of nickel from aqueous solution using synthesized IL/ZnO NPs. *Environ. Sci. Pollut. Res., 27*, 29791–29803, 2020.
62. Pant, B., Ojha, G.P., Kuk, Y.S., Kwon, O.H., Park, Y.W., Park, M., Synthesis and characterization of ZnO-TiO_2/carbon fiber composite with enhanced photocatalytic properties. *Nanomaterials, 10*, 10, 1960, 2020.
63. Parvin, F. and Tareq, S.M., Alleviating environmental pollution with nanoparticles: current advances and future perspectives, in: *Nanoparticles and Plant-Microbe Interactions*, pp. 241–258, 2023.

64. Patel, S. and Gupta, V.K., Cobalt Contamination in Industrial Wastewater: Sources, Impacts, and Remediation. *J. Hazard. Mater.*, 25, 4, 890–907, 2018.
65. Patil, D.S., Chavan, S.M., Oubagaranadin, J.U.K., A review of technologies for manganese removal from wastewaters. *J. Environ. Chem. Eng.*, *4*, 1, 468–487, 2016.
66. Philip, R.S., Aparna, N., Mathew, M., Hexavalent chromium removal using reduced graphene oxide-zinc oxide composite fabricated *via* simple pyrolysis method. *Appl. Surf. Sci. Adv.*, *19*, 100535, 2024.
67. Prajapat, K., Dhonde, M., Sahu, K., Murty, V.V.S., Shirage, P.M., Nanomaterials-Based Sustainable Wastewater Treatment Strategies for a Sustainable Planet, in: *Opportunities and Challenges in Climate-Friendly Clean Water and Energy Technologies*, pp. 15–39, IGI Global, 2023.
68. Primo, J.D.O., Bittencourt, C., Acosta, S., Sierra-Castillo, A., Colomer, J.F., Jaerger, S., Anaissi, F.J., Synthesis of zinc oxide nanoparticles by ecofriendly routes: adsorbent for copper removal from wastewater. *Front. Chem.*, *8*, 571790, 2020.
69. Qu, X., Alvarez, P.J., Li, Q., Applications of nanotechnology in water and wastewater treatment. *Water Res.*, *47*, 12, 3931–3946, 2013.
70. Rad, L.R., Momeni, A., Ghazani, B.F., Irani, M., Mahmoudi, M., Noghreh, B., Removal of Ni^{2+} and Cd^{2+} ions from aqueous solutions using electrospun PVA/zeolite nanofibrous adsorbent. *Chem. Eng. J.*, *256*, 119–127, 2014.
71. Rahimi, M., Zolgharnein, J., Farahani, S.D., Modification of magnetic active carbon nanocomposite with SiO2 and Zincon for efficient removal of Pb (II) and 201Pb (II); multivariate optimization and adsorption characterization. *Inorg. Chem. Commun.*, 112127, 2024.
72. Ren, S., Pang, L., Wang, X., Fang, Z., Zhou, D., Zhao, Y., Li, W., Improvement of high-temperature energy storage properties of polyimide-based nanocomposites with sandwich structure. *J. Appl. Polym. Sci.*, *141*, 4, e54848, 2024.
73. Rengaraj, S., Yeon, K.H., Moon, S.H., Removal of chromium from water and wastewater by ion exchange resins. *J. Hazard. Mater.*, *87*, 1–3, 273–287, 2001.
74. Rodríguez, C., Tapia, C., Leiva-Aravena, E., Leiva, E., Graphene oxide–ZnO nanocomposites for removal of aluminum and copper ions from acid mine drainage wastewater. *Int. J. Environ. Res. Public Health*, *17*, 18, 6911, 2020.
75. Salvador, F., Martin-Sanchez, N., Sanchez-Hernandez, R., Sanchez-Montero, M.J., Izquierdo, C., Regeneration of carbonaceous adsorbents. Part II: chemical, microbiological, and vacuum regeneration. *Microporous Mesoporous Mater.*, *202*, 277–296, 2015.
76. Sethy, N.K., Arif, Z., Mishra, P.K., Kumar, P., Green synthesis of TiO_2 nanoparticles from Syzygium cumini extract for photo-catalytic removal of lead (Pb) in explosive industrial wastewater. *Green Process. Synth*, *9*, 1, 171–181, 2020.

77. Sharifi, M.J., Nouralishahi, A., Hallajisani, A., Fe_3O_4-chitosan nanocomposite as a magnetic biosorbent for removal of nickel and cobalt heavy metals from polluted water. *Int. J. Biol. Macromol.*, *248*, 125984, 2023.
78. Sharon, I., McKay, G.A., Nguyen, D., Schmeing, T.M., Discovery of cyanophycin dipeptide hydrolase enzymes suggests widespread utility of the natural biopolymer cyanophycin. *Proc. Natl. Acad. Sci.*, *120*, 8, e2216547120, 2023.
79. Shim, H.Y., Lee, K.S., Lee, D.S., Jeon, D.S., Park, M.S., Shin, J.S., Chung, D.Y., Application of electrocoagulation and electrolysis on the precipitation of heavy metals and particulate solids in wash water from soil washing. *J. Agric. Chem. Environ.*, *3*, 04, 130, 2014.
80. Smith, A. and Johnson, B., Assessment of Ecological Impacts of Uranium Discharges from Industrial Sources. *Environ. Monit. Assess.*, 40, 7, 567–589, 2020.
81. Sörme, L. and Lagerkvist, R., Sources of heavy metals in urban wastewater in Stockholm. *Sci. Total Environ.*, *298*, 1–3, 131–145, 2002.
82. Stoian, O., Covaliu, C.I., Paraschiv, G., Catrina, G.A., Niță-Lazăr, M., Matei, E., Tudor, P., Magnetite oxide nanomaterial used for lead ions removal from industrial wastewater. *Materials*, *14*, 11, 2831, 2021.
83. Sun, L., Liu, L., Wu, M., Wang, D., Shen, R., Zhao, H., Yao, J., Nanocellulose interface enhanced all-cellulose foam with controllable strength *via* a facile liquid phase exchange route. *Carbohydr. Polym.*, *299*, 120192, 2023.
84. Tiravanti, G., Petruzzelli, D., Passino, R., Pretreatment of tannery wastewaters by an ion exchange process for Cr (III) removal and recovery. *Water Sci. Technol.*, *36*, 2–3, 197–207, 1997.
85. Vakili, M., Deng, S., Cagnetta, G., Wang, W., Meng, P., Liu, D., Yu, G., Regeneration of chitosan-based adsorbents used in heavy metal adsorption: A review. *Sep. Purif. Technol.*, *224*, 373–387, 2019.
86. Vuković, G.D., Marinković, A.D., Čolić, M., Ristić, M.D., Aleksić, R., Perić-Grujić, A.A., Uskoković, P.S., Removal of cadmium from aqueous solutions by oxidized and ethylenediamine-functionalized multi-walled carbon nanotubes. *Chem. Eng. J.*, *157*, 1, 238–248, 2010.
87. Wang, J., Tsuzuki, T., Tang, B., Hou, X., Sun, L., Wang, X., Reduced graphene oxide/ZnO composite: reusable adsorbent for pollutant management. *ACS Appl. Mater. Interfaces*, *4*, 6, 3084–3090, 2012.
88. Wang, L.K., Hung, Y.T., Shammas, N.K. (Eds.), *Advanced Physicochemical Treatment Technologies*, vol. 5, pp. 575–610, Humana Press, Totowa, NJ, 2007.
89. Wang, Q., *et al.*, Assessment of Ecological Impacts of Zinc Discharges from Industrial Sources. *Environ. Monit. Assess.*, 40, 7, 567–589, 2019.
90. Wang, Q., *et al.*, Assessment of Ecological Impacts of Nickel Discharges from Industrial Sources. *Environ. Monit. Assess.*, 40, 7, 567–589, 2020.
91. Wang, Q., Zhang, H., Yu, D., Qin, W., Wu, X., Ultra-sensitive and stable N-doped carbon dots for selective detection of uranium through electron

transfer induced UO2+ (V) sensing mechanism. *Carbon, 198*, 162–170, 2022.

92. Wang, X., Chen, L., Xia, S., Zhao, J., Chovelon, J.M., Renault, N.J., Biosorption of Cu (II) and Pb (II) from aqueous solutions by dried activated sludge. *Miner. Eng., 19*, 9, 968–971, 2006.
93. Wang, X., *et al.*, Lead in Industrial Wastewater: Sources, Impacts, and Treatment Technologies. *J. Environ. Eng.*, 35, 9, 1234–1256, 2018.
94. Yamini, V. and Rajeswari, V.D., Effective bio-mediated nanoparticles for bioremediation of toxic metal ions from wastewater–A review. *J. Environ. Nanotechnol., 12*, 2, 12–33, 2023.
95. Yang, H.L., Huang, C.T., Lin, H.Y., Chen, Y.H., Tsai, H.A., Lee, K.R., Zwitterionic carbon quantum dots incorporated ultrafiltration membrane for efficient removal of copper ion. *Sep. Purif. Technol., 331*, 125709, 2024.
96. Yean, S., Cong, L., Yavuz, C.T., Mayo, J.T., Yu, W.W., Kan, A.T., Tomson, M.B., Effect of magnetite particle size on adsorption and desorption of arsenite and arsenate. *J. Mater. Res., 20*, 12, 3255–3264, 2005.
97. Yu, S., Wang, X., Ai, Y., Liang, Y., Ji, Y., Li, J., Wang, X., Spectroscopic and theoretical studies on the counterion effect of Cu (II) ion and graphene oxide interaction with titanium dioxide. *Environ. Sci.: Nano, 3*, 6, 1361–1368, 2016.
98. Zelmanov, G. and Semiat, R., Iron (Fe^{+3}) oxide/hydroxide nanoparticles-based agglomerates suspension as adsorbent for chromium (Cr^{+6}) removal from water and recovery. *Sep. Purif. Technol., 80*, 2, 330–337, 2011.
99. Zeng, T., Yu, Y., Li, Z., Zuo, J., Kuai, Z., Jin, Y., Peng, C., 3D MnO_2 nanotubes@ reduced graphene oxide hydrogel as reusable adsorbent for the removal of heavy metal ions. *Mater. Chem. Phys., 231*, 105–108, 2019.
100. Zhang, F.S. and Itoh, H., Photocatalytic oxidation and removal of arsenite from water using slag-iron oxide-TiO_2 adsorbent. *Chemosphere, 65*, 1, 125–131, 2006.
101. Zhang, L., Liu, N., Yang, L., Lin, Q., Sorption behavior of nano-TiO_2 for the removal of selenium ions from aqueous solution. *J. Hazard. Mater., 170*, 2–3, 1197–1203, 2009.
102. Zhao, G., Li, J., Ren, X., Chen, C., Wang, X., Few-layered graphene oxide nanosheets as superior sorbents for heavy metal ion pollution management. *Environ. Sci. Technol., 45*, 24, 10454–10462, 2011.
103. Zhu, H., Jia, Y., Wu, X., Wang, H., Removal of arsenic from water by supported nano zero-valent iron on activated carbon. *J. Hazard. Mater., 172*, 2–3, 1591–1596, 2009.
104. Zhu, X., Song, T., Lv, Z., Ji, G., High-efficiency and low-cost α-Fe_2O_3 nanoparticles-coated volcanic rock for Cd (II) removal from wastewater. *Process Saf. Environ. Prot., 104*, 373–381, 2016.
105. Zou, Y., Wang, X., Khan, A., Wang, P., Liu, Y., Alsaedi, A., Wang, X., Environmental remediation and application of nanoscale zero-valent iron and its composites for the removal of heavy metal ions: a review. *Environ. Sci. Technol., 50*, 14, 7290–7304, 2016.

106. Shahriari, T., Mehrdadi, N., Tahmasebi, M., Study of cadmium and nickel removal from battery industry wastewater by Fe2O3 nanoparticles. *Pollution*, 5, 3, 515–524, 2019.

12

Regulatory Concerns for Solid Waste Management

Anuradha Jayaraman, Sandeep Tripathi and Sanjeevi Ramakrishnan*

Department of Environmental Science, Nims Institute of Allied Medical Science and Technology, Nims University Rajasthan, Jaipur, India

Abstract

The regulation of solid waste management faces numerous challenges, requiring a sustainable balance between public health, environmental, and economic considerations. Variances in the enforcement of the regulations within and between nations contribute to inadequate waste management practices, posing major threats to the environment and public health. Inconsistencies in implementing rules, coupled with insufficient oversight, lead to non-compliance by waste producers and disposal sites, exacerbating issues like illegal dumping and unlicensed waste collectors. Financial constraints, limited infrastructure, and inadequate resources hinder the establishment and execution of efficient waste management initiatives. Lagging policies may fail to keep pace with technological advancements, impeding the development of eco-friendly waste management techniques. Transnational waste movements, especially hazardous waste, present regulatory challenges due to their global nature, potentially causing severe environmental and health repercussions. Insufficient public awareness and education about waste management contribute to improper disposal practices. Effective waste management necessitates collaboration between the public and regulatory bodies, emphasizing proper waste sorting and the need for consistent, clear regulations. Long-term planning is vital, but political cycles and inconsistent policymaking can impede robust waste management systems. Addressing these regulatory hurdles requires a multifaceted strategy, involving legislative improvements, enhanced enforcement, public engagement, and investments in sustainable technologies. The shift from a linear to a circular economy, prioritizing reuse and recycling, holds promise. Technological advancements, including advanced sorting methods and data analytics, can optimize waste management efficiency. Extended Producer Responsibility programs

**Corresponding author*: r.sanjeevi@nimsuniversity.org

Priyanka Singh, Pooja Agarwal and V. Vivekanand (eds.) Solid Waste Management: Challenges, Sustainability and Advancements, (397–430) © 2026 Scrivener Publishing LLC

and "zero waste" goals align with sustainable practices, while decentralized waste management systems and global cooperation further contribute to effective solutions. In conclusion, the future of solid waste management necessitates a holistic, technologically advanced, and collaborative approach, emphasizing sustainability and stakeholder involvement to overcome emerging challenges and prioritize environmental well-being and resource conservation.

Keywords: Solid waste, circular economy, waste-to-energy, sustainability, environmental awareness

12.1 Introduction

12.1.1 Solid Waste Generation: A Growing Global Concern

In recent decades, the global proliferation of solid waste has become a pressing concern. Annually, over two billion metric tons of municipal solid waste (MSW) are generated worldwide, a figure projected to surge by roughly 70% by 2050. This surge is largely driven by the rapid growth of population and urbanization, compounding challenges for waste management authorities. The expansion of cities and shifts in consumption patterns intensify the urgency for effective waste treatment and disposal services [1]. To tackle this burgeoning crisis, embracing innovative solutions and implementing sustainable practices is paramount. Such measures could include advanced recycling technologies, composting initiatives, and waste-to-energy schemes. Additionally, promoting awareness and fostering community engagement are vital components in fostering a culture of responsible consumption and waste reduction. Ultimately, concerted efforts on a global scale are necessary to mitigate the environmental, social, and economic impacts of escalating solid waste generation [1, 2].

Regulation plays a crucial role in solid waste management (SWM) by providing guidelines and standards for managing waste, ensuring its proper handling, processing, and disposal. Regulatory bodies are established in many countries to supervise waste management practices, set prices, and provide incentives for efficiency and environmental sustainability. Waste classification is an important aspect of waste management regulations, as it helps differentiate between hazardous and non-hazardous waste, minimizing their negative impact on the environment and human health. Regulations also address the issue of international waste transfer, preventing the transfer of waste to countries with less stringent controls. Economic regulation, such as tariff setting mechanisms, is proposed to ensure financial sustainability, environmental standards, and quality of service in the waste sector. Overall, regulations are necessary to prevent market failures, promote efficiency, and protect the environment and public health.

12.1.2 Historical Context in Solid Waste Management

In the early days of waste management regulation, waste disposal practices were largely unregulated, resulting in significant environmental pollution and public health risks. With the rise of industrialization and urbanization in the 19th century, cities faced mounting challenges in managing the increasing volumes of waste generated by growing populations and industries.

The 19th century witnessed rapid urbanization and industrialization in many parts of the world, leading to the concentration of populations in urban centers and the proliferation of manufacturing activities. The influx of people into cities and the expansion of industries resulted in the generation of large quantities of waste, including household garbage, industrial byproducts, and agricultural residues. During this period, waste disposal was often haphazard and unsanitary. Many cities relied on open dumping, where waste was simply discarded in vacant lots, streets, or water bodies. Open dumping posed significant health risks, attracting vermin and disease vectors, contaminating soil and water sources, and generating foul odors that plagued urban neighborhoods. The late 19th and early 20th centuries saw the emergence of the sanitation reform movement, driven by concerns about public health and hygiene. Reformers advocated for improved sanitation infrastructure, including the construction of sewer systems, the establishment of municipal waste collection services, and the implementation of basic hygiene measures such as street cleaning. Some cities began to implement rudimentary waste management practices, such as organized waste collection by municipal authorities or private contractors. Recycling and resource recovery efforts also emerged during this period, with materials like paper, glass, and metal being collected for reuse or recycling.

12.1.3 Key Legislation and Regulations

Many cities enacted local ordinances to regulate waste collection and disposal activities. These ordinances established standards for waste storage, collection schedules, and designated disposal sites. While often rudimentary, these early regulations laid the groundwork for more comprehensive waste management legislation at the state and federal levels. The Federal

Water Pollution Control Act of 1948, later amended and expanded into the Clean Water Act (CWA), was one of the earliest federal laws addressing water pollution in the United States. While primarily focused on water pollution, the CWA indirectly influenced waste management practices by promoting pollution prevention and control measures that impacted waste disposal into water bodies. The Solid Waste Disposal Act of 1965 was the first federal legislation specifically targeting solid waste management. It provided grants to states for waste management planning and research, marking the federal government's initial involvement in waste management regulation. The Resource Conservation and Recovery Act (RCRA) is one of the most significant pieces of legislation shaping the modern regulatory landscape of waste management. Enacted in 1976, RCRA established a comprehensive framework for the management of hazardous and non-hazardous solid waste. RCRA regulates the generation, transportation, treatment, storage, and disposal of hazardous waste, as well as the management of non-hazardous solid waste. The Comprehensive Environmental Response, Compensation, and Liability Act (CERCLA), commonly known as Superfund, was enacted in 1980 to address the cleanup of hazardous waste sites and spills. CERCLA established liability for parties responsible for hazardous waste contamination and created a trust fund for cleanup activities, shaping the regulatory approach to hazardous waste management and remediation. Enacted in 1995, the MSW Rule regulates emissions from municipal solid waste incinerators, aiming to reduce air pollution and protect public health by establishing emission standards and monitoring requirements for incineration facilities.

The early days of waste management regulation were characterized by unregulated disposal practices and limited infrastructure, leading to environmental degradation and public health hazards. Over time, the introduction of key legislation and regulations such as RCRA and CERCLA marked significant milestones in the development of comprehensive waste management frameworks, establishing standards for waste handling, disposal, and pollution control. These regulatory efforts continue to evolve and adapt to address emerging environmental challenges and promote sustainable waste management practices.

12.1.4 The Differences in Solid Waste Generation across Nations and Regions

The world generates a staggering 2.01 billion tonnes of municipal solid waste (MSW) annually. Shockingly, at least 33% of this waste is not managed in an environmentally safe manner. On average, each person produces 0.74 kilograms of waste per day, but this varies widely from as low as 0.11 kilograms to as high as 4.54 kilograms. High-income countries, accounting for only 16% of the global population, generate approximately 34% (683 million tonnes) of the world's waste. By 2050, global waste is projected to grow to 3.40 billion tonnes, more than double the population growth over the same period.

There exists a positive correlation between waste generation and income level. In high-income countries, daily per capita waste generation is expected to increase by 19% by 2050. In contrast, low- and middle-income countries will experience an increase of approximately 40% or more [3]. The total quantity of waste generated in low-income countries is projected to increase by more than three times by 2050. East Asia and Pacific generate the highest share of the world's waste, accounting for 23%. Conversely, the Middle East and North Africa region produces the least in absolute terms, contributing only 6%. The fastest-growing regions in terms of waste generation are Sub-Saharan Africa, South Asia, and the Middle East and North Africa. By 2050, their waste production is expected to more than triple, double, and double, respectively [4]. Waste collection rates vary significantly by income levels. Upper-middle- and high-income countries provide nearly universal waste collection. Low-income countries collect about 48% of waste in cities but only 26% outside urban areas. Sub-Saharan Africa collects approximately 44%, while Europe and Central Asia and North America collect at least 90%.

Consumption patterns influence waste composition. High-income countries generate relatively less food and green waste (32%) and more dry waste (51%) that can be recycled (e.g., plastic, paper, cardboard, metal, glass). Middle- and low-income countries have a higher fraction of organic waste (53% and 57%, respectively) [5]. Addressing solid waste challenges requires tailored strategies based on income levels, regional disparities, and waste composition. Urgent action is needed to ensure environmental sustainability, health, and prosperity worldwide.

Illegal dumping, also known as fly dumping or fly tipping, refers to the unauthorized disposal of waste instead of using authorized methods such as curbside collection or authorized rubbish dumps. It involves depositing waste onto land without the necessary license [6]. Materials dumped include building materials such as drywall, roofing shingles, lumber, brick, concrete, and siding; discarded vehicle components; and improperly disposed of medical materials. Illegal dumping occurs when there are limited alternatives, such as when legal waste disposal facilities are scarce. Charging per bag of garbage can lead to cost-saving measures, including illegal dumping [7]. Inadequate enforcement of anti-dumping laws contributes to the problem. Public understanding of the environmental and health risks associated with illegal dumping is essential.

Toxins and hazardous materials infiltrate soil and drinking water, posing risks to local residents. Unlike regulated landfills, illegal dumping sites lack proper containment, leading to pollution. Cleaning up illegal dumpsites and addressing health issues incur significant expenses [8]. The people who do this work operate informally, scavenging waste materials for recycling or resale. While some contribute positively to recycling efforts, others engage in illegal activities by collecting waste without proper authorization. Unlicensed rubbish pickers often face economic hardships and lack access to formal waste management systems. Combating illegal dumping requires a combination of strict enforcement, public awareness, and accessible waste disposal options. Responsible waste management benefits both communities and the environment.

12.2 Current Challenges Associated with Solid Waste Management across Different Regions

Solid waste management affects every single person globally. Decisions related to consumption and waste management impact daily health, productivity, and community cleanliness. Poorly managed waste has dire consequences like waste-polluted oceans and the harming of marine ecosystems. Improper waste disposal leads to drainage issues and floods. Breeding of vectors (e.g., mosquitoes) in waste contributes to disease spread. Burning waste releases harmful particles affecting respiratory health. Poor waste management leads to diminished tourism and is a hindrance to economic development [9]. As countries develop, their waste management situations evolve. Urbanization and population growth lead to larger population centers, making waste collection and disposal challenging. Urban waste management is expensive and competes with other priorities like clean water, education, and health care in budget allocation. Local authorities often manage waste with limited resources and capacity [10]. Sustainable waste management is complicated due to planning, contract management, and operational monitoring challenges. Poor waste management disproportionately affects the poor, who often lack influence over waste disposal near their homes. Understanding waste generation patterns, types, and quantities is crucial for policy and planning. Data informs efficient waste collection routes, diversion targets, and adaptation strategies.

Environmental pollution poses significant risks to human health, affecting millions of lives worldwide. Air pollution is a major threat. Nine out of ten people now breathe polluted air, leading to approximately 7 million deaths annually. Air pollution contributes to heart disease and stroke. Long-term exposure increases the risk of lung cancer [11]. Air pollutants worsen respiratory infections. The health impact of air pollution is equivalent to that of smoking tobacco, surpassing the effects of excessive salt consumption.

Meeting climate change goals could save a million lives annually by 2050 through reduced air pollution, which is linked to heart attacks, strokes, diabetes, and respiratory diseases. It also affects the brain, heart, and circulatory system. Air pollution costs more than 6 million premature deaths each year, exceeding the combined toll of AIDS, tuberculosis, and malaria. Pollution, radiation, noise, and land use patterns impact health. Environmental risks represent 6.1% of the global GDP, emphasizing the need for sustainable choices [12].

The lack of regulation and enforcement in solid waste management (SWM) poses significant challenges across various regions. The parliamentary Standing Committee on Urban Development highlighted the need for a comprehensive review of SWM challenges involving all stakeholders. A roadmap must be charted, and adequate funds arranged to address these challenges. Infrastructure gaps in India, compared to European countries and the United States, require urgent attention. As waste composition evolves, planning waste process technology becomes crucial. Accurate data on the amount of waste generated is essential for effective management [13, 14]. The Central Pollution Control Board's plea for accurate data underscores this need. Despite SWM Rules, 2016, and Swachh Bharat Mission (SBM), gaps remain. Inadequate capacity building, weak legislation, and lack of enforcement hinder progress (Table 12.1). Connecting with the private sector and integrating informal waste pickers are essential steps. Developing countries face similar challenges due to rapid urbanization, inadequate resources, and weak organizational structures. Lack of political will, corruption, and low public awareness exacerbate the situation. The Union government accepted only 14 out of 37 recommendations from the Standing Committee on Urban Development. Proper responses are crucial, as incomplete actions hinder effective SWM.

Table 12.1 A comparison of solid waste management practices by income level.

Income level	Waste management practices
Low Income	Limited infrastructure: Often lacks proper waste collection and disposal systems. Open dumping: Common practice, leading to environmental pollution. Informal recycling: Waste pickers salvage recyclables from dumps.
Middle Income	Basic collection services: Some organized waste collection. Landfills: Used for disposal, but may lack proper design and management. Incineration: Increasingly used, but environmental concerns exist.
High Income	Advanced systems: Comprehensive waste collection, recycling, and disposal. Sanitary landfills: Engineered facilities with environmental safeguards. Waste-to-energy: Incineration with energy recovery. Recycling programs: Prominent.

12.3 Challenges

The challenges that have been highlighted—inadequate financial support, insufficient infrastructure, and limited resources—are critical obstacles in achieving efficient waste management. Waste management requires substantial investment in infrastructure, technology, and human resources. However, many municipalities struggle due to insufficient budget allocations. Governments can provide financial support through grants, low-interest loans, and tax exemptions [15]. These mechanisms encourage waste management investments by businesses and specialized producers. Involving private waste operators and companies is crucial. Public-private partnerships (PPPs) can enhance funding and operational efficiency. A major challenge is the lack of adequate collection and disposal infrastructure. Proper waste collection, transportation, and disposal are essential for effective waste management [16, 17]. Insufficient infrastructure leads to illegal dumping and littering, exacerbating environmental pollution. Addressing contaminated sites requires clear liability rules, registers, and prioritization based on environmental and health objectives. As cities expand, waste generation increases. Limited resources strain waste management systems. Balancing waste management with other essential services (clean water, health care) is challenging. Prioritizing waste management within limited resources is crucial for public health and safety. To overcome these challenges, countries must adopt innovative financing mechanisms, integrate waste management strategies into overall policies, and promote circular economy investments. Collaboration between governments, private sector, and communities is essential for building sustainable waste management systems.

Inadequate facilities for garbage collection, recycling, and disposal have far-reaching implications for both public health and the environment. In many rural areas, there is a dearth of recycling facilities. As a result, a higher volume of waste ends up being transferred to landfills. Additionally, several items that could be recycled or composted—such as certain types of plastic—are often rejected by regional recycling facilities. One of the major waste management problems is the lack of adequate collection infrastructure [18]. Proper waste collection is essential for effective waste management. When waste collection systems are inadequate, it leads to serious environmental issues, including littering and illegal dumping. Improper waste disposal can pose health risks to communities. Contaminated water, air pollution, and disease transmission are common consequences. Inadequate facilities contribute to environmental pollution, affecting soil quality, water bodies, and air quality. Poor waste management practices exacerbate climate change due to greenhouse gas emissions from landfills and incineration. Upper-middle and high-income countries provide nearly universal waste collection, with a significant portion of waste being recovered through recycling and composting. In contrast, low-income countries collect only about 48% of waste in cities and even less in rural areas [13]. Only a small fraction is recycled. It is essential to invest in better infrastructure, promote recycling awareness, and encourage responsible waste disposal practices. By building sustainable waste management systems, we can create a cleaner and healthier future for all.

The challenges posed by insufficient public knowledge and lack of instruction regarding waste management techniques can indeed lead to inappropriate disposal practices and non-adherence to rules. Public knowledge about waste management is crucial (Figure 12.1). When people understand the impact of their actions, they are more likely to adopt responsible waste disposal practices. Educating the public about source segregation—separating waste at its origin—ensures that recyclables, organic waste, and hazardous materials are appropriately handled. Awareness campaigns can encourage behavioral change, promoting recycling, composting, and proper disposal [19]. Governments play a vital role in creating and enforcing waste

Figure 12.1 Challenges for solid waste management.

management regulations. These rules guide waste collection, disposal, and recycling practices. Effective waste management requires active participation from the public. Citizens must adhere to rules, sort waste correctly, and use designated bins. Community-based initiatives, workshops, and educational programs foster cooperation and responsible waste management. The results are well worth the effort: For example, smart bins, equipped with sensors and IoT technology, optimize waste collection routes and monitor fill levels; AI-based sorting systems automate waste segregation, reducing human error; waste level sensors help track bin capacity, ensuring timely collection, and recycling apps provide information on recycling centers, schedules, and guidelines. Public-private partnerships, NGOs, and local communities must collaborate for effective waste management [20, 21]. Regulatory bodies should communicate guidelines clearly to the public through campaigns, signage, and educational materials. Public awareness, regulatory support, and technological innovations are essential for sustainable waste management. By working together, we can create cleaner, healthier environments for all.

Long-term planning is essential for creating and implementing efficient waste management programs.

Long-term planning allows for strategic investment in waste collection, recycling, and disposal infrastructure. Properly planned facilities—such as landfills, recycling centers, and waste-to-energy plants—can be strategically located to minimize environmental impact and optimize efficiency. Long-term programs can focus on public awareness campaigns, educating citizens about responsible waste disposal practices. Encouraging households and businesses to segregate waste at the source ensures better recycling and composting outcomes [22, 23]. Long-term planning allows for research into cutting-edge technologies. Smart bins, AI-based sorting systems, and waste-to-energy solutions can be integrated. As technology evolves, long-term programs can adapt to incorporate new tools and methodologies. Long-term planning enables the development of robust waste management policies and regulations. Effective enforcement of rules requires sustained efforts over time. Long-term waste management strategies contribute to climate change mitigation by reducing greenhouse gas emissions. Proper waste management improves public health by minimizing disease transmission and environmental pollution. Long-term planning ensures continuity, efficiency, and sustainability in waste management [24, 25]. By thinking ahead, we can create cleaner, healthier communities for generations to come. Strong waste management systems are essential for environmental sustainability and public health. However, they often face challenges due to transient political cycles and inconsistent policymaking. Political leaders often operate within short-term election cycles. Their focus may prioritize immediate gains over long-term planning. When new administrations take office, they may alter existing waste management policies, leading to disruptions in continuity. Inconsistent policies arise from shifting priorities. What was a priority for one administration may not be for the next. Without sustained commitment, waste management programs may suffer from inadequate funding and implementation [26]. Inconsistent policies affect the development of waste management infrastructure. Long-term planning is crucial for building robust systems. Public awareness campaigns require consistent messaging over time to drive behavioral change. Cutting-edge technologies need stable policies to thrive.

The EU has issued policies on waste management over the last 50 years. Policymaking on waste management and circularity in agroecosystems differs from other sectors. Agricultural waste management has been excluded from the European Waste Framework Directive. The growing interest in circular economy has potential, but agroecosystems lack a general framework compared to other sectors. Policymakers should adopt a long-term vision for waste management, transcending political cycles [27]. They must engage stakeholders, scientists, and decision-makers to create robust, sustainable policies and use innovative tools like policy briefs and scientific advice mechanisms to address complexity. Overcoming the challenges posed by political cycles and inconsistent policymaking requires commitment, collaboration, and a forward-looking approach [28]. The ongoing development of consumer goods and technologies introduces new waste streams and contaminants, posing challenges for waste management.

As technology advances, electronic waste (e-waste) grows. Discarded gadgets, obsolete devices, and electronic components contribute to this stream. The proliferation of consumer goods leads to increased plastic packaging waste [29, 30]. Single-use plastics are a major concern. The rise of portable devices and electric vehicles results in more battery waste. Fast fashion and changing trends contribute to textile waste, including synthetic fabrics that are hard to recycle. New consumer products often contain chemicals that can leach into the environment when disposed of improperly. Synthetic materials break down into microplastics, affecting soil, water, and wildlife. Some technologies contain heavy metals, toxic substances, or rare earth elements that require specialized handling. AI-based bins sort recyclables automatically, improving recycling rates. They monitor fill levels, optimize collection routes, and reduce operational costs. Automated sorting processes in recycling facilities and optimized collection routes based on weight are necessary [15]. Underground systems transport waste efficiently. There is a need to reduce volume and improve efficiency, encourage proper disposal of electronic devices, and provide information on recycling centers and schedules. As technology evolves, waste management must adapt to handle new waste streams and contaminants. Smart technologies play a crucial role in optimizing collection, recycling, and disposal processes.

The circular economy emphasizes reuse, recycling, and resource recovery. It aims to minimize waste and maximize the value of materials. This shift is gaining momentum globally, driven by factors such as technological advancements, consumer preferences, and a growing focus on sustainability. By embracing circular principles, we can create a more resilient, resource-efficient world for future generations.

12.3.1 Evolution of SWM Rules in India

The evolution of Solid Waste Management (SWM) rules in India has an interesting history. The problem of waste management, especially in urban areas, has been a major administrative and environmental concern in India. The issue was further highlighted by the fact that urban solid waste not only contains household waste but also hazardous waste, construction and demolition waste, plastic waste, and electronic waste.

The Supreme Court decision in Almitra H. Patel (2000) played a pivotal role in the evolution and development of SWM Rules in India. In this case, a writ petition was filed regarding the management of Municipal Solid Waste (MSW) in four metropolitan cities of India, namely Chennai, Calcutta, Delhi, and Mumbai. The Central Pollution Control Board (CPCB) had issued guidelines regarding MSW handling, but due to the municipalities being under the administrative control of the States, CPCB expressed its inability to enforce these guidelines effectively. This led the Supreme Court to appoint a committee under the chairmanship of Mr. Asim Burman.

These guidelines were incorporated in the judgment of the Supreme Court in the Almitra H. Patel case. This decision prompted the Union Government to enact the rules for the management of solid waste in India, leading to the notification of the Municipal Solid Waste (Management and Handling) Rules, 2000.

The Municipal Solid Waste (Management and Handling) Rules, 2000, were notified in the year 2000 and came into force on September 25, 2000. However, post the suggestions of the public, the set of rules were updated in the areas of plastic, e-waste, biomedical, hazardous, and construction and demolition waste management rules. Later, the government notified the Solid Waste Management Rules, 2016.

The new set of rules are now not only applicable in municipal areas but have also included urban agglomerations, census towns, notified industrial townships, and also areas which fall under the control and direction of Indian Railways, Airports, Special economic zones, Places of Pilgrimage, Religious and Historical importance, and Organizations under the conduct of State and Central Government.

This evolution of SWM rules in India demonstrates the country's commitment to addressing the challenges of waste management and creating a cleaner and healthier environment for its citizens.

12.3.2 The Future Holds Exciting Possibilities for Solid Waste Management

Among the possibilities for the future are AI-powered bins that automatically sort recyclables, monitor fill levels, and optimize collection routes; transparent waste tracking, ensuring proper disposal and recycling; advanced waste-to-energy technologies that generate clean electricity from waste; and robots for sorting, dismantling e-waste, and handling hazardous materials. Consumers increasingly favor products designed for reuse, repair, and recycling. A shift toward owning fewer possessions reduces waste generation, as does demand for sustainable packaging materials and reduced plastic use [31]. There must be ambitious goals to minimize waste generation and maximize recycling. Manufacturers must be responsible for product end-of-life management. There must also be grassroots efforts for waste reduction and community composting, as well as education campaigns on waste reduction, proper disposal, and recycling. Businesses, governments, and NGOs must work together for sustainable waste management [32, 33]. Research and innovation must drive waste management solutions. The convergence of technology, consumer consciousness, and sustainability will transform waste management into a more efficient, eco-friendly, and circular process.

The continued advancements in recycling technology hold immense promise for sustainable waste management. High-speed cameras and sensors identify different materials based on color, shape, and composition. This enables precise sorting of recyclables. Powerful magnets extract ferrous metals (such as iron and steel) from waste streams. These devices use magnetic fields to separate non-ferrous metals (like aluminum and copper) from other materials [34]. Chemical recycling, also known as advanced recycling, breaks down plastics into their molecular components. Plastics are heated in the absence of oxygen, converting them into oil or gas. These can be used as feedstock for new plastics or fuels. Polymers are broken down into monomers, which can be used to create new plastics. Derived from renewable resources (such as corn or sugarcane), bioplastics offer an eco-friendly alternative to traditional plastics. Advancements include recovering carbon fiber from discarded composite materials (used in aerospace and automotive industries) for reuse and recycling electronic waste to extract valuable rare earth metals used in electronics and renewable energy technologies. These advancements not only reduce waste but also contribute to resource conservation and a more circular economy [35]. By harnessing these technologies, we can extract valuable resources from waste streams and create a more sustainable future.

The integration of data and technology is transforming waste management practices, with smart waste bins leading the charge. Equipped with sensors, these bins measure fill levels and send notifications to collection services when they need emptying. This real-time data enables optimized collection routes, cutting fuel consumption and operational costs while minimizing carbon emissions. Moreover, monitoring environmental conditions through sensors helps prevent issues like odor and pest infestation, maintaining cleanliness and hygiene in waste management processes. By analyzing historical data alongside machine learning algorithms, waste management authorities can predict waste generation patterns, facilitating better resource allocation and planning. Data analytics play a crucial role in assessing waste composition, distinguishing recyclables from hazardous materials. This insight enables more efficient sorting and processing of waste, enhancing recycling rates and reducing environmental harm.

Furthermore, the availability of real-time data allows for dynamic adjustments to collection schedules based on demand fluctuations, ensuring timely waste removal while avoiding overburdening collection resources.

Automation of processes reduces the need for manual intervention, streamlining operations and improving overall efficiency in waste management. This increased efficiency not only saves time and labor costs but also contributes to a more sustainable waste management system by minimizing human error and maximizing resource utilization. The integration of data and technology in waste management brings numerous benefits, including optimized collection routes, better resource allocation, enhanced recycling rates, and improved operational efficiency. By leveraging these advancements, waste management systems can become more sustainable, cost-effective, and environmentally friendly, paving the way for a cleaner and greener future [36]. Data-driven decisions play a pivotal role in optimizing resource allocation within waste management systems, leading to reduced waste and costs. Smart systems integrated with advanced sensors contribute to enhanced sustainability by minimizing fuel consumption and emissions. These sensors monitor fill levels of waste containers in real-time, enabling timely and efficient waste collection processes.

Furthermore, by analyzing historical data, waste management authorities can identify usage patterns and trends, thereby informing strategic decision-making and resource allocation. Machine learning algorithms can be leveraged to predict optimal collection times and routes based on historical data and real-time inputs from sensors.

The synergy between data, sensors, and analytics empowers waste management systems to operate in a smarter, greener, and more efficient manner. By embracing these technologies, stakeholders in waste management can not only improve operational efficiency but also contribute significantly to environmental sustainability efforts. Ultimately, these advancements pave the way for a more sustainable future by minimizing waste generation, reducing environmental impact, and optimizing resource utilization within waste management processes.

The integration of ecologically friendly materials and recyclable product design aligns with a sustainable waste management approach. Additionally, setting "zero waste" goals is a powerful strategy. Using materials that have minimal environmental impact throughout their life cycle is crucial. These break down naturally, reducing long-term waste. Materials like organic cotton, hemp, and bamboo are renewable and biodegradable [37]. Derived from plant sources, they offer an alternative to traditional plastics. Products should be easy to take apart for recycling. Combining materials that are difficult to separate during recycling should be avoided, and consumers should be informed about recyclability and proper disposal.

Zero waste entails minimizing waste sent to landfills or incineration by designing products for reuse, recycling, and composting. The focus is on avoiding waste generation whenever possible, prioritizing reuse, and minimizing consumption. This approach emphasizes the importance of reducing the environmental impact of waste by promoting sustainable practices and mindful consumption habits [38]. To mitigate the burden on landfills, it's crucial to divert waste through recycling and composting initiatives. Businesses play a pivotal role by embracing circular practices that prioritize resource efficiency and waste reduction throughout their operations. Grassroots movements advocate for zero waste living through educational campaigns, lobbying for local policies, and fostering community engagement. By collectively embracing ecologically friendly materials, designing products for recyclability, and setting ambitious zero waste goals, we can pave the way for a more sustainable future. This holistic approach addresses waste management challenges while promoting environmental stewardship and fostering a culture of sustainability across society.

Incorporating ecologically friendly materials and designing products for recyclability are integral components of sustainable waste management strategies. Setting "zero waste" goals serves as a powerful framework for minimizing environmental impact. Utilizing materials with minimal ecological footprint throughout their life cycle is essential for reducing long-term waste accumulation. Renewable and biodegradable materials such as organic cotton, hemp, and bamboo offer sustainable alternatives that align with the principles of waste reduction and environmental stewardship. By prioritizing the use of these materials and adopting zero waste principles, businesses and consumers can contribute to a more sustainable future with reduced waste generation and enhanced environmental resilience [20, 39]. Derived from renewable plant sources, plant-based plastics offer a sustainable alternative to traditional petroleum-based plastics. To maximize their environmental benefits, products made from plant-based plastics should be designed with recyclability in mind. This entails ensuring that products are easy to disassemble and that they do not incorporate materials that are difficult to separate during recycling processes.

Informing consumers about the recyclability and proper disposal of plant-based plastic products is crucial for encouraging environmentally responsible behavior. Zero waste principles advocate for minimizing waste sent to landfills or incineration by designing products for reuse, recycling, and composting. This involves prioritizing reuse over single-use items, minimizing consumption, and diverting waste from landfills through recycling and composting initiatives.

Businesses play a key role in advancing zero waste practices by adopting circular economy principles that emphasize resource efficiency and waste reduction throughout the product life cycle. Grassroots movements also contribute to promoting zero waste living through education, advocating for local policies that support waste reduction, and fostering community engagement in sustainable practices.

By prioritizing design for recyclability, informing consumers about proper disposal practices, and embracing circular economy principles, plant-based plastics can play a significant role in advancing zero waste goals and promoting a more sustainable approach to plastic use and disposal. Collaboration between businesses, consumers, and community organizations is essential for driving meaningful progress towards a zero waste future.

Campaigns play a pivotal role in fostering responsible consumer behavior and promoting sustainable choices. Media coverage admitting climate change or emphasizing sustainability positively impacts consumer behavior [40]. When climate change becomes an implicit social norm, consumers are more likely to engage in sustainable actions. Consumers are swayed by the behaviors of their social groups. Highlighting sustainable actions within communities encourages others to follow suit.

Companies wield significant influence in encouraging waste reduction, often through subtle means. For instance, informing consumers that their coffee cups are compostable can effectively increase composting rates. Studies suggest that consumers are motivated by the behavior of out-group members, individuals they don't wish to be associated with. When out-group members exhibit sustainable practices, such as improved composting, it inspires in-group members to follow suit.

Effective waste reduction initiatives require collaboration among businesses, communities, and government entities. Public-private partnerships play a crucial role in stimulating innovation, investing in infrastructure, and developing comprehensive waste management solutions. Collaborative alliances can facilitate shared advertising campaigns promoting sustainable behaviors, like recycling plastic beverage bottles.

Consumer Social Responsibility (CSR) initiatives exert economic pressure on companies through boycotts and conscious purchasing. This encourages responsible consumer behavior while holding companies accountable for their practices. Campaigns emphasizing social norms, peer influence, and responsible choices can drive positive behavioral changes.

Ultimately, by fostering collaboration across sectors, we can pave the way towards a more sustainable future. By leveraging the influence of businesses, harnessing the power of public-private partnerships, and promoting consumer engagement through CSR initiatives, we can collectively work towards reducing waste and building a more environmentally responsible society.

The adoption of biodegradable and compostable materials, particularly in single-use items, represents a significant stride in mitigating environmental impact. These materials are designed to degrade into natural components like water, carbon dioxide, and biomass through the action of microorganisms over time. Specifically engineered to break down under composting conditions, they yield nutrient-rich compost, contributing to soil health and fertility.

While some biodegradable and compostable items can be composted in home composters, widespread implementation necessitates robust infrastructure for organic waste treatment. Industrial composting facilities are essential for handling larger volumes of organic waste efficiently. These facilities create optimal conditions for composting on a larger scale, accelerating the decomposition process and producing high-quality compost suitable for agricultural and landscaping purposes.

However, successful integration of biodegradable and compostable materials relies heavily on establishing and maintaining proper waste segregation practices, along with effective management systems. Public awareness campaigns can play a crucial role in educating consumers about the importance of separating compostable materials from other waste streams to ensure their proper treatment and diversion from landfills. By investing in infrastructure and fostering collective efforts, communities can harness the environmental benefits of biodegradable and compostable materials, advancing towards more sustainable waste management practices [41]. Efficient waste collection systems necessitate the separation of organic waste from other waste types. Composting plays a crucial role in diverting organic waste from landfills, mitigating methane emissions, and enriching soil fertility and structure. This process completes the loop by returning organic matter to the earth. Public awareness campaigns are vital to educate citizens about the proper disposal of compostable items. Governments and businesses must invest in composting infrastructure, supported by clear labeling to help consumers identify compostable products. While biodegradable and compostable materials offer environmental advantages, their effective integration hinges on the development of robust infrastructure and collective efforts from various stakeholders.

12.3.3 Major Regulatory Challenges

The definition of solid waste holds pivotal importance in the regulation of hazardous waste under the Resource Conservation and Recovery Act (RCRA). Ongoing debates revolve around the breadth of this definition and whether certain materials merit exclusion from it. The regulation of hazardous waste constitutes a multifaceted legal landscape, characterized by diverse requirements and standards. Debates persist regarding the efficacy of performance-based standards versus more prescriptive regulations in ensuring effective hazardous waste management.

Waste-to-energy facilities, which convert waste into energy through processes like incineration, are subject to various regulatory requirements. Discussions continue regarding the appropriate level of regulation for these facilities, emphasizing the need to balance potential benefits with environmental risks.

Many communities have implemented recycling and waste diversion programs to curtail the volume of waste destined for landfills. However, debates persist regarding the most effective regulatory approach for these programs. Issues encompass ensuring proper sorting and processing of recyclables and addressing environmental impacts associated with recycling activities.

Furthermore, there is a growing recognition of the imperative to address the climate change impacts of waste management. Discussions focus on determining the most effective regulatory mechanisms for controlling greenhouse gas emissions from waste management activities. Considerations include exploring carbon pricing and other market-based approaches to incentivize emissions reduction.

These regulatory challenges are intricate and multifaceted, demanding careful deliberation of the legal and policy intricacies at hand. Ultimately, the overarching goal of regulation in this domain is to ensure that solid waste is managed in a safe and responsible manner, simultaneously promoting environmental sustainability and safeguarding public health.

By engaging in ongoing discussions and deliberations, policymakers and stakeholders can work towards refining regulatory frameworks that effectively address the complexities of waste management while aligning with overarching goals of environmental protection and sustainability. As technologies evolve and societal expectations change, regulatory approaches must remain dynamic, adaptable, and responsive to emerging challenges and opportunities in waste management.

12.3.4 Current Regulatory Framework

The Resource Conservation and Recovery Act (RCRA) of 1976 stands as the cornerstone of the current regulatory framework governing solid waste management in the United States. RCRA encompasses two main programs: the Solid Waste program, detailed in Subtitle D, and the Hazardous Waste program, outlined in Subtitle C. Under Subtitle D, states are encouraged to devise comprehensive plans for managing nonhazardous industrial and municipal solid waste. Meanwhile, Subtitle C establishes a structured system for controlling hazardous waste from its generation to its ultimate disposal.

Beyond federal regulations, many states have been granted authorization to operate their own RCRA programs, tailoring regulations to meet specific regional needs. Consequently, consulting state waste management programs or regional authorities becomes essential for understanding specific regulations and policies. Solid waste management thus becomes a collaborative effort involving federal, state, regional, and local entities.

The regulation of solid waste management extends to various federal agencies and departments. For instance, the Department of Transportation's Pipeline and Hazardous Materials Safety Administration oversees the transport of hazardous materials. Moreover, the Environmental Protection Agency (EPA) regulates specific waste types like medical and biomedical waste under different RCRA subtitles and regulations.

State and local governments also wield significant authority in waste management regulation. They possess the power to establish and enforce regulations such as criteria for landfill siting and operation and the implementation of solid waste management programs. Local governments often take charge of managing and operating solid waste facilities while spearheading recycling and waste reduction initiatives.

However, the regulatory landscape is not without its complexities and debates. Legal and policy discussions revolve around matters such as the definition of solid waste, hazardous waste regulation, waste-to-energy facility oversight, recycling and diversion program management, and the intersection of waste management with climate change concerns. Effective solid waste management necessitates the involvement of stakeholders from diverse backgrounds, including citizens, businesses, and government agencies, to develop proper mechanisms that balance environmental protection, public health, and economic considerations.

12.3.5 Emerging Regulatory Trends

Several new and emerging regulatory issues are shaping the landscape of solid waste management. These issues reflect evolving environmental concerns, technological advancements, and changing societal attitudes towards waste.

The environmental impact of plastic waste, especially single-use plastics, has spurred regulatory action in many jurisdictions. Some are implementing bans or restrictions on certain single-use plastics like bags, straws, and Styrofoam containers. Additionally, efforts are focused on promoting alternatives to plastics, encouraging recycling, and implementing Extended Producer Responsibility (EPR) programs, which hold producers accountable for the disposal of their products.

Recognizing the importance of diverting organic waste from landfills, regulations promoting composting and anaerobic digestion are on the rise. Mandatory organic waste recycling programs, food waste bans, and incentives for organics diversion are becoming more common as well.

Electronic waste (e-waste) poses unique challenges due to its complex composition and hazardous components. Emerging regulations aim to ensure proper recycling and disposal of e-waste to prevent environmental contamination and recover valuable materials.

Governments are increasingly exploring policies to promote a circular economy approach, which minimizes waste generation, maximizes resource efficiency, and encourages product reuse, repair, and recycling. This may involve regulations incentivizing eco-design, product stewardship, and sustainable consumption practices.

Solid waste management is a significant source of greenhouse gas emissions, primarily through methane generation in landfills. Regulatory efforts target methane emissions reduction through landfill gas capture and utilization, waste-to-energy technologies, and renewable energy production from waste.

Overall, regulatory efforts are multifaceted and evolving, addressing various aspects of waste management to mitigate environmental impact, promote resource efficiency, and transition towards a more sustainable waste management system. These regulations aim to tackle specific waste streams like plastics and e-waste, while also promoting broader strategies like composting, recycling, and circular economy principles to achieve comprehensive and long-term sustainability goals.

12.3.6 Potential Future Regulatory Developments

Extended Producer Responsibility (EPR) programs are poised to undergo expansion to encompass a broader array of products and materials, thereby shifting the onus of waste management and recycling onto producers. This evolution holds the potential to foster more robust product take-back schemes and elevate recycling rates across various product categories. Leveraging emerging technologies like blockchain and digital tracking systems can revolutionize waste management data collection, tracking, and reporting, enhancing the efficiency and transparency of the process. Regulatory frameworks are anticipated to adapt to integrate digital solutions, facilitating improved monitoring of waste streams, compliance verification, and transparency.

Furthermore, there's a growing impetus for greater standardization and harmonization of waste management regulations across jurisdictions. This harmonization is critical for facilitating interstate and international trade in recyclable materials and promoting consistency in waste management practices. Such regulatory alignment can streamline regulatory compliance for businesses operating across multiple regions while fostering more efficient and sustainable waste management practices globally.

As regulatory frameworks evolve, there's likely to be a concerted push to incentivize the adoption of innovative waste-to-resource technologies. This may involve offering incentives, simplifying permitting processes, and addressing regulatory hurdles to facilitate the deployment of advanced recycling, chemical recycling, and conversion technologies. By encouraging the adoption of these technologies, regulators aim to unlock new avenues for resource recovery and minimize waste generation, thereby advancing sustainability goals.

Moreover, future regulatory developments are expected to place a heightened emphasis on addressing environmental justice concerns associated with waste management. This entails ensuring equitable distribution of environmental benefits and burdens and actively engaging marginalized communities in decision-making processes. By prioritizing environmental justice, regulators seek to redress historical disparities in waste management practices and foster inclusive and participatory approaches to waste management governance.

In conclusion, the evolving regulatory landscape of solid waste management is characterized by a concerted effort to address emerging challenges and opportunities. With a focus on waste reduction, recycling promotion, and environmental impact mitigation, future regulatory developments are poised to advance sustainability, innovation, and social responsibility in waste management practices. By embracing these changes, stakeholders can contribute to building a more resilient, resource-efficient, and equitable waste management system for the future.

12.3.7 Decentralized Waste Management

Decentralized waste management systems offer numerous advantages, particularly in emerging economies like India, where they have established a functioning network of informal waste collectors, kabadiwalas (scrap dealers), and aggregators. These systems inherently embody the principles of the circular economy, ensuring efficient resource utilization and environmental sustainability.

Informal waste pickers, integral to decentralized systems, often surpass machines in waste sorting efficiency. Their hands-on approach and intimate knowledge of materials result in higher-quality outputs compared to centralized systems. Moreover, decentralized systems involve small-scale entrepreneurs who play a significant role in waste collection and recycling efforts. By building their capabilities, providing access to capital, and ensuring reliable waste streams, their performance can be further enhanced.

Efficiency in resource recovery is another hallmark of decentralized systems. Integration of the informal sector expands waste collection and recycling capacity, maximizing resource utilization and minimizing environmental impact. Additionally, decentralized systems mitigate the risks associated with corruption often prevalent in centralized waste management systems. By distributing responsibilities across various stakeholders and empowering local communities, decentralized systems foster resilience against monopolistic practices and corrupt influences.

Furthermore, decentralized waste management systems align closely with the principles of the circular economy, emphasizing resource conservation, reuse, and recycling. By empowering small-scale entrepreneurs and promoting collaboration among stakeholders, these systems contribute to the development of more efficient and sustainable waste management solutions.

Investing in decentralized waste management infrastructure and supporting small-scale entrepreneurs is crucial for realizing the full potential of these systems. By providing training, access to financing, and technological support, governments and organizations can enhance the capabilities of local stakeholders and optimize resource recovery efforts. Moreover, fostering collaboration among government agencies, businesses, civil society organizations, and communities is essential for creating an enabling environment for decentralized waste management initiatives.

In conclusion, decentralized waste management systems offer a range of benefits, including efficient resource recovery, resilience against corruption, and alignment with circular economy principles as shown in Figure 12.2. By investing in and supporting these systems, we can create more inclusive, sustainable, and resilient waste management solutions that benefit both local communities and the environment.

Hazardous waste management in today's globalized trade environment presents multifaceted challenges. The Basel Convention addresses the trans-boundary movement of hazardous waste, aiming to safeguard human health and the environment. However, the global trade in hazardous waste involves complex interactions between environmental regulations and international trade agreements.

The World Trade Organization (WTO) recognizes the importance of sustainable waste management, emphasizing the need to harmonize rules under the Basel, Rotterdam, and Stockholm Conventions with global trade regulations. Achieving this balance is crucial to ensure effective hazardous waste management while facilitating international trade.

The globalization of hazardous waste trade has led to hidden externalities, prompting studies to analyze the effectiveness of the Basel Convention and its impact on international trade. Understanding these dynamics is essential for developing policies that address environmental concerns without stifling economic growth.

In emerging economies, decentralized waste management systems, often involving informal waste collectors, play a significant role. These systems highlight the importance of community involvement and collaboration across various stakeholders, including communities, businesses, and

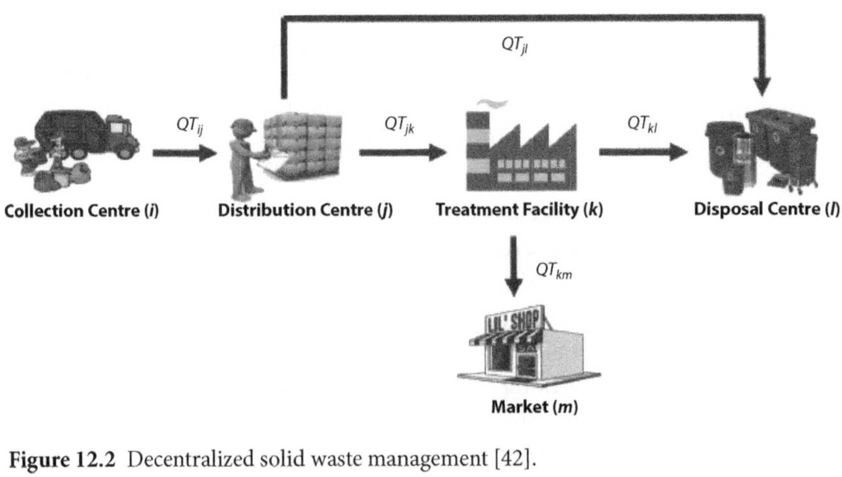

Figure 12.2 Decentralized solid waste management [42].

governments. By working together, comprehensive waste management solutions can be developed to mitigate the risks associated with hazardous waste while promoting sustainable development.

Overall, addressing the challenges of hazardous waste management in a globalized trade environment requires a coordinated effort that considers environmental, economic, and social dimensions. By fostering collaboration and harmonizing regulations, countries can work towards more effective and sustainable management of hazardous waste on a global scale.

International agreements serve as vital tools in regulating the transport and disposal of hazardous materials, ensuring responsible waste management practices and environmental protection on a global scale. One prominent agreement is the Basel Convention on the Control of Transboundary Movements of Hazardous Wastes and their Disposal, negotiated under the United Nations Environment Program (UNEP). This multilateral treaty aims to promote environmentally sound management of exported and imported waste, with a particular focus on developing countries.

The Basel Convention establishes standards for the transboundary movement of hazardous waste, solid waste, and municipal incinerator ash. Key provisions include requirements for prior notification to and consent from the receiving country, the issuance of international movement documents, and compliance with packaging and labeling regulations. By setting these standards, the Convention seeks to prevent the dumping of hazardous waste in countries with inadequate infrastructure for safe disposal.

Additionally, the Organization for Economic Cooperation and Development (OECD) Council Decision addresses the transboundary movement of hazardous waste among its member countries. This decision aims to harmonize waste management practices, providing guidelines for environmentally sound waste management and promoting cooperation in waste trade.

Furthermore, bilateral agreements between countries further regulate hazardous waste trade, ensuring adherence to international standards and fostering collaboration in waste management. Examples include agreements between the United States and Canada, Mexico, Costa Rica, Malaysia, and the Philippines.

These international frameworks play a crucial role in ensuring responsible waste management, protecting human health, and safeguarding the environment. By adhering to these agreements and implementing their provisions, countries can effectively manage hazardous materials, prevent environmental pollution, and promote sustainable development worldwide. Cooperation and compliance with international agreements are essential for addressing the challenges of hazardous waste management in an increasingly interconnected global community.

Encouraging sustainable practices is paramount for safeguarding the well-being of our planet. Recognizing this urgency, governments and businesses worldwide have implemented various incentives to promote sustainability. Governments offer tax credits and grants to businesses embracing sustainable practices, encouraging investments in renewable energy systems like solar panels or wind turbines. Businesses adopting plug-in hybrid or electric vehicles can also benefit from these incentives.

Sustainable businesses not only enjoy financial benefits but also attract environmentally conscious consumers. People prefer spending money at establishments that prioritize environmental responsibility, leading to increased customer loyalty and brand reputation. Embracing sustainability not only benefits businesses but also contributes to cleaner air, water, and overall public health. By reducing pollution, businesses positively impact their communities and ecosystems.

Although initial investments in sustainable practices may seem costly, the long-term benefits far outweigh the expenses. Sustainable initiatives often result in lower utility bills, tax advantages, and enhanced brand image. Additionally, businesses play a crucial role in mitigating climate change and preserving natural resources for future generations. By embracing sustainability, businesses not only contribute to a healthier planet but also ensure their own long-term success and resilience in an increasingly environmentally conscious market.

12.4 Waste to Energy

Ongoing advancements in waste-to-energy technologies are crucial for diversifying and enhancing sustainability in waste management. These innovations enable efficient conversion of waste into usable energy, reducing reliance on landfills and fossil fuels while contributing to a more circular economy.

Waste-to-energy conversion technologies play a crucial role in addressing waste management challenges while simultaneously contributing to energy production and resource recovery (Figure 12.3). These technologies utilize various processes to transform different types of waste into usable energy forms.

In conventional waste-to-energy facilities, solid waste undergoes combustion at high temperatures, typically in modern incinerators designed to minimize emissions. The heat generated from this process can be harnessed

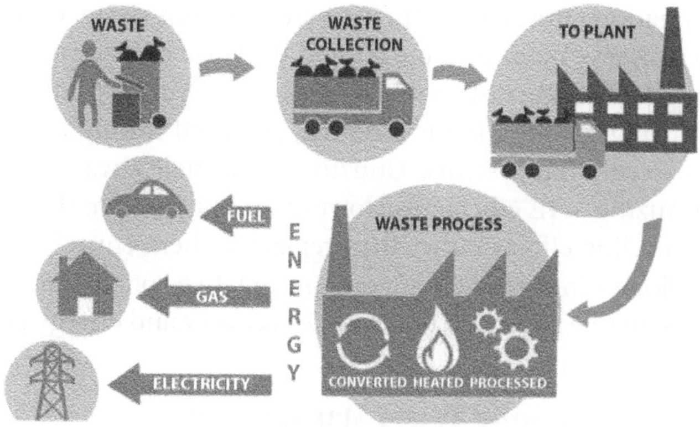

Figure 12.3 Waste-to-energy generation [43].

to produce steam, driving turbines to generate electricity or providing heat for district heating systems. Advanced technologies ensure efficient energy recovery while controlling air pollutants, such as particulate matter and nitrogen oxides.

For organic waste streams like food scraps and agricultural residues, anaerobic digestion is employed. In this process, microorganisms break down organic matter in the absence of oxygen, producing biogas primarily composed of methane. Biogas can be utilized directly as a fuel for electricity generation or heating purposes, offering a renewable and sustainable energy source.

Additionally, thermal processes can convert waste into syngas or biochar. Syngas, a mixture of hydrogen, carbon monoxide, and other gases, can be used as a fuel for electricity generation or as a feedstock for the production of chemicals and fuels. On the other hand, biochar, a stable carbon-rich material, can improve soil quality and sequester carbon, contributing to climate change mitigation efforts.

Overall, waste-to-energy conversion technologies offer a multifaceted approach to waste management and energy production, facilitating the transition towards a more sustainable and circular economy. By harnessing the energy potential of waste streams and minimizing environmental impacts, these technologies play a vital role in driving towards a more resource-efficient and resilient future.

Waste-to-energy (WtE) conversion technologies play a vital role in diverting waste from landfills, thus alleviating the strain on available landfill space. Managed properly, WtE processes emit fewer greenhouse gases compared to uncontrolled landfill decomposition, contributing to climate change mitigation efforts. By extracting energy from waste, these technologies help conserve natural resources and decrease reliance on fossil fuels, contributing to a more sustainable energy mix. Crucially, minimizing air pollution during incineration is essential, achieved through advanced filtration and scrubbing systems. Utilizing high-quality waste feedstock is key to optimizing WtE efficiency. However, it's imperative that WtE complements recycling efforts rather than replacing them entirely. Integrating WtE technologies into waste management practices offers a more diversified and sustainable approach to resource recovery and energy generation.

12.4.1 Policy for Solid Waste Management

The National Policy on Waste Management in India represents a comprehensive approach to address the growing challenges of waste management in the country. With the overarching goal of managing waste sustainably for a healthier life and a cleaner environment, the policy covers solid, liquid, and gaseous waste, recognizing the need for a holistic approach to waste management. Central to the policy's objectives is the waste hierarchy principle, emphasizing the importance of waste reduction, reuse, and recycling to minimize environmental impact and resource depletion.

One of the key strategies outlined in the policy is extended producer responsibility (EPR), which holds producers accountable for the environmental impact of their products throughout their life cycle [44]. By shifting the burden of waste management to producers, the policy incentivizes the adoption of eco-friendly production processes and product designs, ultimately promoting a circular economy where resources are kept in circulation for as long as possible.

Real-time data monitoring mechanisms are integrated into the policy framework to ensure efficient waste management practices. By leveraging technology and data analytics, authorities can track waste generation, collection, and disposal processes, enabling timely interventions and informed decision-making to optimize resource utilization and minimize waste generation.

The overarching goal of the policy is to transform waste into valuable resources, aligning with the principles of sustainable development and resource conservation [45]. By promoting innovative waste management technologies and practices, such as composting, biogas generation, and material recovery facilities, the policy aims to maximize the recovery of resources from waste streams, reducing the burden on landfills and mitigating environmental pollution.

However, the policy also acknowledges the challenges posed by rising waste generation rates driven by economic growth and urbanization. As India's economy continues to expand and urban areas become more densely populated, the pressure on existing waste management infrastructure and resources increases. Addressing this challenge requires concerted efforts from both the government and the private sector to invest in modernizing waste management infrastructure, enhancing public awareness, and promoting sustainable consumption and production patterns.

In the Philippines, the Environmental Management Bureau (EMB) oversees waste management regulations aimed at addressing environmental concerns related to waste, pollution, and sustainable practices. These regulations encompass various aspects of waste management, including waste segregation, disposal methods, recycling initiatives, and compliance monitoring. By enforcing these regulations, the EMB plays a crucial role in safeguarding the environment and promoting responsible waste management practices across the country.

Additionally, state-specific solid waste management policies and programs in India play a crucial role in addressing waste challenges at the local level. Tailored to the unique needs and contexts of individual states, these policies encompass waste segregation, collection, transportation, treatment, and disposal methods [46]. State governments collaborate with local bodies, NGOs, and communities to promote awareness, implement sustainable practices, and enhance waste infrastructure, contributing to cleaner environments, improved public health, and resource conservation.

12.5 Conclusion

Solid waste management regulation faces a myriad of challenges that necessitate a delicate balance between public health, environmental preservation, and economic viability. Inadequate waste management practices, exacerbated by disparities in enforcement within and among nations, pose serious threats to ecosystems and human well-being. The inconsistent implementation of rules, coupled with insufficient oversight, fosters non-compliance by waste producers and disposal sites, resulting in issues like illegal dumping and unlicensed waste collection.

Financial limitations, deficient infrastructure, and scarce resources impede the establishment and execution of efficient waste management initiatives. Lagging policies may struggle to keep up with technological advancements, hindering the adoption of eco-friendly waste management techniques. Moreover, the trans-boundary movement of waste, particularly hazardous materials, presents regulatory complexities due to their global reach, potentially causing severe environmental and health ramifications.

Insufficient public awareness and education about waste management further contribute to improper disposal practices. Effective waste management necessitates collaborative efforts between the public and regulatory bodies, promoting proper waste sorting and the enforcement of consistent, transparent regulations. However, long-term planning may be thwarted by political cycles and inconsistent policymaking, impeding the development of robust waste management systems.

Addressing these regulatory hurdles demands a multifaceted strategy encompassing legislative enhancements, strengthened enforcement mechanisms, public engagement initiatives, and investments in sustainable technologies. Transitioning from a linear to a circular economy, prioritizing reuse and recycling, shows promise in minimizing waste generation and maximizing resource efficiency.

Technological innovations such as advanced sorting techniques and data analytics can optimize waste management processes, improving efficiency and resource utilization. Implementing Extended Producer Responsibility programs and striving for "zero waste" goals align with sustainable practices, encouraging product design for longevity and recyclability.

Decentralized waste management systems, complemented by global cooperation frameworks, offer additional avenues for effective waste solutions. By decentralizing waste processing and encouraging community involvement, localities can tailor solutions to their unique needs while fostering a sense of responsibility and ownership.

In conclusion, the future of solid waste management hinges on a comprehensive, technologically driven, and collaborative approach. Emphasizing sustainability and stakeholder engagement is paramount to overcoming emerging challenges and safeguarding environmental well-being and resource conservation for future generations. By addressing regulatory complexities, embracing innovative solutions, and fostering global cooperation, we can pave the way for a more sustainable and resilient waste management landscape.

References

1. World Bank, Solid Waste Management, Accessed: Feb. 27, 2024, [Online]. Available: https://www.worldbank.org/en/topic/urbandevelopment/brief/solid-waste-management.

2. World Bank, Trends in Solid Waste Management, Accessed: Feb. 27, 2024, [Online]. Available: https://datatopics.worldbank.org/what-a-waste/trends_in_solid_waste_management.html.
3. Chen, C.C., Spatial inequality in municipal solid waste disposal across regions in developing countries. *Int. J. Environ. Sci. Technol.*, 7, 3, 447–456, Jun. 2010, doi: 10.1007/BF03326154.
4. Zambrano-Monserrate, M.A., Ruano, M.A., Ormeño-Candelario, V., Determinants of municipal solid waste: a global analysis by countries' income level. *Environ. Sci. Pollut. Res.*, 28, 44, 62421–62430, Nov. 2021, doi: 10.1007/s11356-021-15167-9.
5. Karak, T., Bhagat, R.M., Bhattacharyya, P., Municipal Solid Waste Generation, Composition, and Management: The World Scenario. *Crit. Rev. Environ. Sci. Technol.*, 42, 15, 1509–1630, Aug. 2012, doi: 10.1080/10643389.2011.569871.
6. Kawai, K. and Tasaki, T., Revisiting estimates of municipal solid waste generation per capita and their reliability. *J. Mater. Cycles Waste Manage.*, 18, 1, 1–13, Jan. 2016, doi: 10.1007/s10163-015-0355-1.
7. Jayasinghe, P.A., Jalilzadeh, H., Hettiaratchi, P., The Impact of COVID-19 on Waste Infrastructure: Lessons Learned and Opportunities for a Sustainable Future. *Int. J. Environ. Res. Public Health*, 20, 5, 5, Jan. 2023, doi: 10.3390/ijerph20054310.
8. Zhou, A., Wu, S., Chu, Z., Huang, W.-C., Regional differences in municipal solid waste collection quantities in China. *Sustainability*, 11, 15, 4113, 2019.
9. Perkoulidis, G., Karagiannidis, A., Kontogianni, S., Diaz, L.F., Solid waste management in developing countries. Present problems and future perspectives. *J. Environ. Prot. Ecol.*, 12, 2, 570, 2011.
10. Cheema, S.M., Hannan, A., Pires, I.M., Smart Waste Management and Classification Systems Using Cutting Edge Approach. *Sustainability*, 14, 16, 16, Jan. 2022, doi: 10.3390/su141610226.
11. Khan, S., Anjum, R., Raza, S.T., Bazai, N.A., Ihtisham, M., Technologies for municipal solid waste management: Current status, challenges, and future perspectives. *Chemosphere*, 288, 132403, 2022.
12. Kumar, A. and Agrawal, A., Recent trends in solid waste management status, challenges, and potential for the future Indian cities–A review. *Curr. Res. Environ. Sustainability*, 2, 100011, 2020.
13. Guerrero, L.A., Maas, G., Hogland, W., Solid waste management challenges for cities in developing countries. *Waste Manage. (Oxford)*, 33, 1, 220–232, 2013.
14. Sanjeevi, R., Haruna, M., Tripathi, S., Singh, B., Jayaraman, A., Impacts of Global Carbon Foot Print on the Marine Environment. *Int. J. Eng. Res. Technol.*, 5, 51–54, May 2017.
15. Khatib, I.A., Municipal solid waste management in developing countries: Future challenges and possible opportunities. *Integr. Waste Manage.*, 2, 35–48, 2011.

16. Abbasi, T., Sanjeevi, R., Makhija, M., Abbasi, S.A., Role of Vitamins B-3 and C in the Fashioning of Granules in UASB Reactor Sludge. *Appl. Biochem. Biotechnol.*, 167, 2, 348–357, May 2012, doi: 10.1007/s12010-012-9691-y.
17. Joshi, R. and Ahmed, S., Status and challenges of municipal solid waste management in India: A review. *Cogent Environ. Sci.*, 2, 1, 1139434, Dec. 2016, doi: 10.1080/23311843.2016.1139434.
18. Henry, R.K., Yongsheng, Z., Jun, D., Municipal solid waste management challenges in developing countries–Kenyan case study. *Waste Manage. (Oxford)*, 26, 1, 92–100, 2006.
19. Das, S., Lee, S.-H., Kumar, P., Kim, K.-H., Lee, S.S., Bhattacharya, S.S., Solid waste management: Scope and the challenge of sustainability. *J. Clean. Prod.*, 228, 658–678, 2019.
20. Zohoori, M. and Ghani, A., Municipal solid waste management challenges and problems for cities in low-income and developing countries. *Int. J. Sci. Eng. Appl.*, 6, 2, 39–48, 2017.
21. Abbasi, T., Sanjeevi, R., Anuradha, J., Abbasi, S.A., Impact of Al 3+ on sludge granulation in UASB reactor. *Indian J. Biotechnol.*, 12, 2, 254–259, 2013.
22. Ravichandran, C. and Venkatesan, G., Toward sustainable solid waste management–challenges and opportunities, in: *Handbook of Advanced Approaches towards Pollution Prevention and Control*, pp. 67–103, 2021.
23. Ranjan, H., Sanjeevi, R., Vardhini, S., Tripathi, S., Anuradha, J., Biogenic Production of Silver Nanoparticles Utilizing an Arid Weed (Saccharum munja Roxb.) and Evaluation of its Antioxidant and Antimicrobial Activities.
24. Otchere, A.F., Sarpong, K.O., Okyere, S., Assessing the challenges affecting solid waste management system in the Kumasi Metropolis. *J. Arts Humanit.*, 3, 2, 50–63, 2014.
25. Sathvara, P.B., Anuradha, J., Sanjeevi, R., Tripathi, S., Rathod, A.B., Spatial Analysis of Carbon Sequestration Mapping Using Remote Sensing and Satellite Image Processing, in: *Multimodal Biometric and Machine Learning Technologies: Applications for Computer Vision*, pp. 71–83, 2023.
26. Jin, J., Wang, Z., Ran, S., Solid waste management in Macao: practices and challenges. *Waste Manage. (Oxford)*, 26, 9, 1045–1051, 2006.
27. Chang, N.-B., Pires, A., Martinho, G., Empowering Systems Analysis for Solid Waste Management: Challenges, Trends, and Perspectives. *Crit. Rev. Environ. Sci. Technol.*, 41, 16, 1449–1530, Aug. 2011, doi: 10.1080/10643381003608326.
28. Kibria, M.G., Masuk, N.I., Safayet, R., Nguyen, H.Q., Mourshed, M., Plastic Waste: Challenges and Opportunities to Mitigate Pollution and Effective Management. *Int. J. Environ. Res.*, 17, 1, 20, Jan. 2023, doi: 10.1007/s41742-023-00507-z.
29. Abarca-Guerrero, L., Maas, G., Hogland, W., Solid waste management challenges for cities in developing countries. *Rev. Tecnol. Marcha*, 28, 2, 141–168, 2015.
30. Mishra, S., Anuradha, J., Tripathi, S., Kumar, S., *In vitro* antioxidant and antimicrobial efficacy of Triphala constituents: Emblica officinalis, Terminalia

belerica and Terminalia chebula. *J. Pharmacogn. Phytochem.*, 5, 6, 273–277, 2016.
31. Sinthiya, N.J., Chowdhury, T.A., Haque, A.K.M.B., Artificial Intelligence Based Smart Waste Management—A Systematic Review, in: *Computational Intelligence Techniques for Green Smart Cities*, Green Energy and Technology, M. Lahby, A. Al-Fuqaha, Y. Maleh (Eds.), pp. 67–92, Springer International Publishing, Cham, 2022, doi: 10.1007/978-3-030-96429-0_3.
32. Soni, A., Das, P.K., Kumar, P., A review on the municipal solid waste management status, challenges and potential for the future Indian cities. *Environ. Dev. Sustain.*, 25, 12, 13755–13803, Dec. 2023, doi: 10.1007/s10668-022-02688-7.
33. Aly, A.A., Al-Omran, A.M., Sallam, A.S., Al-Wabel, M.I., Al-Shayaa, M.S., Vegetation cover change detection and assessment in arid environment using multi-temporal remote sensing images and ecosystem management approach. *Solid Earth*, 7, 2, 713–725, Apr. 2016, doi: 10.5194/se-7-713-2016.
34. Bazrbachi, A., Sidique, S.F., Adam, S.U., Hamdan, A., Smart Technology and Waste Management, a Review on Current and Potential Applications in Malaysia, in: *Explore Business, Technology Opportunities and Challenges After the Covid-19 Pandemic*, Lecture Notes in Networks and Systems, B. Alareeni, A. Hamdan (Eds.), pp. 1105–1117, Springer International Publishing, Cham, 2023, doi: 10.1007/978-3-031-08954-1_93.
35. McCarty, J.A. and Shrum, L.J., The recycling of solid wastes: Personal values, value orientations, and attitudes about recycling as antecedents of recycling behavior. *J. Bus. Res.*, 30, 1, 53–62, 1994.
36. Duquennoi, C. and Martinez, J., European Union's policymaking on sustainable waste management and circularity in agroecosystems: The potential for innovative interactions between science and decision-making. *Front. Sustain. Food Syst.*, 6, 2022, Accessed: Feb. 27, 2024. [Online]. Available: https://www.frontiersin.org/articles/10.3389/fsufs.2022.937802.
37. Zaman, A., Zero-Waste: A New Sustainability Paradigm for Addressing the Global Waste Problem, in: *The Vision Zero Handbook: Theory, Technology and Management for a Zero Casualty Policy*, K. Edvardsson Björnberg, M.-Å. Belin, S.O. Hansson, C. Tingvall (Eds.), pp. 1–24, Springer International Publishing, Cham, 2020, doi: 10.1007/978-3-030-23176-7_46-1.
38. Clark, J.H., Solid Acids for Green Chemistry. *Acc. Chem. Res.*, 35, 9, 791–797, Sep. 2002, doi: 10.1021/ar010072a.
39. Sanjeevi, R., Rathod, A.B., Sathvara, P.B., Tripathi, A., Anuradha, J., Tripathi, S., Vegetational Cartography Analysis Utilizing Multi-Temporal Ndvi Data Series: A Case Study from Rajkot District (Gujarat), India. *Tianjin Daxue Xuebao (Ziran Kexue Gongcheng Jishu Ban)/J. Tianjin Univ. Sci. Technol.*, 55, 04, 490–497, 2022, doi: 10.17605/OSF.IO/UGJYM.
40. Bonnet, F., Lahboubi, N., Habchi, S., El Bari, H., Waste Management Institutional and Legislation Aspects in Developing Countries, in: *Waste Management in Developing Countries*, Waste as a Resource, H. El Bari and

C. Trois (Eds.), pp. 91–109, Springer International Publishing, Cham, 2023, doi: 10.1007/978-3-031-28001-6_6.

41. Teshome, F.B., Municipal solid waste management in Ethiopia; the gaps and ways for improvement. *J. Mater. Cycles Waste Manage.*, 23, 1, 18–31, Jan. 2021, doi: 10.1007/s10163-020-01118-y.
42. Muneeb, S.M., Adhami, A.Y., Jalil, S.A., Asim, Z., Decentralized bi-level decision planning model for municipal solid waste recycling and management with cost reliability under uncertain environment. *Sustain. Prod. Consum.*, 16, 33–44, Oct. 2018, doi: 10.1016/j.spc.2018.05.009.
43. Nag, A., Waste to Energy May Not Be a Brilliant Idea, Yet. *Eng. Rev.*, Jan. 21, 2020, Accessed: Feb. 28, 2024. [Online]. Available: https://engmag.in/waste-to-energy-may-not-brilliant-idea-yet/.
44. CPCB, Central Pollution Control Board, [Online]. Available: https://cpcb.nic.in/rules-2/
45. Hassan, S.H., Halim, A.A., Yusoff, M.S., Wang, L.K., Wang, M.-H.S., Legislation for Solid Waste Management, in: *Solid Waste Engineering and Management: Vol. 1, Handbook of Environmental Engineering*, L.K. Wang, M.-H.S. Wang, Y.-T. Hung (Eds.), pp. 85–141, Springer International Publishing, Cham, 2021, doi: 10.1007/978-3-030-84180-5_2.
46. Singh, V.K., Solanki, P., Ghosh, A., Pal, A., Solid Waste Management and Policies Toward Sustainable Agriculture, in: *Handbook of Solid Waste Management: Sustainability through Circular Economy*, C. Baskar, S. Ramakrishna, S. Baskar, R. Sharma, A. Chinnappan, R. Sehrawat (Eds.), pp. 523–544, Springer Nature, Singapore, 2022, doi: 10.1007/978-981-16-4230-2_27.

About the Editors

Priyanka Singh, PhD is an Associate Professor at the Institute of Allied Medical Science and Technology at the National Institute of Medical Sciences University. She has published 31 research articles in peer-reviewed journals and ten book chapters in international book publications, and filed patent with the Government of India. She is actively engaged in research related to nanotechnology, bioprocess technology, and enzyme engineering.

Pooja Agarwal, PhD is a Professor in the Department of Chemistry in the School of Basic and Applied Sciences at Galgotias University with more than 13 years of teaching experience. She has more than 18 research publications in reputed journals and book chapters and one edited book to her credit. Her research focuses on material science, biomaterials, nanomaterials, and composite materials.

V. Vivekanand, PhD is an Assistant Professor and the Head of the Centre for Energy and Environment at the Malaviya National Institute of Technology. He has published more than 90 international research articles in journals of high repute and more than 30 articles in international conferences, and authored more than seven book chapters. He has a keen research interest in biomass pretreatment, bioprocessing, and its conversion for bioenergy applications.

Index

Also of Interest

Check out these related titles from Scrivener Publishing

Waste Management and Utilization for a Sustainable Environment, Edited by Amit Srivastava and Suman Lata Tripathi, ISBN: 9781394292974. *Waste Management and Utilization for a Sustainable Environment* is an essential guide to understanding how cutting-edge technologies and innovative engineering solutions can transform waste management into a cornerstone of a sustainable future.

Sustainable Production Innovations: Bioremediation and Other Biotechnologies, Edited by Alok Patel and Amit Kumar Sharma, ISBN: 9781119791904. Presenting the latest technologies and practices in this ever-changing field, this groundbreaking new volume covers the gambit for providing solutions and practical applications of smart and efficient energy systems.

Ecosystem Management: Climate Change and Sustainability, Edited by Arnab Banerjee, Manoj Kumar Jhariya, Abhishek Raj, and Taher Mechergui, ISBN: 9781394231218. This book is essential for anyone who wants to understand the challenges of environmental degradation and learn about the sustainable solutions needed to address these critical issues.

Sewage and Biomass from Wastewater to Energy: Possibilities and Technology, Edited by Inamuddin, Tariq Altalhi, Mohammad Luqman, and Joseph K. Bwapwa, ISBN: 9781394204311. Written and edited by a team of industry experts, this exciting new volume covers clean energy production from sewage and biomass while achieving a zero-waste strategy.

Biotechnological Processes for Industrial and Biomedical Waste Management, Edited by Naveen Dwivedi, Shubha Dwivedi, Ranjeet Kumar Mishra and Charles Oluwaseun Adetunji, ISBN: 9781394272006.

LAND AND ENVIRONMENTAL MANAGEMENT THROUGH FORESTRY, Edited by Abhishek Raj, Manoj Kumar Jhariya, Arnab Banerjee, Sharad Nema, and Kiran Bargali, ISBN: 9781119910404. Written and edited by a group of experts in the field, this groundbreaking reference work sets the standard for engineers, students, and professionals working in forestry, agriculture, ecology and environmental science, offering the scientific community a way towards combating climate change.

GREEN CHEMISTRY FOR ENVIRONMENTAL REMEDIATION, Edited by Rashmi Sanghi and Vandana Singh, ISBN: 9780470943083. Presents high quality research papers as well as in depth review articles on the new emerging green face of multidimensional environmental chemistry.

BIOREMEDIATION AND SUSTAINABILITY: Research and Applications, Edited by Romeela Mohee and Ackmez Mudhoo, ISBN: 9781118062845. Bioremediation and Sustainability is an up-to-date and comprehensive treatment of research and applications for some of the most important low-cost, "green," emerging technologies in chemical and environmental engineering.

EMERGENCY RESPONSE MANAGEMENT OF OFFSHORE OIL SPILLS: Guidelines for Emergency Responders, By Nicholas P. Cheremisinoff, PhD, and Anton Davletshin, ISBN: 9780470927120. The first book to examine the Deepwater Horizon disaster and offer processes for safety and environmental protection.

COMPLEX BIOLOGICAL SYSTEMS: Adaptation and Tolerance to Extreme Environments, By Irina R. Fomina, Karl Y. Biel, and Vladislav G. Soukhovolsky, ISBN: 9781119510406. Written and edited by some of the most well-respected authors in the area of the adaptation of plants and animals to climate change, this groundbreaking new work is an extremely important scientific contribution to the study of global warming.

GLOBAL CLIMATE CHANGE DEMYSTIFIED, By James G. Speight, ISBN: 9781119653851. Tackling one of the most controversial subjects of our time, one of the world's foremost environmental and petroleum engineers explores the potential causes and ramifications of global climate change.

DEATH OF THE CHESAPEAKE: A History of the Military Role in Polluting the Bay, By Richard D. Albright, ISBN: 9781118686270. This unique book focuses attention on the failure of current efforts to cleanup the Chesapeake Bay and suggests an approach often used in cleaning up environmentally damaged sites.

Printed and bound by CPI Group (UK) Ltd, Croydon, CR0 4YY

28/07/2026

14931338-0004